The Theory of Evolution

The Theory of Evolution

Principles, Concepts, and Assumptions

SAMUEL M. SCHEINER AND
DAVID P. MINDELL

The University of Chicago Press
Chicago and London

The University of Chicago Press, Chicago 60637
The University of Chicago Press, Ltd., London
© 2020 by The University of Chicago
Published 2020
Printed in the United States of America

29 28 27 26 25 24 23 22 21 20   1 2 3 4 5

ISBN-13: 978-0-226-67102-4 (cloth)
ISBN-13: 978-0-226-67116-1 (paper)
ISBN-13: 978-0-226-67133-8 (e-book)
DOI: https://doi.org/10.7208/chicago/9780226671338.001.0001

Library of Congress Cataloging-in-Publication Data

Names: Scheiner, Samuel M., 1956– editor. | Mindell, David P., editor.
Title: The theory of evolution : principles, concepts, and assumptions /
    [edited by] Samuel M. Scheiner, David P. Mindell.
Description: Chicago ; London : The University of Chicago Press, 2019. |
    Includes bibliographical references and index.
Identifiers: LCCN 2019024351 | ISBN 978022671024 (cloth) |
    ISBN 9780226671161 (paperback) | ISBN 9780226671338 (ebook)
Subjects: LCSH: Evolution (Biology)
Classification: LCC QH366.2 .T525 2019 | DDC 576.8—dc23
LC record available at https://lccn.loc.gov/2019024351

♾ This paper meets the requirements of ANSI/NISO Z39.48-1992
(Permanence of Paper).

CONTENTS

For all of my colleagues at the National Science Foundation
—Samuel M. Scheiner

For my family —David P. Mindell

CONTRIBUTORS

Joel Cracraft
  Department of Ornithology
  American Museum of Natural History

Scott V. Edwards
  Department of Organismic and Evolutionary Biology
  Harvard University

Patrick Forber
  Department of Philosophy
  Tufts University

Gordon A. Fox
  Department of Integrative Biology
  University of South Florida
  Department of Biology
  University of New Mexico

Steven A. Frank
  Department of Ecology and Evolutionary Biology
  University of California, Irvine

Rosemary G. Gillespie
  Department of Environmental Science, Policy, and Management
  University of California, Berkeley

Charles Goodnight
  Department of Biology
  University of Vermont

Robin Hopkins
    Department of Organismic and Evolutionary Biology
    Harvard University

David Jablonski
    Department of Geophysical Sciences
    University of Chicago

Maureen Kearney
    American Association for the Advancement of Science

Jun Y. Lim
    Department of Integrative Biology
    University of California, Berkeley

Alan C. Love
    Department of Philosophy
    University of Minnesota

James Mallet
    Department of Organismic and Evolutionary Biology
    Harvard University

James O. McInerney
    Faculty of Biology, Medicine, and Health
    University of Manchester
    School of Life Sciences
    University of Nottingham

David P. Mindell
    Museum of Vertebrate Zoology
    University of California

Marco J. Nathan
    Department of Philosophy
    University of Denver

Maria E. Orive
    Department of Ecology and Evolutionary Biology
    University of Kansas

Patrick C. Phillips
    Department of Biology
    University of Oregon

Timothée Poisot
    Département de Sciences Biologiques
    Université de Montréal

Andrew J. Rominger
    Department of Environmental Science, Policy, and Management
    University of California, Berkeley

Samuel M. Scheiner
    Arlington, VA

Vassiliki Betty Smocovitis
    Departments of Biology and History
    University of Florida

ONE

The Theory of Evolution

DAVID P. MINDELL AND SAMUEL M. SCHEINER

The Goals and Organization of This Book

The goal of this book is to explore and develop the current framework for evolutionary theory. Theory organizes our ideas for both research and teaching. It can help us see which questions are most important and how models and hypotheses can be constructed in seeking answers. We use the term framework to mean the principles, concepts, and assumptions inherent in evolutionary theory, and the logic employed in using them to advance ideas.

The chapters in this book address the current state of affairs across a range of fields within evolutionary biology, recognizing that such a book cannot possibly be comprehensive. The book is divided into two sections. The first section (chapters 2–8) examines topics that cross multiple specific theories. The second section (chapters 9–17) examines the strengths, weaknesses, and gaps in particular constitutive theories. Where appropriate, we asked the authors to discuss how their components of evolutionary theory integrate with other components and how a more explicit framework for evolutionary theory could change the thinking and approach of investigators.

Our discussion of evolutionary theory is in the context of a more general theory of biology that has five constituent theories: genetics, cells, organisms, ecology, and evolution (fig. 1.1, Scheiner 2010). A consequence of this approach is that some key aspects of evolution and the corresponding principles (tables 1.1–1.3) fall within other constitutive theories. For example, processes of genetic mutation and organismal development are in the theories of genetics and organisms, respectively, though they remain very much a part of evolution.

In chapter 1, we discuss the general theory of evolution in terms of its content and structure, including background on its general principles. We also discuss how the general theory of evolution relates to its constitutive theories and to other general theories within biology. Other chapters in the first section examine the historical, philosophical, and mathematical underpinnings of evolutionary theory (Smocovitis, chap. 2; Forber, chap. 3; Phillips, chap. 4; Love, chap. 8) or explore key evolutionary concepts such as homology (McInerney, chap. 5), species (Nathan and Cracraft, chap. 6), and higher-order relationships (Kearney, chap. 7). The second section considers nine constitutive theories. Those theories are not intended or claimed to provide comprehensive coverage of evolutionary biology and theory. Rather they are meant to be representative, including natural selection (Frank and Fox, chap. 9); multilevel selection (Goodnight, chap. 10); life history (Fox and Scheiner, chap. 11); specialization (Poisot, chap. 12); plasticity (Scheiner, chap. 13); sex (Orive, chap. 14); speciation (Edwards et al., chap. 15); biogeography (R. G. Gillespie et al., chap. 16); and macroevolution (Jablonski, chap. 17).

This book is intended for students and practitioners across the broad field of evolutionary biology. It should have particular appeal to those interested in identifying explicit components of evolutionary theory and how those components—the constitutive theories—function separately and in combination to provide evolutionary science with predictions and testable hypotheses. The past two decades have seen extensive growth and debate in evolutionary biology, and it is our hope that an examination of evolution's theoretical foundations will enhance future research and understanding.

The Role of Evolutionary Theory

Any claim that new research findings provide new or more general understanding is based on the theoretical framework within which that research is embedded. Take, for example, the descriptive study of mutation rates via mutation accumulation experiments. The motivation for studying mutation accumulation patterns in *Drosophila* (Keightley 1994; Keightley et al. 2009), *Arabidopsis* (Ossowski et al. 2010; Rutter et al. 2012), or *Caenorhabditis* (Denver et al. 2000; Estes et al. 2004) is that these data will also be relevant to other insects and plants or even mammals and fungi. It is because of the general principle of common descent that we can assume such relevance. The reason for wanting to measure mutation rates is that they are a basic parameter in many evolutionary models (Turelli 1984). By measuring those parameters in a disparate set of species, we ground our models in the empirical world and gain assurance that the parameter values are widely applicable. The utility of models is well known (e.g., Huson and Bryant 2006; Beaulieu et al. 2012; Servedio et al. 2014). But what about the utility of constitutive and general theories, the focus of this book? Constitutive theories are the workhorses of theory. They are the wellsprings of models, because they provide the conceptual definitions for model components, guide model development, and link seemingly disparate models and concepts in a clear framework.

Evolutionary theory is dynamic and responsive, incorporating new data about organisms and the environment as they become available. In this way, evolutionary theory plays an important role in its own improvement. Ideally, all research should make explicit reference to underlying theory or theories. Even better, the research should include an explanation of how its various findings will alter or support our understanding of those theories. Will the research improve model parameterization? Will it help shape the structure of models by demonstrating the existence of phenomena that the models fail to include? Will it force us to reconsider the origins of particular traits or lineages? Will it provide additional confirmation of a proposition in a constitutive theory or a general principle in a general theory? Will it test the relative importance or validity of alternative models or propositions? Will it make us rethink the meaning of a general principle or show that the list of principles needs to be substantially reconsidered? Although those last two activities are the ones that we hold up in class as the scientific ideal, rarely do we attempt to do those things in our day-to-day research. One reason is the lack of an explicit set of propositions and principles to which we can relate our research. Well articulated and mature theories let us make those connections.

Evolutionary theory also has powerful practical applications, ranging from agriculture to medicine, infectious disease studies, public health, and the adjudication of justice within the courts (Metzker et al. 2002; Mindell 2006; Losos et al. 2013; Hoberg and Brooks 2015). A classic example is the interplay of animal breeding and evolutionary theory. Darwin used the results of artificial selection both as a guide in developing the concept of natural selection and as proof of its efficacy. In turn, evolutionary models in the twentieth century gave us the breeder's equation that became a basis for designing artificial selection regimes for improving product yields of farm animals and plants. More recently, genetic algorithms (actually selection processes) are used in software development to allow the program to find the optimal set of instructions for a computational task. The programmers decide on some fitness function (e.g., minimizing computational time), some mutational process for altering the computer code, a selection process for choosing among variants, and then just let evolution happen (e.g., Soule and Foster 1998; Fortin et al. 2012; M.-A.Gardner et al. 2015). The development of genetic algorithms has been informed by models of biological evolution (Beyer and Schwefel 2002). They have even been used for evolving physical objects such as robots (e.g., Soule and Heckendorn 2013).

One of the most important uses of evolutionary theory may be in helping us understand the origin or origins of life on Earth and its possible existence elsewhere in the universe. If the theory of evolution presented here is truly a general theory, it should apply at all times and places. It should tell us something about how deterministic and random processes operated in the assemblage of abiotic molecules into the first living cells.

The Content of Evolutionary Theory

We differentiate between the content of evolutionary theory and its structure. Content includes the components of evolutionary theory such as concepts, facts, and assumptions (table 1.1), while structure includes the functional relationships among components, making a dynamic, working theory. Although the content and structure of any scientific theory are deeply entwined, making this general distinction promotes the utility of theory in research and understanding. One may disagree with part or all of the content or structure while still accepting other aspects. When we are explicit about both and their separate roles, debate over one aspect need not hinder development of the other. Both the content and the structure may be revised. For example, if we compare the list of general principles presented here with that of Scheiner (2010), we see that none have survived intact. Those changes were the result of scrutiny by a second person with a very different perspective. We hope that others will take what we present here, subject it to equal scrutiny, and possibly further refine what we have done.

Darwin's (1859) and Wallace's (1858) theory of natural selection, proposed as the mechanism for evolutionary change, was a watershed event in the development of evolutionary theory. It introduced and applied natural selection to all features of living creatures. Despite the limitations of the data then available, Darwin was correct in stipulating an important role for natural selection in effecting change among organisms. He was also right in saying that all life forms share common ancestry, that much of evolution is a population-level process, and that changes in frequencies of traits among individuals within populations are important in driving evolution.

Table 1.1.   The Content of Theory (from Pickett et al. 2007)

Component	Definition
Assumptions	Conditions or structures needed to build a theory or model
Concepts	Labeled regularities in phenomena
Confirmed generalizations	Condensations and abstractions from a body of facts that have been tested
Definitions	Conventions and prescriptions necessary for a theory or model to work with clarity
Domain	The scope in space, time, and phenomena addressed by a theory or model
Facts	Confirmable records of phenomena
Framework	Nested causal or logical structure of a theory or model
General principle	A concept or confirmed generalization that is a component of a general theory
Hypotheses	Testable statements derived from or representing various components of a theory or model
Laws	Conditional statements of relationship or causation, or statements of process that hold within a domain of discourse
Model	Conceptual construct that represents or simplifies the natural world
Translation modes	Procedures and concepts needed to move from the abstractions of a theory to the specifics of model, application, or test

However, much of today's evolutionary theory content was unknown to Darwin, including plate tectonics, particulate evolution of traits, chromosomes and DNA as the material of inheritance, processes of molecular replication and recombination, endosymbioses, lateral gene transfer, polyploidy, rarity or absence of inheritance of acquired characters, epigenetics, and the extent of rapid organismal change and species diversification. These topics and many others have been working their way into evolutionary theory, changing it and improving its correspondence with evolution in nature. The broad scope of evolution, its operation across billions of years and all aspects of life, including numerous levels of biological organization, from genes to clades and ecosystems, presents a significant challenge for those seeking to integrate new data and content into a general theory of evolution.

The Structure of Evolutionary Theory

The structure of evolutionary theory should organize its content in an intellectual framework that reflects operational relationships among the components (table 1.2). An analogous approach to ecological theory structure was provided by Scheiner and Willig (2008) and has recently been extended to other biological domains (Scheiner 2010; Zamer and Scheiner 2014).

Scientific theories and their components exist along a continuum from general to specific. Our structure incorporates that continuum by recognizing general theories, constitutive theories, and models. While some components play similar roles at different levels (e.g., domain definitions), others differ. For example, general principles at the level of a general theory primarily act as background assumptions for the constitutive theories. Propositions at the level of a constitutive theory serve as guidelines or rules for building models.

We emphasize that these categories (table 1.2) are denoted for convenience only; a particular theory can contain aspects of more than one category. Although these categories compose a logical hierarchy (models within constitutive theories within general theories), any given theory need not be solely nested within another. For example, a constitutive theory about the evolution of development (Love, chap. 8) may draw on general principles from other domains, such as the theory of organisms (Zamer and Scheiner 2014), so that it sits at the overlap between the domains of two general theories. The theory of evolution is presented here as one of

Table 1.2.   A Continuum of Types of Theories Including Their Components (from Scheiner and Willig 2011)

General Theory

Background: domain, assumptions, framework, definitions
General principles: concepts, confirmed generalizations
Outputs: constitutive theories

Constitutive Theory

Background: domain, assumptions, framework, definitions
Propositions: concepts, confirmed generalizations, laws
Outputs: models

Model

Background: domain, assumptions, framework, definitions, propositions
Construction: translation modes
Outputs: hypotheses
Tests: facts

Theory of Biology

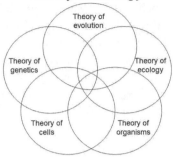

Figure 1.1. The theory of evolution, as conceived here, is contained within a more general theory of biology and intersects four other general theories of cells, organisms, genetics, and ecology (Scheiner 2010). Alternative scenarios, with differing relationships among the various general theories, are certainly possible. The intersections among the theories occur at two levels. First, the principles of each theory can make reference to the other theories, as when the principle "The characteristics of organismal lineages change over generations" from evolutionary theory (table 1.3) references variation in genetic "characteristics," which is a component of the theory of genetics. Correspondingly, a constitutive theory can include components from more than one general theory.

five general theories of biology (fig. 1.1)—and relies on the general principles of those theories to provide context and deeper meaning for its principles. Similarly, constitutive theories can have overlapping domains or be nested within other constitutive theories. The relationships and feedback among evolution's constitutive theories presented in this book are discussed below.

Alternative arrangements for the biological theories in figure 1.1, with differing relationships among them, are possible and valid depending on the intentions and explanations of investigators (Love 2013), in keeping with our pluralist approach as explained in the next section. The relationships (structure) among these theories and their components are created, by choice, to promote research and understanding; there is no single correct structure. One of us (DPM) could readily justify a partially hierarchical alternative to figure 1.1, with the general theory of evolution encompassing all four of the others, and ecology then encompassing the theories of genetics, cells, and organisms, which are arranged nonhierarchically, although the content of such theories would have to be altered from those of Scheiner (2010). Justification for this configuration stems from the fact that evolution and ecology are processes and that organisms, cells, and genomes are material entities subject to those processes. Evolution, in this structural conception, would be the most inclusive theory within biology with the most inclusive domain: the causes and consequences of change from the present back to the origin of life.

Our Philosophy and Approach

Designating an explicit structure for evolutionary theory, or any scientific theory, entails making some pragmatic decisions about the approach for evaluating theory and content. For biologists, these decisions are largely driven by the nature of the questions they ask and especially the data they analyze (Forber, chap. 3). Someone analyzing speciation processes in vertebrates using DNA may employ different methods and draw on different content than someone using organismal mate-choice behaviors. And the methods of someone analyzing speciation in asexual plants or fungi will differ from those of someone studying vertebrates.

The diversity of explanatory aims among evolutionary biologists, in seeking to understand evolution across all taxa, at multiple levels of biological organization, and as a result of many different causes, requires a pluralistic approach to evolutionary theory. In this context, pluralism means that there is no single correct definition of various concepts or configuration of models for universal application across the domain of evolution. For example, species are often defined or identified differently by those working on different organisms or organismal attributes (Nathan and Cracraft, chap. 6). Similarly, there is no single correct view of how evolutionary theory, and its components, should be structured or presented. Those features depend on the particular aims and methods of researchers. In addition, the general theory of evolution and its many constitutive theories draw on content from other biological theories (fig. 1.1), as is to be expected of any theory with diverse explanatory aims.

Of three generally recognized philosophical views regarding the structure of scientific theory, known as syntactic, semantic, and pragmatic (Winther 2015), our approach is most closely aligned with the pragmatic view. The syntactic and semantic approaches can be characterized as emphasizing, respectively, mathematical axioms and models for those mathematical axioms. The pragmatic approach, in contrast, combines components of both the syntactic and the semantic views with less formal, nonmathematical components including metaphor and analogy. Rice (2004) provides an example of the treatment of evolutionary theory from the standpoint of a semantic approach emphasizing models (as in population genetics and game theory); however, his is a narrow representation of the theory, missing important aspects of evolutionary biology related to paleobiology, systematics, speciation, evolutionary ecology, and biogeography. In describing the pragmatic view, Winther (2015) says, *"The emphasis is on internal diversity, and on the external pluralism of models and theories"* and *"As when studying an organism, the structure of theory cannot be understood independently of its* [the theory's] *history and function."* By incorporating a pluralistic and utilitarian view of scientific theories, the pragmatic view recognizes that scientists in different disciplines or subdisciplines may employ particular theories, models, and concepts in different contexts and with differences in meaning and intention.

The pragmatic approach is consistent with a pluralist and pragmatic structure for evolutionary theory in which theory content is organized according to the research questions being asked (Love 2010). Research questions often originate in different disciplines or beyond the domain of a given theory. Examples of relatively new research questions that originated in part in other domains are: How do developmental processes evolve? How do genomes evolve? A theory of the evolution of developmental processes would draw on components from the theory of organisms (Love, chap. 8), while a theory of genome evolution would draw on components of the theories of genetics and cells. Allowing flexibility of theory structure facilitates syntheses across disciplines and counters any tendency toward an overly reductionist or proscribed general theory, criticisms made of the modern synthesis (Gould 1977; Woese 2004).

Concepts and Metaphor

Concepts are key elements of theory and require careful description. This is particularly important when theory is being used to bridge disciplines or subdisciplines where connotation for the same concepts may differ. Within the domain of evolution, some concepts, like species, population, gene, exon, or specific morphological traits, are used to discretize units of evolution or targets of selection. Other concepts such as phylogeny, homology, and convergence correspond to relationships of one kind or another for taxa or traits. Many other concepts, such as developmental constraint, contingency, and adaptation, denote phenomena whose relative strength, frequency, causes, or consequences are being evaluated. Because of their central importance, and because of ongoing expansion in types of data and methods of analysis, the meaning and scope of evolutionary concepts need to be continually reassessed and revised.

Evolutionary theory has a long tradition of using metaphors as informal descriptors for its concepts. For example, phylogeneticists describe genealogies of both species and genes as trees, networks, webs, or radiations. Evolutionary biologists refer to adaptive landscapes and genome architectures, as well as conceiving of various phenomena as arms races, canalization, coalescence, conversion, cost/benefit ratios, drift, drive, explosions, fitness, hitchhiking, invasions, plasticity, random walks, silent substitutions, sweeps, and trade-offs. This pervasive and continuing use of metaphor in evolutionary theory is necessary, given the important role of metaphor in helping us think about complex ideas and mechanisms and turn those thoughts into concepts. Indeed, metaphors are not simply linguistic expressions—they are concepts themselves. Lakoff and Johnson (2008) make the case that conceptual metaphors are central to human cognition, grounded in our experiences and unavoidable. Metaphors can help us reason and synthesize linked phenomena into an overarching view of evolutionary process and pattern. For example, the metaphor of an arms race, whether between predator and prey or between pathogens and hosts, synthesizes processes of reproduction, heritability, mutation, selection, adaptation, co-evolution, and so on, all linked over time, into a single phrase drawn from the context of human warfare. However, because the brevity of metaphor invites imprecision and misunderstanding, its use requires constant vigilance.

The Domain of the Theory of Evolution

Evolutionary theory seeks to describe and explain the patterns, causes, and consequences of change in the characteristics and diversity of life across generations and geologic eras (table 1.3). Thus, the domain of the theory includes all of life's lineages and their characteristics at all levels of biological organization, from molecules and genes to morphology and behavior, for the entire 3.8 billion years of life's existence on Earth.

As part of placing evolutionary theory's domain within a broader theory of biology, we distinguish within-generational change from that which occurs across generations. Change that occurs within a generation is part of other domains, such as the developmental unfolding of an organism's

Table 1.3. The Domain and General Principles of the Theory of Evolution

Domain

Patterns, causes, and consequences of change in the characteristics and diversity of life across generations and geologic eras

General principles

1. The characteristics of organismal lineages change over generations.
2. Change in characteristics can lead to diversification of lineages.
3. All organisms and their lineages are linked through common descent.
4. Variation among individual organisms in genotype and phenotype is necessary for evolutionary change.
5. Variation arises from the genetic properties of organisms.
6. Evolutionary change is caused by deterministic and random processes.
7. Evolutionary processes depend on the properties of organisms.
8. Heterogeneity in the environment and evolutionary processes yield variable rates of evolutionary change.

form (theory of organisms) or mutational change in a genome (theory of genetics). Changes that occur within an individual organism's lifetime remain relevant to an individual's fitness and among-generation evolutionary dynamics, and we recognize that some versions of the theory of evolution include at least some of those types of changes within its domain (e.g., Hall 1999). We separate them here from across-generational change for epistemological and practical reasons. We could lump all of the general principles of the various general theories into a single theory of biology or a single theory of evolution. But doing so would obscure the domain-specific roles of a given set of principles in structuring and applying various constitutive theories and models.

In the next section, we identify eight general principles of the general theory of evolution and their context. The general principles are rooted in Darwin's *The Origin of Species* (1859) but expand his original formulation. They would be recognizable by the research community of the modern synthesis sixty years ago; however, the interpretation of some of those principles, as described below, has changed. The content of a theory evolves along with differences in the meanings given to various terms and concepts, and it is important that any theory be as explicit as possible about meanings. In keeping with a pragmatic approach to theory structure, the principles as articulated in table 1.3 cannot be understood in isolation. A much fuller explanation is necessary to give meaning to the single sentence that embodies each principle. Those explanations help define the terms used and the range of phenomena being referenced.

Evolution's General Principles

1. Evolutionary Change Occurs over Generations

Describing and explaining change in the characteristics of organismal lineages is the central goal of evolution science and, thus, is the first general principle of its theory. As noted above, this principle refers to change across generations, both from one generation to the next and across geologic eras. Such changes are distinct from changes occurring within a single lifetime. The recognition of cross-generational change preceded Darwin (e.g., Lamarck 1809) and was one impetus for his development of a theory of evolution. As defined, these changes include those from nongenetic causes, including phenotypic plasticity or maternal and other environmental effects. Such changes are included because they affect phenotypes that then can affect fitness or other evolutionary phenomena.

2. Lineages Diversify

Darwin was motivated in developing evolutionary theory by the desire to explain not only incremental organismal change over time but also the diversity of all life. Thus, the second general principle is that change leads to diversification of lineages. In turn, diversification can lead to greater change in the characteristics of the newly independent lineages. Darwin expressed this principle as species giving rise to other species. This common formulation was articulated by the architects of the modern synthesis and is often used in textbooks and reviews of evolutionary biology and theory (including Scheiner 2010). We express the principle here, though, as a statement about lineages rather than species. Species concepts are variable and difficult to apply across all life forms. For broader relevance within the domain of evolution, especially across all life forms, we refer to lineage divergence.

By diversification we mean increases in both the number of and variation among independent lineages. By lineage we mean a genealogical sequence of individual organisms, species, or clades linked by common descent (Hull 1980; Nathan and Cracraft, chap. 6). In this context, lineage is a more general term than species, making no assumptions about reproductive isolation or mechanisms of inheritance. This generality makes lineage more widely applicable across different life forms, but also requires additional description by investigators regarding the intended level of lineage inclusiveness in their particular research. For example, separately evolving lineages of populations are often considered species (e.g., G. G. Simpson

1961; de Queiroz 2007), and separate, closely related species lineages can compose a monophyletic group or clade. For the purposes of articulating a general principle, it is sufficient to recognize that distinct lineages exist and let exact definitions be operational in the context of specific models and taxa. The concept also applies to genes, as in gene genealogies, emphasizing the importance of clarity in using the term.

3. All Life Forms Share a Common Descent

Common descent for all life forms is now accepted as scientific fact, though a great deal of their diversity and their detailed phylogenetic relationships remain to be determined. Arguably, it was the discovery of the universality of the genetic code in the 1960s that provided the material basis for confirming this principle. Although common descent of living forms is a fact, the details of life's origin or origins from inanimate sources over 3.8 billion years ago are less certain (e.g., Oparin 1938; Pace 1991; Orgel 1998), as are the characteristics of the hypothetical last universal common ancestor of today's organisms (Penny and Poole 1999; Koonin 2003; Glansdorff et al. 2008). Common descent is at the heart of many evolutionary applications such as the use of *Arabidopsis* to discover genes important for crops, and the use of mouse, rat, and pig models in human medicine. This principle is a corollary of the second principle, as relatedness implies that lineage divergence occurred.

4. Variation in Genotypes and Phenotypes

The focus on variation was one of Darwin's great insights. Ernst Mayr (1982) noted that what distinguished Darwin's ideas from what had come before was a switch from typological to populational thinking. While Darwin would recognize this general principle as a component of his theory, he did not know about genetics. It was the modern synthesis that fully articulated the meaning of this principle (Smocovitis, chap. 2). Even today, however, what we mean by "genotype" and "phenotype" continues to change as new kinds of data are discovered and compared. Many kinds of changes at the molecular level that were previously unknown or unimagined now inform our understanding of evolutionary mechanisms, including changes in karyotypes and chromosome structure, gene duplication and loss, the movement of transposable elements, the gain of extrachromosomal elements, lateral gene transfer, regulation of gene expression, and epigenetic change. As with "lineage," we do not attempt to provide exact definitions of "genotype" and "phenotype"; rather we leave those definitions to be operationalized by constitutive theories and models. We also resist the temptation to "modernize" our terminology by adding "omics" anywhere.

5. Variation Arises from Genetics

This principle is grounded in one of the other general theories, the theory of genetics. It may seem odd to evolutionary biologists that the genetic properties of organisms are not placed immediately within evolutionary theory, but placement in the complementary theory of genetics yields a more tractable general theory of biology (Scheiner 2010). The theory of genetics defines the genetic properties of organisms and encompasses all of the different sources of heritable variation that arise from various kinds of genetic mutation, including events like chromosome, gene, and genome duplication. It also includes the effects of gene shuffling and recombination that can affect the phenotypic expression of genes through pleiotropic, dominance, and epistatic relationships. Constitutive theories within the domain of genetic theory provide information about rates for these processes, which may be critical for evolutionary models. Thus, many evolutionary models draw on principles from the domains of both general theories. Because of this tight coupling of the sources of variation with evolutionary change, principles about the sources of variation were historically included within the theory of evolution (e.g., Kutschera and Niklas 2004). We place these processes in their own domain so as to clearly separate those that occur within a single lifetime (e.g., the appearance of a new mutation) from those that change across generations (e.g., the increase in the frequency of that mutation).

6. Evolutionary Change Consists of Both
Deterministic and Random Processes

Evolutionary change stems from both deterministic and random factors. Random events can include the appearance of particular mutations in particular individuals, whereas more deterministic processes include aspects of natural selection, such as differential reproductive success linked to heritable differences among individuals. Selection, however, has random components (e.g., which of two identical individuals happens to get caught by a predator). Conversely, mutational processes can include deterministic components and constraints such as differences in mutation rates along a DNA sequence or the rates of different types of mutations.

The operation of both deterministic and random processes in evolutionary change has been recognized since Darwin's time. However, their relative importance has been the subject of some of the most heated debates in evolutionary biology (Mayr and Provine 1980). As with other general principles, we state this one generally, to acknowledge that all such processes can play a role while being agnostic about their relative importance (contra Scheiner 2010). Deterministic and random processes are context dependent and usually discussed at the level of constitutive theories (chaps. 9–17). Influences of the environment link this principle to the theory of ecology, and influences of organismal and cellular conditions link it to theories of cells and organisms.

This principle plays an important role in helping to make explicit the assumptions inherent in most models. Many evolutionary models are strictly deterministic or entail equilibrial conditions. However, those models often fail to acknowledge that they assume that random processes are either absent or minor relative to deterministic processes. Failing to acknowledge this assumption can result in applying that model to situations where the assumption does not hold, leading to incorrect inferences about the processes responsible for a given evolutionary pattern (e.g., J. H. Gillespie 1974).

7. Evolution Is Influenced by the Properties of Organisms

This general principle provides an acknowledgment of the significant constraints and influences on evolution that stem from the organisms themselves. For example, the rate and direction of evolutionary change can be constrained by genomes, developmental processes, and phenotypes acting as integrated systems. One of the complaints about many of the evolutionary models that came out of the modern synthesis (e.g., gene frequency models) was that they neglected organismal biology. Recent claims about the need for a new evolutionary theory (e.g., Pigliucci and Müller 2010b) often reduce to an assertion of this principle. While it may be that many aspects of organismal biology were not captured in those earlier models, it seems unlikely that Darwin and the architects of the synthesis would find problems with inclusion of such a principle. As with the previous principle, the relative importance of particular organismal properties (e.g., development) can be debated without doubting the validity of the principle. Thus we differ from those who claim that any current theory of evolution is a break from past theories. Rather, we see current efforts as expanding or shifting content at the level of constitutive theories, resulting in either the development of new constitutive theories (e.g., Love, chap. 8) or a new recognition of theories that previously had been considered of minor importance (e.g., Poisot, chap. 12; Futuyma 2015).

8. Rates of Change Are Heterogeneous

One key characteristic of evolutionary change is variation in rates of change of organismal characteristics and lineage diversification. Debates about those patterns of variation and the processes underlying them is long-standing. Darwin's version of the theory would have explicated a version of this principle about gradual rates of change. G. G. Simpson (1944), one of the architects of the modern synthesis, would have amended that to a statement about heterogeneity of rates. The theory of punctuated equilibrium (Eldredge and Gould 1972) would have further amended that statement, and debate about that amendment invigorated research on rates of evolution across levels of biological organization. As with the previous principles, in our version of the theory we simply acknowledge heterogeneity without indicating the relative distribution of rates. Those details remain at the levels of constitutive theories and models. Again, this is a reminder that any assumptions about constancy of processes or patterns should be made explicit.

Models

Models are an aspect of theory that formalizes relationships among entities. While this formalization can take many forms (e.g., Watson and Crick's ball and wire model of the structure of DNA), one of the most common is a mathematical equation. Such mathematical models formed the backbone of the modern synthesis (Phillips, chap. 4). Models exist within the context of specific constitutive theories. Those theories serve as guidance for model building and as nexus points for connecting apparently disparate models. The start of the modern synthesis is sometimes linked to the publication of Fisher (1918), in which he demonstrated how quantitative genetic models could be reconciled with Mendelian genetics (Provine 1971).

Alongside such analytic models is the growing importance of computer simulation. Simulation models allow the exploration of evolutionary scenarios that are analytically intractable, and increases in computer speed allow analysis of complex, high-dimensional problems. Another type of model that is central to evolutionary explanations is diagrams, including simulations, of phylogenetic relationships (Kearney, chap. 7). Phylogenies are often referred to as hypotheses, which is another way of saying that they are models. While relational models are not unique to evolution within biology, they play a large role here because evolutionary theory is often based on understanding patterns of descent. The development of phylogenetic models has also greatly benefited from the growth of computational power, which makes reliable phylogenetic hypotheses easier to build and test.

Evolutionary biologists use models in several ways. First, models provide heuristics. Much of the theory that came out of the modern synthesis was not directly connected to data because at that time it was not possible to measure many model parameters. Yet those models were influential in shaping conceptual understanding. For example, the way we think of evolution as changes in gene frequencies was largely based on the models of R. A. Fisher, Sewall Wright, and J. B. S. Haldane. As heuristics, models often serve to place boundaries on the potentialities of the world. For example, using a computer simulation model, Scheiner (2013a) explored the effects of various amounts and patterns of temporal heterogeneity on the evolution of phenotypic plasticity (Scheiner, chap. 13). The parameter space explored likely exceeded any found in the empirical world, but by showing general trends for when plasticity was favored and disfavored, the model provides general guidance for expectations of real systems. Servedio et al. (2014) refer to this use of mathematical models as providing proof-of-concept tests of verbal models.

Second, models provide predictions in at least two contexts. First, they can be used to solve a specific problem. For example, evolutionary models have been applied to fisheries management to show how setting minimal size requirements for a catch can result in selection for smaller, faster-developing adults (Kuparinen and Merilä 2007). Or they have shown how minimizing the use of antibiotics can minimize the evolution of drug resistant pathogens (e.g., Atkins et al. 2013). The second context, more relevant for this book, is in theory testing. The general principles of evolutionary theory and the propositions of its constitutive theories must be grounded in the empirical world, and models can link theory to data. Phylogenetic models are often used in testing aspects of selection theory (Gharib and Robinson-Rechavi 2013; however, see Maddison and FitzJohn 2015) or hypothesized differences in rates of evolution for different traits or organismal lineages (G. J. Slater et al. 2012; Pyron and Burbrink 2013; Forber, chap. 3). As noted above in our descriptions of the general principles, central questions in evolution are often about the relative importance of different phenomena (e.g., positive selection versus genetic drift in shaping evolutionary change). In such cases, we are not attempting falsification in the sense that philosophers of science mean (Popper 1959). Instead, we are determining the likelihood that a specific process is important in a particular situation.

A third use for models is in providing parameters. For example, various likelihood and Bayesian models of DNA sequence change are used in building models of phylogeny, which, in turn, tell us about patterns of descent among species, rates of change for different characters, rates of speciation, and evolutionary processes such as natural selection and genetic drift. Models denoting patterns of relationship, increasingly accompanied by estimates of time since divergence events, are valuable in estimating rates of character evolution and speciation. The results of one set of models can become the input for others that are used for theory testing, applied predictions, and general heuristics.

Theory Growth

Evolutionary theory is growing as new datasets, methods, and ideas are explored and integrated with existing theory (Smocovitis, chap. 2). Larger and different datasets from more taxa can yield greater understanding of the mechanisms influencing evolutionary change with a potential decrease in reliance on statistical correlation alone. Questions can be addressed now, at new levels of resolution, that could not be addressed ten years ago. Multiple causation can be better discerned, and with the advent of more knowledge about phenomena such as development, epigenetics, and mutational variation, traditional population genetic and quantitative genetic models can be made more mechanistic, with less reliance on gene-centric and selection-centric explanations. Usually the integration of new knowledge is relatively easy; however, occasionally it is disruptive of prevailing views. These situations, if lasting, generate great interest in the growth of evolutionary theory and can take many years or even decades to resolve (e.g., Tax 1960a; Grene and Depew 2004; Smocovitis 2005; Rose and Oakley 2007; M. A. Bell et al. 2010; Pigliucci and Müller 2010b; P. Bateson 2014; Futuyma 2015). For example, one of the oldest and most general debates within evolutionary theory concerns the relative contributions of adaptation via positive selection and chance events as explanations for observed features and their variation among organisms (e.g., ENCODE Project Consortium 2012; Doolittle 2013; Graur et al. 2013).

Efforts to expand, reconfigure, and move beyond the modern synthesis have been going on for decades (e.g., Gould 1977; Endler 1986; West-Eberhard 2003; Pigliucci and Müller 2010b; Gilbert et al. 2012; Jablonka et al. 2014). In a recent review, Laland et al. (2015) make a case for an expanded perspective on the workings of natural selection, in which organisms are seen as significantly influencing their own evolution. One of the ways this can happen is via developmental bias, in which some phenotypic variants are more likely to arise in the course of development than others (Arthur 2004; Davidson and Erwin 2006; G. P. Wagner and Draghi 2010). Another way is through inclusive inheritance, which accommodates cultural, epigenetic, physiological, and ecological inheritance (Cavalli-Sforza and Feldman 1981; Danchin et al. 2011; Jablonka et al. 2014). A third is habitat construction, in which organisms modify their environments, as with nest, burrow, or dam building by vertebrates and nutrient cycling by fungi and bacteria, as this provides a guiding influence on selection of their own lineages (Lewontin 1983; Odling-Smee et al. 2003). All of these effects call for more complex, dynamic evolutionary models, and it is unclear whether such models can fit within current theories of adaptation (e.g., Goodnight, chap. 10; Poisot, chap. 12; Orive, chap. 14) or whether new theories are needed (Love, chap. 8).

Another area of recent growth within evolutionary theory is the study and elucidation of reticulate evolution. Reticulate evolution denotes patterns of inheritance of traits that are not acquired directly from parents. Rather, they are acquired from different species or lineages. Reticulation includes horizontal or lateral gene transfer (LGT), hybridization among species, and endosymbiosis in which an initial invasion or ingestion of one individual by another leads over time to symbiosis and genetic integration of two disparate lineages. As an indication of the importance of reticulate evolution, over 80 percent of all bacterial and archaeal genes are estimated to have been laterally transferred among distinct lineages at some time in the past (Dagan et al. 2008), and endosymbioses are widely understood as key events in the evolution of eukaryotic cells, mitochondria, and chloroplasts (Margulis 1991; Gontier 2015; Koonin 2015). Reticulation events indicate coevolutionary processes, and there is increasing appreciation for the importance and frequency of coevolution and symbioses over time. This is particularly the case for the assembly and maintenance of diverse communities of microbial organisms that colonize and sustain multicellular eukaryotes, including humans (e.g., Gilbert et al. 2012; Douglas 2014). While models of pairwise coevolution are well established (J. N. Thompson 2005), we currently lack models for the simultaneous coevolution of many species that may have a complex set of positive and negative interactions. It is worth noting that LGT, endosymbioses, cultural inheritance, and aspects of immune system evolution and development (e.g., Koonin and Wolf 2009; Bhaya et al. 2011) entail inheritance of acquired character-istics, demonstrating the potential for evolution to be significantly influ-enced by inheritance beyond that of parental genes.

Recognizing that host-symbiont systems (holobionts) and their associ-ated genomes (hologenomes) can be studied as integrated units (Margu-lis 1991; Mindell 1992; E. Rosenberg and Zilber-Rosenberg 2013; Gilbert 2014) broadens our conception of what we recognize as individuals and provides new avenues for investigating life's evolution (Zilber-Rosenberg and Rosenberg 2008; McFall-Ngai et al. 2013; Bordenstein and Theis 2015). The extent of coevolution and shared selection pressure, if any, for holobionts and hologenomes remains to be determined (Moran and Sloan 2015). Theis et al. (2016) note that coevolution and selection on holobi-onts and hologenomes need not be assumed and that multilevel selection (Goodnight, chap. 10) accommodates selection across levels of symbionts, hosts, holobionts, and hologenomes (Shapira 2016). Doolittle and Inkpen (2018) even suggest that natural selection may act on the collective func-tion of microbial communities. While divergence leading to speciation has long been considered the primary force driving organismal diversification, the relative importance of reticulation among lineages in driving diversifi-cation is less well known.

Systematics, including both phylogenetics and taxonomy, has helped to expand our understanding of many aspects of evolution in recent years. This has been accomplished by applying phylogeny-based inferences about genealogy, homology, lineage divergence dates, and divergence rates to studies of evolutionary processes such as selection, speciation, and lateral gene transfer (table 1.4). We mention systematics here as emblematic of

Table 1.4. **Kinds of Phylogenetic Inference and Their Application to Understanding the Process and Theory of Evolution**

Kinds of phylogenetic inference	Applications to evolutionary processes
Genealogy for lineages of organisms (including fossils) and genes	Understanding speciation, selection, constraint, convergence, parallelism, LGT, endosymbiosis, hybridization, genome evolution
Relationships of homology for traits	Understanding origins of evolutionary novelty, adaptation
Absolute dates for nodes in phylogeny	Determining age and timing of diversification and extinction for lineages and traits
Rates of lineage diversification	Correlating diversification rates with events of earth history and environmental and ecological change

recent growth within the body of evolutionary theory for three reasons. First, theory and models for phylogenetic analyses (Kearney, chap. 7) have become increasingly sophisticated in distinguishing phylogenetic signal from noise, as evident in current software packages (e.g., Drummond et al. 2012; Höhna et al. 2014; Stamatakis 2014; Maddison and Maddison 2015; Darriba et al. 2018). Second, ongoing discovery of taxa and phylogenetic analyses are substantively enhancing our understanding of the patterns of life's common descent, including reticulation. Third, incorporation of better phylogenetic information leads to a better understanding of evolutionary process, where the latter is based on comparisons across more than two lineages, as in studies of biogeography (Gillespie et al., chap. 16), speciation (Edwards et al., chap. 15), and macroevolution (Jablonski, chap. 17).

Linkages and Feedback among Constitutive Theories

The constitutive theories presented in this book are linked in a variety of ways. The theory of natural selection (Frank and Fox, chap. 9) is linked to many of the other theories that also focus on selective processes. Theories of multilevel selection (Goodnight, chap. 10), life history evolution (Fox and Scheiner, chap. 11), the evolution of plasticity (Scheiner, chap. 13), and the evolution of sex (Orive, chap. 14) each focus on a specific type of trait, trait property, or population structure while still being primarily about natural selection. Each of those theories could be considered a subset of the theory of natural selection. However, recognizing each as a separate constitutive theory allows more specific propositions to be articulated while still acknowledging their linkages to the broader constitutive theory through the inclusion of propositions about those linkages (e.g., table 13.1, proposition 4; table 12.1, proposition 1.3).

Random events as a topic are not traditionally considered in a single theory. However, elements of chance are present in many or most constitutive theories, in a manner similar to selection. Any theories invoking selection on variation in genotype and phenotype, such as theories of multilevel selection (Goodnight, chap. 10), and life history evolution (Fox and Scheiner, chap. 11), are linked to chance events involved in the origins and distribution of that variation. Biotic and abiotic environmental changes involved in theories of biogeography (Gillespie et al., chap. 16) and macroevolution (Jablonski, chap. 17) also include an element of chance, such as which species colonize an island first or which lineage escapes a mass extinction event. Constitutive theories considering processes of genetic mutation and drift, which involve random events, fall within the theory of genetics (fig. 1.1). Constitutive theories considering development (Love, chap. 8), which may also entail chance events in getting from genotype to phenotype, fall within the theory of organisms and more specifically the subtheory of multicellular organisms (Zamer and Scheiner 2014).

The constitutive theories also feed back upon and provide meaning to the general theory. For example, the theory of the evolution of developmental systems (Love, chap. 8) informs us about the ways in which organismal properties affect evolutionary processes (general principle 7). Speciation theory (Edwards et al., chap. 15) provides definition to our concept of independently evolving lineages (general principle 2). Biogeography theory (Gillespie et al., chap. 16) and macroevolution theory (Jablonski, chap. 17) expand our consideration of the types of events that cause evolutionary change (general principle 6).

Acknowledgments

We thank Rob Pennock and Mike Willig for valuable comments on an earlier draft of this chapter. This chapter is based on work done by DPM and SMS while serving at the US National Science Foundation and by DPM at the University of California, Berkeley. The views expressed in this chapter do not necessarily reflect those of the National Science Foundation or the United States Government.

PART I

Overarching Issues

TWO

Historicizing the Synthesis

Critical Insights and Pivotal Moments in the
Long History of Evolutionary Theory

VASSILIKI BETTY SMOCOVITIS

Elsewhere, I have examined how evolutionary theory as well as practice (a distinction often blurred in the everyday language of evolutionary science) worked together in a number of biological disciplines to lead to the emergence of a new science and discipline of evolutionary biology (Smocovitis 1994, 1996, 1999). This new science was accompanied by standard disciplinary apparatus including an agreed-upon set of problems and a community infrastructure composed of professional societies and journals, textbooks and courses of instruction, and rituals of celebration and commemoration. Here, I offer a brief narrative focusing on pivotal moments in the long history of evolutionary biology, conveying a sense of its conceptual richness and its heterogeneity, as well as the many twists and turns in what has become an increasingly complex narrative. I then close with some critical remarks about the conceptual status of the field, recent attempts to extend or expand its domains of inquiry, and the need for a more pluralistic account, one that is properly interdisciplinary and integrative and that takes into account not just a range of organismal systems and differing methodologies but also differing views of the history, philosophy, and sociology of science.

From our present vantage point, this is a difficult task. We know a great deal about the history of evolutionary science, but some of this literature gives us a dissonant if not a contradictory series of perspectives. This chapter draws on multiple historical sources. Several involve works by Ernst Mayr, including *The Growth of Biological Thought* (1982) and the 1980 collection (revised with a new introduction in 1998), co-edited with historian William B. Provine, titled *The Evolutionary Synthesis: Perspectives on the Unification of Biology*. It comprised a disparate set of perspectives from a 1974 conference that showed the lack of agreement among scientists, historians, and philosophers about what constituted this event. Stephen J. Gould, in *The Structure of Evolutionary Theory* (2002), attempted to find the same points of agreement and had a hard time doing so. Other works that I have drawn on include Peter J. Bowler's *Evolution: The History of an Idea* (2009), the work of philosophers like Jean Gayon (1998), but especially the work of William B. Provine, including *The Origins of Theoretical Population Genetics* (1971) and his biography of Sewall Wright (1986).

In short, a great deal of work has been done to explore the history of evolutionary biology and its theoretical foundations; indeed, this is but a sampler of the abundant work to date that gives us a more nuanced understanding of the twists and turns in the development of the field. What have we actually learned from all this study, and what useful insights can we glean? Where can we hope evolutionary biology will take us, especially as calls for an extended or expanded synthesis of the field are being made? Are such reforms or amendments even necessary, and are we in actual need of some new "paradigm"?

Pivotal Moment I: The Long History of Evolutionary Theory and Charles Darwin

First and foremost, we must recall that as an area of inquiry, the history of evolutionary thought began well before 1959, the year of the "Darwin Centennial" that celebrated the centenary of the publication of Darwin's 1859 *On the Origin of Species* and the 150th anniversary of Darwin's birth and that followed the establishment of the new or "modern" synthesis of evolution between Mendelian genetics and Darwinian selection theory. That history includes hundreds, if not thousands, of workers contributing some small brick of knowledge to the growing edifice of understanding or to the rapidly changing design plan (a kind of intellectual *Bauplan*) of what became a complex science only in the middle decades of the twentieth century. This fact needs underscoring and makes for an important starting point for anyone seeking historical perspectives, but especially for all of us engaged in outreach or the teaching of evolution, because it is the first line of defense against evolution's many critics. The idea of evolution did not emerge like an intact unit particle from the brain of a proverbial Zeus or a scientific genius working in isolation, in this case Charles Darwin, as is sometimes thought or as is depicted in some biology textbooks. Instead, evolutionary thinking grew out of concerns with the problem of change, and especially organic change, beginning in antiquity and continuing through the Middle Ages and the early modern period, making an appearance in thinkers such as the misunderstood Charles Bonnet and his "great chain of being" and in the brilliant but oftentimes confused ramblings of naturalists like the comte de Buffon (1749, 1753). It is of course recognizable, though frequently depicted as wrong, in the writing of Jean-Baptiste Lamarck (1809), who remains one of the most underappreciated and least understood individuals in the history of science, but whose adaptationist and transformationist thinking undergirded early nineteenth-century views of organic change. Lamarck was but one of a score of individuals who held transformationist or transmutationist beliefs, meaning that he upheld a view of organic change, in contrast to a belief in the fixity of species. He was joined by others who wrote enormously popular works of natural history, including the sensational but much reviled anonymous author of the 1844 *Vestiges of the Natural History of Creation*, Robert Chambers, who along with a number of other individuals flirted with transformationism (Secord 2000). Still others not only engaged transmutation but also came up with precursor-like views of natural selection before Charles Darwin.

Darwin's own celebrated theory, which he called "descent with modification" primarily, though not solely, by means of natural selection, did not therefore emerge in a vacuum, nor did it suddenly spring up in an "aha" moment while its author was observing the finches of the Galapagos during his celebrated voyage on HMS *Beagle* (Sulloway 1982). Recent scholarship has given us a more nuanced and complex portrait of how Darwin's theoretical understanding of descent with modification was formulated, not only from his keen insights, but also as a result of serious engagement with the work of others including geologists and naturalists like Charles Lyell and Edward Grant, along with scores of naturalists he corresponded with all over the world, like Fritz Müller who worked on a staggering diversity of living forms and geographic contexts (Browne 1995, 2002; D. A. West 2016). It was with good reason, therefore, that it took over twenty years from its first rough incarnation to its published expression in the famous account of his theory in 1859. It took that long for him to sharpen his thinking, earn his credentials (hence his studies on the systematics of barnacles), and garner support as well as evidence to buttress a theory he rightly suspected would be controversial (watching the critical reception of *Vestiges* had made Darwin squirm). As is evident in the second half of his magnum opus, he also spent a considerable amount of time anticipating criticism and responding to it in an effective manner. Had Darwin not been nudged by the appearance of the younger Alfred Russel Wallace, an intrepid traveler and gifted naturalist, who independently formulated a view of species change resembling that of Darwin, he probably would not have published *On the Origin of Species*, which was intended to be merely an "abstract" of 490 pages of what he was ambitiously planning to be a multivolume work.

Some scholars rightly refer to the Darwin-Wallace theory of evolution, which is inclusive of Wallace's many contributions to evolutionary theory, especially his writings on island biogeography that are now informing contemporary understanding of conservation biology (Quammen 1996; Fichman 2004; Slotten 2004). "Evolution" as a term for the general theory of descent with modification was not used by either individual initially but gained currency shortly after the publication of Darwin's work in 1859, thanks to usage by Darwin's contemporaries like the social evolutionist Herbert Spencer and Darwin's notorious "bulldog," Thomas Henry Huxley. Only at the very end of *On the Origin* did the word *evolve* appear; until then the term *evolution*, which meant "unfolding" or "unrolling," was mostly associated with embryology or astronomy. Largely as a result of his keen insights, his many examples, and his rhetorical flourish in crafting *On the Origin of Species*, Darwin is now regarded as the giant of modern evolutionary theory, his name virtually synonymous with evolution by means of natural selection.

Darwin's many studies after 1859 were not, however, the whimsical forays of a Victorian naturalist reluctant to leave his backyard; they were designed to lend additional support for his theory. His work on orchids, for example, not only capitalized on the availability of the plants in an era obsessed with them but was also designed to examine the adaptive functions of the many parts or "contrivances" in a group known for its stunning morphological diversity (Edens-Meier and Bernhardt 2014; Endersby 2016). His attempt to apply his theory to humans and human social evolution in his 1871 *Descent of Man* was also a prolonged discussion of sexual selection, a kind of amendment to natural selection. (Darwin had only one cryptic reference to what his theory meant for humans in 1859.) And knowing that his theory of descent with modification rested on the shaky foundations of a theory of heredity he did not clearly understand, Darwin also embarked on an examination of the study of variation and its origins in his 1868 *The Variation of Animals and Plants under Domestication*. It resulted in his celebrated "provisional hypothesis of pangenesis," a theory of heredity that, while appropriate to his thinking, turned out to be dead wrong.

Darwin did not, therefore, have a viable theory of heredity, one that included a particulate form of inheritance as opposed to a blending. That remained one of a number of holes in his theory that fanned the fires of criticism from scientists. Nor did he have adequate direct support for natural selection; his examples in *On the Origin of Species*, for example, were either indirect or "imaginary" in nature. Thus, though Darwin—and others—convinced the scientific world that species change took place in some kind of orderly mechanistic fashion, they lacked a deep knowledge of the precise means by which that took place. By the end of the nineteenth century, a number of amendments, interpretive spins, and alternative theories had been proposed. So many of these extended, appropriated, reinterpreted, or distorted Darwin's views, especially as they applied to humans and to human social evolution, inventing entire areas of science and pseudoscience such as eugenics or racial theories of anthropology, that scholars have justly drawn the distinction between Darwin, Darwinism, and Darwinisticism in order to capture the permutations and distortion at the hands of scientists, philosophers, and theologians, as well as social and political theorists (Peckham 1959).

Pivotal Moment II: The "Eclipse of Darwin" and the "Modern Synthesis of Evolution"

The turn of the twentieth century was thus a historical moment characterized by widespread confusion, if not outright dissent, concerning evolution. It saw a proliferation of theories that accounted for evolution, including some that appeared to be non-Darwinian, or at times even anti-Darwinian; it was for this reason that Julian Huxley in 1942 designated this interval of time as the "eclipse" of Darwin (Bowler 1983). Such was the case with mutation theory (or *Mutationstheorie*) associated with the Dutch botanist Hugo de Vries (1903), which argued for rapid or saltationist species change due to mutation pressure, giving selection merely an eliminative instead of a creative role. This was in contrast to Darwin's postulated slower, gradual process of species change through natural selection of small, individual differences. Ironically, de Vries was also one of the three co-discoverers of Mendel's theory of heredity in 1900, along with Eric von Tschermak and Carl Correns, so that Mendelism, which supplied the particulate theory of heredity needed by classical Darwinian evolution, was instead seen to bolster the rival de Vriesian mutation theory. Equally important was the fact that mutation theory, and what came to be known as Mendelian genetics (W. Bateson 1902), was largely experimental in methodology, while Darwinism was seen as mostly descriptive and came to be associated with statistics and the school of biometricians (Provine 1971), who for methodological, intellectual, and personal reasons were at odds with Mendelians like William Bateson. Thus ensued the celebrated controversy between the Mendelians and the biometricians, one of the most counterproductive episodes in the history of science (Provine 1971).

Not until a number of the protagonists in the debate were dead, and not until natural selection, genetic drift, and mutation were seen as evolutionary variables by mathematicians who formulated what came to be called the "Hardy-Weinberg equilibrium principle" that determined the conditions under which there would be no evolutionary change, did a synthesis between Darwinian selection theory and Mendelian genetics begin to take place (Provine 1971). The work of mathematical population geneticists R. A. Fisher, Sewall Wright, and J. B. S. Haldane, in concert with the work of field-oriented naturalists-systematists such as E. B. Ford and Theodosius Dobzhansky, who were keen to understand the interplay of theory and practice in evolution, laid the foundations of what came to be called the synthetic theory of evolution or sometimes "neo-Darwinism." (The latter term is more appropriately reserved for late nineteenth-century theorizing by individuals such as George Romanes who meant it to mean Darwin without his Lamarckian leanings [Provine 1971].) Though there remained some differences of opinion about the relative importance of those variables within varied population structures, all agreed that Darwinian natural selection was the primary mechanism of evolution and that alternatives or rivals (such as *Mutationstheorie*, orthogenesis/aristogenesis or directed evolution, and neo-Lamarckism) were diminished if not eliminated outright; indeed, they became "alternative" once natural selection was established as the primary mechanism of evolution. This is why Provine (1988) justly referred to the "evolutionary constriction" that took place at this time when he attempted to find the precise core elements that defined the historical event.

Although historians and scientists have tended to favor the contributions of one or more individuals (especially depending on their field of interest), most agree that Theodosius Dobzhansky, a Ukrainian-born émigré to the United States, did the most to establish the modern synthesis of evolution, especially as embodied in his important book *Genetics and the Origin of Species* (1937b; Provine 1981, 1986). Associated with the prestigious Jesup Lectures at Columbia University, the book drew on a number of experimental studies Dobzhansky had conducted in natural populations of *Drosophila pseudoobscura* that had been informed by his celebrated collaboration with mathematical population geneticist Sewall Wright. He was predisposed to such collaborative ventures by his training in Russia, where he had been influenced by Russian naturalists-systematists who, following figures like Sergei Chetverikov, sought to assimilate insights from genetics into a populational approach to evolution (Adams 1968, 1980, 1990). Immigrating to the United States in 1927, Dobzhansky brought with him what was to remain of the legacy of those ecumenical approaches to genetics and evolution that were common in Russia but that were tragically eliminated under Stalin and Lysenko. That ecumenical, integrative vision was responsible for his synthesis of evolution in his 1937 book. As its title suggested, the book was meant to redress the absence of a theory of heredity in Darwin's original work and to explore the genetic basis of evolutionary change. It also opened inquiry into the mechanisms of speciation as well as species definitions left unresolved in Darwin's formulations of 1859 (Nathan and Cracraft, chap. 6; Edwards et al., chap. 15). Besides linking the new understanding of genetics with Darwinian selection theory, the book demonstrated how the mathematical theories of Fisher, Haldane, but especially Sewall Wright, Dobzhansky's long-standing collaborator, could be put to use in the field. As such, it was a foundational work in the new area of evolutionary genetics and is generally regarded as the single most influential book of twentieth-century evolution, functioning as a kind of textbook for a generation of workers (Provine 1986; Levine 1995).

Pivotal Moment III: The Convergence of Evolutionary Disciplines and the Emergence of Evolutionary Biology

The publication of *Genetics and the Origin of Species* in 1937 is thus regarded as one of the pivotal moments in the history of evolution, in part because of its synthesis between genetics and evolution and its functioning as a kind of bridge or "trading zone" (Galison 1997) between theory and practice, but also because it served as a kind of catalyst for the publication of a number of other books, associated with the Jesup Lectures at Columbia University, that brought a number of related disciplines to consensus, or at least appeared to bring a number of disciplines to conversation with each other (Mayr and Provine 1980). These other books included Ernst Mayr's (1942) *Systematics and the Origin of Species*, which brought systematics into the synthesis and set forth the modern-day biological species concept; George Gaylord Simpson's (1944) *Tempo and Mode in Evolution*, which brought the paleontological record, long thought to be problematic, in line with Darwinian evolution, and G. Ledyard Stebbins's (1950) *Variation and Evolution in Plants*, which offered a comprehensive synthesis of plant genetics, evolution, and the plant fossil record that was especially consistent with Dobzhansky's view of evolution.

These books were also broadly synthetic in themselves and drew heavily on a disparate body of literature that included original insights by additional figures who have too often been overlooked. These include the "Carnegie team" of Jens Clausen, David Keck, and William Hiesey (Clausen et al. 1940, 1948; Clausen and Hiesey 1958), who provided crucial experimental insights into distinguishing phenotype from genotype, thereby illuminating the origin and maintenance of variation, elucidating the action of natural selection, and dealing a body blow to neo-Lamarckism, especially in the plant world (Sandage et al. 2004). Also important was E. B. Babcock, the Berkeley-based plant geneticist who produced the first comprehensive phylogenetic treatment of any group, namely, the genus *Crepis*, which integrated methods and insights from genetics, cytology, systematics, biogeography, and fossil history (Smocovitis 2009). Yet another overlooked figure was Bernhard Rensch, the German systematist whose 1947 book *Neuere Probleme der Abstammungslehre* was translated into English only after the period of synthesis but whose insights into the pattern and process of speciation nonetheless were formative, shaping the thinking of systematists like Mayr, and whose conclusions on macroevolution paralleled those of Simpson to a remarkable degree (Rensch 1947; Mayr and Provine 1980).

But the most underappreciated if not misunderstood figure was perhaps Julian Huxley (Smocovitis 1996). His efforts at synthesis began with a number of organizational activities, especially in Britain, spearheading what came to be called the new systematics, which used genetics, ecology, and experimentation to solve taxonomic problems (Huxley 1940). His edited collection, and other such efforts to reform systematics so that it reflected evolutionary processes, began to focus on the patterns and process of speciation and helped to establish the biological species concept as articulated by Dobzhansky, Mayr, and others. His other notable contribution was his ambitious 1942 book *Evolution: The Modern Synthesis*. More than any other, this work heralded the new synthesis of Mendelian genetics and Darwinian selection theory and celebrated the emergence of a new unifying discipline of evolutionary biology within a political context that enabled a secular, progressivist, and liberal worldview (Huxley 1942). Though it drew some criticism from his peers who thought the work too general in nature as well as lacking in coherence, it was nonetheless widely read and very influential in intellectual circles at the time; it also gave the name "modern synthesis" to the new understanding of Darwinian selection theory that incorporated Mendelian genetics and attempted to understand the origins of biological diversity (Smocovitis 1996).

Though most of the key participants—later called "architects"—of the historical event known as the "evolutionary synthesis" were active in the United States or, like Huxley, Fisher, and Haldane, in Great Britain, they belonged to an international community as a whole (Mayr and Provine 1980). Some, like Dobzhansky, had bridged Russian and American perspectives, while others like Ernst Mayr, a German immigrant to the United States, brought with them insights from continental workers and gave the evolutionary synthesis an international perspective. Being world travelers, furthermore, many of the architects of the synthesis rapidly disseminated the view that a consensus had emerged. Theodosius Dobzhansky, for example, took his path-breaking insights and methodologies to Latin America, and especially Brazil, creating entire centers of evolutionary genetics with many students, research associates, and visitors (Glass 1980; Levine 1995; Araújo 2004), while Sewall Wright collaborated with Brazilian geneticist Warwick Estevam Kerr (Kerr and Wright 1954b, 1954a; Wright and Kerr 1954).

The growing consensus on common problems of evolution also fueled a number of organizational activities, starting as early as the mid-1930s when an informal group of workers in the San Francisco Bay area founded a group initially called the Linnaean Club to discuss the new insights coming from genetics and ecology and how they could inform systematics. Echoing Huxley's call for a new systematics that integrated approaches from genetics with systematics, they renamed themselves shortly thereafter the Biosystematists, a group that included every major student of evolution on the West Coast, among them Dobzhansky, whose field sites were in Mather, California, and who was a frequent visitor. In 1939 at a meeting of the American Association for the Advancement of Science, a wider community of workers came together to discuss new evolutionary perspectives in systematics; they called their organization the Society for the Study of Speciation. Soon after, in 1943, the National Research Council-backed Committee on Common Problems of Genetics and Paleontology formed in the Northeast, later changing its name to the Committee on Common Problems of Genetics, Paleontology, and Systematics at the request of Ernst Mayr, who was then at the American Museum of Natural History. During the war years when meetings and communication were especially challenging, members circulated mimeographed bulletins that included substantive queries and discussion on the special problems of evolution. Appropriately enough, the very last bulletin included formal notice from G. G. Simpson, newly returned from war service, that a field common to the disciplines of genetics, paleontology, and systematics had come into existence and was "beginning to be clearly defined" (G. G. Simpson 1944a). In 1946, shortly after the war, members of these groups came together in St. Louis, Missouri, to found the first formal scientific society for the study of evolutionary biology, the Society for the Study of Evolution, which sponsored the first international publication venue, the journal *Evolution* (Smocovitis 1994).

As expected, many of the same authors and architects of the synthesis also served as organizers and leaders of the new organization. Ernst Mayr, for example, was instrumental in laying down the initial infrastructure of the society and journal, serving as its first editor, while G. G. Simpson served as the first president. Most of the membership were aware that a new synthetic field had emerged at some point during the war years, but it was not until 1947, on the occasion of Princeton University's bicentennial, that it was officially recognized by an international assemblage of evolutionists who began to explicitly reidentify themselves as evolutionary biologists. The proceedings of that meeting, edited by Glenn L. Jepsen, G. G. Simpson, and Ernst Mayr, and titled *Genetics, Paleontology and Evolution*, made explicit reference to the new synthetic field of evolution (Jepson et al. 1949). In one well-known essay by geneticist Hermann J. Muller (1942), for example, reference was made to the "convergence of evolutionary disciplines." Drawing an analogy with an evolutionary convergence of types, Muller noted that there had been a convergence of evolutionary types between paleontologists and geneticists to form a synthetic type of evolutionist. Most important, what had made this possible was a "common ground of theory" that served to unite formerly disparate areas. According to Muller (1949), consensus—agreement—had been reached on these basic tenets:

1. Natural selection was the primary mechanism of evolutionary change.
2. It operated on the level of small, individual differences, making evolution a slow, gradual process.
3. The same processes that operated at lower levels (for example beneath the species) also accounted for higher-order phenomena; in other words, there was a continuum between microevolution and macroevolution.

Echoing Julian Huxley's 1942 insights, the new discipline of evolutionary biology indicated not only that a more unified science of evolution had emerged but that it served as the unifying principle of the whole of biology (hence the name "evolutionary biology").

Muller's brief paper is important for understanding the theoretical foundations of the modern synthesis. It is one of the few such brief and clear declarations of the core elements of what would eventually be known as "the synthetic theory of evolution," associated with the new synthetic type of evolutionist. These core elements were hardly new, revolutionary, or conceptually profound, however. Indeed, taking a broad historical view, what was notable in this "newer" synthetic theory was the absence of the alternative theories that were popular at the beginning of the twentieth century, once again supporting the view that the synthetic theory that emerged by the middle decades involved not so much building on any one strikingly new development as slowly reworking and eliminating rival or alternative theories. Provine's "evolutionary constriction" therefore confirms Muller's characterization of 1949, and little argument could be made for the establishment of any new, bold paradigm (Provine 1988). In many respects it was a genetically informed selectionist theory closely resembling Darwin's as originally stated in 1859. What was notable in the late 1940s, however, was the spirit of consensus and agreement that now existed, especially centering on the efficacy of natural selection as a primary mechanism of evolution, and a growing community of practitioners who redefined themselves as evolutionary biologists and who began to function as the unifiers of biology.

That agreement over the synthetic theory that emerged was not an end in itself, however. Only a few years later, developmental biologist Conrad Waddington pointed out the need to continue the process of synthesis so as to include embryology and the insights of embryologically inclined evolutionists such as Richard Goldschmidt (1940) and Ivan Schmalhausen (1949), who continued to pose big questions not answered in part because of the overwhelming dominance given to population genetics (Waddington 1953b). Echoing Waddington, J. B. S. Haldane (1953), himself one of the mathematical population geneticists, noted that the synthesis was but a kind of stage, a developmental "instar" in the path toward the maturation of the theory.

Pivotal Moment IV: The Watershed of the Darwin Centennial of 1959

Thus, the end of the war saw a number of developments leading to a new scientific discipline of research that culminated with the Princeton meetings of 1947. It was surely one of the pivotal moments in the long history of evolutionary thought, and though it may not necessarily have resulted in any revolutionary new insights, or integrated the whole of biology, the argument could be made that it was part of a unifying process that was bringing more and more disciplines into consensus with genetics, systematics, paleontology, and the botanical sciences.

That unifying process could be seen most explicitly in 1959 on the occasion of the hundredth anniversary of the publication of Darwin's magnum opus and the 150th anniversary of his birth. Until then a number of developments had been taking place within evolutionary studies to resolve other long-standing concerns, one of the most obvious being the integration of anthropology with evolutionary studies. Human evolution was barely included in Darwin's 1859 account, and though social evolution appeared in his 1871 *Descent of Man*, and though human evolution and especially human social evolution were of interest to some like R. A. Fisher, who were attempting extensions of the synthetic theory to humans in support of eugenics (Smocovitis 2012), it did not figure prominently until the period of the synthesis.

It fell to Dobzhansky to integrate human evolution into the newly emerging consensus. Drawing on the notion of the gene pool (a concept derived from his Russian mentors) as a replacement for race or racial type, Dobzhansky (1941) moved the study of humans away from the racial theories of anthropology that had dominated the study of human evolution toward a populational or genic view. It was a view of humanity (and species) that diminished, if not entirely eliminated, the crude essentialism that characterized the union of biology and anthropology, a view that had been actively opposed by Franz Boaz and his influential school of cultural anthropology, especially in the United States. Beginning in the early 1940s, a process of integration took place between biology and anthropology, thanks to the efforts of Dobzhansky working in collaboration with anthropologists like Sherwood Washburn and Ashley Montagu (Dobzhansky and Montagu 1947). What emerged by the next decade was a new kind of biological anthropology, a kind of revamped physical anthropology, that enabled the incorporation of evolutionary insights into human evolution without reinforcing a rigid biological concept of race (Washburn 1951; Haraway 1988; Armelagos 2008; Marks 2008; Smocovitis 2012).

By the 1959 Darwin centennial, a broad range of biologists and anthropologists were therefore drawn to evolutionary study. After the biochemical basis for the origins of life was established by the experiments of Stanley Miller and Harold Urey (1959), and after James Watson, Francis Crick, Rosalind Franklin, and Maurice Wilkins articulated the structure of the DNA molecule in the same year, an even larger number of people were also drawn. This was evident at the many formal events dedicated to evolution during the Darwin centennial. At the Cold Spring Harbor Symposium for the Darwin centennial, for example, some two hundred biologists, anthropologists, and physical scientists were in attendance. Organized by geneticist Milislav Demerec, the symposium took as its theme "Genetics and Darwinism in the Twentieth Century." It was clearly slanted toward geneticists but also included systematists, paleontologists, and anthropologists keen on biology like Ashley Montagu and the more controversial Carlton S. Coon, who persisted in his support of racial theories, though he was increasingly losing ground (Demerec 1960).

The emphasis at that event was mostly on genetics and the contributions of geneticists and mathematical theorists such as Fisher, Haldane, and especially Sewall Wright. This focus was not left unnoticed by Ernst Mayr, who challenged the growing historical narrative that was placing population genetics and theoretical developments at the core of evolutionary study in a paper titled "Where Are We?" (Mayr 1959). After giving a good share of the credit to genetics, he then took the opportunity of pointing out its limitations, describing the approach of mathematical population geneticists as mere bean-bag genetics for its lack of consideration of interactive gene effects and for ignoring the organism and its natural environment. Beyond the fact that this became a notorious critique of population genetics, it also led to a heated series of exchanges especially with Haldane, who took umbrage at Mayr's oversimplification and mischaracterization of the work (Haldane 1964; Dronamraju 2011; Rao and Nanjundiah 2011).

An even larger event took place at the University of Chicago at the end of the year to coincide with the date that Darwin's book was actually published. Thanksgiving Day brought a special note of thanks for Darwin from the community of evolutionists assembled there. Organized by Sol Tax (Tax 1960a, 1960b; Tax and Callender 1960; Smocovitis 1999), the five-day event was designed both to bring all of the scientific disciplines together in a spirit of unity made possible by the modern synthesis of evolution, and also to enable the process of unification to continue, especially bringing anthropology into the fold of evolutionary biology. It also served to open discussion on the evolution of mind, culture, and the physical world. The highlight of the formal events was five panels that featured the distinguished representatives offering summaries of their areas as a way of looking forward. In all five panels—"The Origin of Life," which brought chemistry, astronomy, and physics into the fold, "The Evolution of Life," "Man as an Organism," "The Evolution of Mind," and "Social and Cultural Evolution"—the centrality of evolution by means of natural selection was established. Indeed, what began to emerge from the very organization of the panels was a kind of unified, evolutionary cosmology—a modern progressive and secular worldview, locating the "human's place in nature" and dominated by a view of evolution.

For the most part consensus dominated the discussions, though few new insights were gleaned. Some critics, however, viewed the agreement among the many panel members and some of the celebrants as an unhealthy sign of a new growing orthodoxy (Goudge 1961; Smocovitis 1999). Many years later, while launching the critique of the adaptationist program, Stephen J. Gould highlighted the 1959 Darwin Centennial as the historical moment that saw the hardening of the synthesis around a selectionist core (Gould 1980, 1983; Tattersall 2000). But while consensus may have appeared to exist 1959, not all the areas represented in the panels were fully integrated. Disciplines like anthropology were only starting to be integrated into the synthetic theory (Smocovitis 1999; Tattersall 2000; Delisle 2007; Smocovitis 2012), as was embryology or developmental biology (Waddington 1953b; Amundson 2005). Individuals like George Ledyard Stebbins were moving into areas of developmental genetics, though this work was entirely new (Crawford and Smocovitis 2004). Even parts of paleontology remained unassimilated as shown by Everett Olson's dissent in the panels (Olson 1960).

Whatever the criticism or the deficits that remained, however, the Darwin Centennial revealed the extent to which the consensus that had emerged in the late 1940s continued well into the late 1950s. It served to draw together a staggering assortment of workers who agreed, for the most part, that evolution served as a unifying or organizing principle for the biological sciences, within an increasingly unified view of knowledge. Certainly, within the long history of evolutionary thought, the centennial represented a kind of watershed demarcating the long struggle to establish Darwin's theory of evolution by means of natural selection, as well as determining what the core disciplines informing the subject would be; but perhaps equally important the centennial celebrations with the many papers and conferences that ensued also revealed what remained to be integrated.

Pivotal Moment V: Biology in the Post-Sputnik Period and Challenges to the Synthetic Theory

Yet another measure of success of the evolutionary synthesis was the number of new members recruited to the study of evolution. Membership in the Society for the Study of Evolution spiked immediately after the Darwin centennial of 1959 and, despite a backlash from creationist movements galvanized by the lavish attention heaped on Darwin, continued to grow in the 1960s (Smocovitis 1994, 1999). Indeed, the study of evolution was officially made part of the US high school curriculum following efforts by some of the same architects to make it the core subject of a unified biology curriculum that became the Biological Sciences and Curriculum Study (Auffenberg 1959; Mayer 1986; Smocovitis 1996; Rudolph 2002). The curriculum gave rise to a series of popular—and very influential—textbooks of biology that drew on the expertise of some of the architects of the evolutionary synthesis.

The reorganization and fueling of federal research following the cold war, and the successful launching of Sputnik, the Soviet satellite, had energizing effects on the biological sciences as a whole. But by the late 1950s and early 1960s, a number of new sciences began to emerge and become institutionalized. They provided new opportunities, especially in methodologies and philosophies of science, but they also began to pose new challenges, especially to evolutionary biology.

The turn to molecular biology, especially in the late 1950s and early 1960s, for example, caused an initial tumult in the community of architects who were building and promoting evolutionary biology (Smocovitis 1996). Beyond the fact that it led to a kind of crude reductionism to physics and chemistry that threatened the very autonomy of biology, and to the celebrated "molecular wars" at places like Harvard (E. O. Wilson 1996), it also challenged some of the very foundations of the modern synthetic theory. Data from the new technique of polyacrilamide gel electrophoresis revealed that much more genetic variation existed in populations than had been thought (Hubby and Lewontin 1966; Lewontin and Hubby 1966; Lewontin 1974) or than could be accounted for by current theories. Subsequent work with protein sequence data suggested that a kind of molecular clock was operating in evolution since proteins were shown to evolve in a process roughly linear to the time of divergence between species (Zuckerkandl and Pauling 1965). The turmoil all came to a head in the late 1960s when Japanese geneticist Motoo Kimura noted that the observed rate of protein sequence evolution was much higher than selective deaths, casting doubt on the primacy of natural selection as a driver of evolutionary change (Kimura 1968); shortly thereafter, Jack L. King and Thomas Jukes (1969) declared the existence of a "non-Darwinian evolution." By 1983, Kimura (1983) had articulated the "neutral theory of evolution," arguing that the vast majority of variation seen at the molecular level in natural populations is the result of mutation and random genetic drift, while selection acted in a negative or eliminative capacity. In his book of 1983, *The Neutral Theory of Molecular Evolution*, Kimura even went so far as to claim that the synthetic theory had become so well entrenched or overdeveloped as to represent the orthodox view (Kimura 1983, especially chap. 2). Kimura additionally charged that advocates of the synthetic theory had been so critical of his views on molecular evolution that they had actually held back understanding of evolution at the molecular level. Debates between Kimura supporters, or neutralists, and those who favor a more positive and creative role for selection, especially at the phenotypic level of evolution, continue to the present with little resolution.

A far more heated debate that drew the attention of the public began in the 1970s with the appearance of two articles that introduced the theory of punctuated equilibrium (Eldredge and Gould 1972; Gould and Eldredge 1977). This involved a revision of the standard view of the rates of evolution (as especially established by G. G. Simpson) as being slow and gradual. Examining the fossil record, proponents of this theory instead saw long periods of stasis over geological time, punctuated by periods of rapid phenotypic evolution. The publication of these two papers led to a series of debates about the precise mechanisms of evolution that were responsible for such a pattern. In 1980, for example, Stephen Jay Gould suggested that such change might involve radical reorganization at the genic level, but this was rapidly countered with the view that stabilizing selection could also account for long periods of stasis (B. Charlesworth et al. 1982). Gould and others, however, continued to call for a new emphasis on such genic alterations and for exploration of the role played by developmental constraints. With a group of "young Turks" in paleobiology (a new area of research), furthermore, Gould and others continued to challenge the synthetic theory by calling for a decoupling of macro- from microevolution, postulating different mechanisms responsible at different levels of evolution (Smocovitis 1996; D. Sepkoski and Ruse 2009; D. Sepkoski 2012; Jablonski, chap. 17). Along with the critique of the adaptationist program, which he launched with R. C. Lewontin to confront selectionist rhetoric and its application in biology (Gould and Lewontin 1979), Gould was instrumental in launching a series of challenges to the synthetic theory, many of which garnered both negative and positive attention in the national press (Gould 2002). The extent to which this challenge altered the synthetic theory remains unclear and largely dependent on disciplinary background.

Other challenges, amendments, or revisions that precipitated debates, controversies, or prolonged discussions at the time included the application of nonequilibrium thermodynamics and information theory (Wiley and Brooks 1982, 1986) and Willi Hennig's systematic manifesto and its application (Hull 1988; Kearney, chap. 7). Even discussions of Lamarckism were revived under the guise of somatic selection (Steele 1981). Adding more fuel to the fires were debates over group selection, sociobiology, and selfish genes, which started in the late 1970s and peaked in the 1980s (E. O. Wilson 1975; Dawkins 1976; Goodnight, chap. 10). By the 1990s, long-standing concerns over developmental biology in animals, which had largely not been part of the formulation of the synthetic theory of the 1940s as Waddington had noted, began to take center stage. Concerns centering on deep homology, the recognition that distantly related organisms may share a kind of toolkit for developmental genes (McInerney, chap. 5), have proven especially challenging, but for the most part one may now commonly speak of "evo-devo" and even "eco-evo-devo," terms for fields that integrate these insights from developmental genetics and even ecology into the synthetic theory of evolution (Sultan 2007; Futuyma 2013; Gilbert 2013; Love, chap. 8).

Such optimism toward the integration of development with what the proponents term the standard evolutionary theory is not shared by all researchers, however. Building on what those proponents perceive as new ideas, new phenomena being studied, and new fields of inquiry that include niche construction and ecology (Poisot, chap. 12), epigenetic inheritance, genomics and network theory, plasticity and accommodation (Scheiner, chap. 13), modularity and evolvability, among other things, calls continue to be made for an "overhaul" or an extension of the synthesis (Müller 2007; Pigliucci 2009; Pigliucci and Müller 2010b; Laland et al. 2015), or even an expansion of the synthesis that may both widen and thicken the domains of inquiry (Kutschera and Niklas 2004; S. B. Carroll 2008). The extent to which some of these arguments trace their point of origin to, or mirror arguments made by, Stephen J. Gould and others in the early 1980s, or point to the restrictive or even hegemonic use of the notion of a standard theory and its exclusion of areas such as development, is currently the topic of active debate. Such challenges to the synthetic theory are once more receiving a great deal of attention, mirroring similar discussions in the 1980s (see Zimmer 2016 for one popular account of the recent debate). Other researchers strongly maintain that population genetics must remain the core area of interest in evolutionary biology (M. Lynch 2007), or that the synthesis itself is an evolving theory that has been and continues to be sufficiently accommodating to these newer insights, phenomena, or areas of research (Wray et al. 2014; Futuyma 2017; Gupta et al. 2017; Svensson 2018).

Historicizing the Synthesis: Analysis and Closing Thoughts

In this chapter, I have tried to capture the richer, more nuanced, more dissonant twists and turns in the history of evolutionary thinking, at times even demythologizing existing textbook narratives. I have highlighted particular moments in that history that I view as pivotal, in that they represent a series of benchmarks in the history of the field. They do have an arbitrary component since history is continuous, ongoing, and weblike; it is in fact not as well structured or defined as I, and others, have made it out to be (such periodization and narrative flattening are the foibles of the historian). My goal has been to capture the sense of the long history of evolution and the many settings and contexts against which conceptual advances have been made in order to stress the provisional and contingent nature of our understanding of both the history and the theory of evolution, despite all the many insights we may have attained.

What can we learn from such historical reckoning? First, it may help shed light on some contemporary debates especially pertaining to the conceptual structure of the standard evolutionary theory and whether or not it excluded individuals or entire fields of inquiry, such as developmental biology. Indeed, at heart in nearly all of the literature for or against some of these contemporary debates centering on the extension of the synthetic theory or what has been called the "extended evolutionary synthesis" is the question of whether or not the standard evolutionary theory or the modern synthesis has exerted a constraining, restrictive, or even exclusionary hegemonic role. In other words, has it actually been more of a hindrance than a help to some fields of inquiry? The historical record, I believe, can help us with this. As shown here and elsewhere (Smocovitis 1996), there was little in the way of a monolithic, standard theory in the late 1940s or even the late 1950s during the period of the Darwin centennial when some complained of a new orthodoxy. What did exist was an emerging consensus that Darwinian selection theory combined with Mendelian genetics within a populational view of evolution could explain the origins of biological diversity, a consensus that accompanied the elimination of alternative mechanisms responsible for evolutionary change and one that enabled a conversation between a number of domains of scientific inquiry previously thought to be incompatible. This agreement enabled a community of researchers (a scientific discipline, if you will) to come together around common problems in evolution and redefine themselves as evolutionary biologists. Their shared belief was that evolution unified biology within a unified theory of knowledge, or at least they shared a common goal of unifying evolution within a unified biology. If they existed at all, the core elements of the theory (core itself being a philosophical belief) as articulated by individuals such as Hermann J. Muller were minimal, hardly revolutionary, and probably not agreed upon by all individuals keen to align themselves with this new evolutionary biology, Richard Goldschmidt being one of the best-known dissidents. There were of course noisy people, some of whom grandstanded for a particular point of view, in the process stressing their own contribution or discipline, but at the disciplinary level such perspectives were distinctly their own. That is why the collection edited by Ernst Mayr and William B. Provine titled *The Evolutionary Synthesis: Perspectives on the Unification of Biology* (1980) provided only a set of perspectives on the unification of biology. There was no one unified perspective, nor even any real shared agreement on what happened during the synthesis period, what key discoveries or insights had been made, or even who may have played a central role. Stunningly, even Mayr (1993) backpedaled on his own belief in the unification of biology, once it became apparent to him what precisely that entailed.

As noted in that 1980 volume and elsewhere, disciplines that may appear to have been "left out" of the synthesis of the 1930s and 1940s, such as anthropology, were in fact already becoming integrated by 1950, one of the end points of the historic event. This happened, however, only after they had undergone significant amendment so as to redefine race in terms of the populational nonessentialized gene pool. This integration was also true, though perhaps less successful or visible, for developmental biology, as calls were made to integrate it within the newer discipline of evolutionary biology within only a few years of the Princeton conference of 1947. Even a mathematical population geneticist like J. B. S. Haldane recognized an ongoing process of development in the theory, which needed to continue incorporating areas such as development biology; and let us recall that Haldane was such an avid proponent of population genetics that he was duking it out with Mayr in the notorious "bean-bag" dispute in the late 1950s. Mayr himself was vociferous as well as persistent about the centrality of systematics in evolutionary theory and about the importance of his area to the wider evolutionary synthesis as a historical event. He coined the term "evolutionary synthesis" as a way to distinguish the wider synthesis of interest to historians from the original "modern synthesis" coined by Huxley, and he created the periodization that ended in 1947, leaving out the traditional last book of the synthesis by the botanist G. Ledyard Stebbins. (An earlier account by Edgar Anderson who partnered with Mayr to produce the "viewpoint of a botanist" never made it to print.) Lesser known by proponents of the extended evolutionary synthesis, who share mostly zoological, philosophical, and continental theoretical commitments, is that by 1959 or so, right at the time that the Darwin centennial with its "new" orthodoxy was about to be held, plant evolutionary biologists such as G. Ledyard Stebbins recognized that the gene-to-character transformation was the next great step in evolutionary biology (see especially his popular article in *American Scientist*, Stebbins 1965). His own research program from that time until his retirement drew on the insights of Conrad Waddington (1953b) and John Tyler Bonner (1952) to understand development in plants.

My point here is that even a brief historical reckoning gives us a sense of an enormously complex period in a very long history that defies any easy characterization. The actual points of agreement were minimal and involved so many diverse tools and methods, research traditions, organismic systems, and different levels of analysis that only a multivocal, pluralistic view of the synthesis—and the synthetic theory—does it any justice. Thus, the view of the modern synthesis or the synthetic theory, as described by advocates of the extended evolutionary synthesis, fundamentally paints a very simplistic picture, one that is dominated by philosophical or theoretical commitments instead of the historical or sociological commitments that are also needed to get at a richer and more nuanced understanding.

Finally, it may also be useful to distinguish the evolutionary synthesis, which refers to a historical event, from evolutionary theory, which is an undergirding explanatory framework (Mindell and Scheiner, chap. 1) and evolutionary biology, which refers to the scientific discipline of research concentrating on evolution that informs biology. Though they have notable overlap especially in the years between 1930 and 1950 during the period of synthesis, and are often used interchangeably, they each refer to a quite different aspect of the long history of evolutionary thinking. Given the breadth and the fundamental diversity of the areas it potentially unites, it seems to me that the question is not so much what the theory comprised or comprises, but what counts as evolutionary theory and to whom.

To conclude on a presentist note: even from a cursory historical examination of evolutionary biology from about 1980 to the present, it would appear that the synthetic theory has been and continues to be actively challenged from within, by evolutionary biologists themselves. In the 1980s this was so much the case that it appeared as though the synthetic theory had been dissolved, disassembled, or rendered inadequate, if not invalid. While some aided and abetted these views, others worked at amendment, revision, or reintegration; at present, what counts as the synthetic theory, its central tenets and first principles (e.g., Kutschera and Niklas 2004; Mindell and Scheiner, chap. 1, table 1.3), might appear to vary depending on field, experimental system, methodology, or generational and even individual preference, but it nonetheless has abundant instantiations, provides rich resources for further investigation, and retains a large measure of its explanatory power (see the constitutive theories in section 2 of this book). Its unifying properties are still upheld, especially by integrative biologists, the successors to the synthesis, and it has perhaps been most productive in extending its reach to areas like medicine and agriculture, computer science, and even historical and cultural modeling. It is for good reason that some speak of a theoretical versus an applied evolutionary biology. And as noted by at least several recent commentators, it might be useful to speak of multiple evolutionary theories that work productively for biological investigation instead of one monolithic evolutionary theory (Scheiner 2010; Love 2013; Mindell and Scheiner, chap. 1; Love, chap. 8).

Acknowledgments

I thank Alan Love, Erik Svensson, and Doug Futuyma for sharing critical insights into recent evolutionary theory as well as drafts of manuscripts in production, and especially Sam Scheiner for guidance, encouragement, and for editorial assistance on this manuscript.

THREE

Philosophy of Evolutionary Theory

Risky Inferences of Process from Pattern

PATRICK FORBER

Evolutionary theory provides the resources for reconstructing the deep past and identifying the complex processes that explain how, as Darwin (1859, 396) put it, "from so simple a beginning endless forms most beautiful and most wonderful have been, and are being, evolved." Yet gaining access to the deep past is particularly challenging—the very processes of evolution we aim to uncover can destroy the traces of the past in ways that hide their operation and confound the signals of ancestry. A primary task for the philosophy of evolutionary theory is evaluating the conceptual and evidential challenges that confront the application of the theory. Here, as an instance of this task, I compare two broad sets of methods that evolutionary biologists deploy to test hypotheses of natural selection: phenotypic tests of adaptive hypotheses versus molecular tests for selection. In the language of Scheiner (2010), the comparison involves looking at how precise models of evolution are articulated from the constitutive theories to detect the signatures of evolutionary change in real biological systems—such as the theories describing natural selection (Frank and Fox, chap. 9), drift, group selection (Goodnight, chap. 10), or life history evolution (Fox and Scheiner, chap. 11). When we take a closer look at the methods deployed to detect natural selection we uncover a striking contrast. Despite using the same constitutive theories, phenotypic and molecular methods carry out the process of articulation and testing in different ways. While phenotypic tests tend to focus on the morphological or behavioral features of organisms, and the developmental and ecological context of those structures and behaviors, molecular tests focus on comparative sequence data. The same evolutionary processes leave different signatures at the phenotypic and molecular levels.

These differences have philosophical implications for how we should understand evidence. They also support a general point, namely, that articulating our constitutive theories may proceed differently depending on features of the real biological systems we attempt to explain. I conclude by arguing that this flexibility contributes to a diversity that is a strength of thinking explicitly about the role of theory in evolution. The point of this chapter is not to convince empirical researchers to test theory, or to make the case that this philosophical perspective is necessary for this task, but instead to raise awareness that different methods can be developed and deployed to work in synergy together. In short, the point is to open minds to new possibilities.

To analyze the differences between phenotypic and molecular methods for testing selection, I focus on the variety of signatures that these tests aim to detect. After describing a broad contrast, I analyze an older and well-known test for selection, Tajima's D test, in detail. The analysis of Tajima's D test helps reveal the peculiar aspects of molecular tests, provides an illustration of the contrast with phenotypic tests, and motivates some philosophical claims about the nature of evidence.

Drawing out these implications involves focusing on ways scientists confront the problem of *testing holism* (Quine 1951; Duhem 1954). The problem concerns how testing depends on a network of theoretical commitments and complementary studies. While this problem is not explicitly discussed in the day-to-day practice of science, it is no mere philosophical worry. For instance, articulating a model of evolution to apply to Galapagos finches involves drawing upon multiple constitutive theories (e.g., Frank and Fox, chap. 9; Poisot, chap. 12; Gillespie et al., chap. 16) and a variety of background facts about finch ecology, development, and biogeography. No test of theory occurs in isolation. Scientists must decide which studies to trust, which models to use, how to design their experiments and studies to exclude confounding factors, and so on. All of these concerns involve mitigating the problem of testing holism. Even after the test results are in, the problem does not go away. When the test goes wrong, which component do we blame? Is our focal hypothesis mistaken that (say) finch beaks evolved in response to available food resources? Or is some other element of our model to blame? And when a test goes right, does the focal hypothesis count as confirmed? To formulate any test, and thus to obtain evidence for any hypothesis, we require what philosophers of science have (unhelpfully) termed "background theory"—the host of commitments, information, and theoretical bets necessary to bring some focal hypothesis in contact with the data. The contrast between phenotypic and molecular tests for selection brings out a difference in how this body of background theory interacts with the focal tests to provide evidence for selection. The tests face different risks, and recognizing this difference should inform how we think about the nature of evidence.

Signatures of Selection

I define a *signature* as an extant trace of a past causal process. A *signature of selection* is a trace left by a process of evolution by natural selection. Such traces must be detectable, at least in principle, for if they were systematically destroyed by subsequent evolutionary processes, they would cease to be signatures. Also, such traces must fit the model of evolution associated with a specific selection hypothesis, for selection can operate in a variety of ways, and different models of selection will predict different sorts of traces.

Evolution by natural selection occurs when (i) some trait varies; (ii) variation in the trait is responsible for difference in reproductive success (i.e., evolutionary fitness); and (iii) the trait is transmitted from parent to offspring to some degree. This recipe formulation of evolution by natural selection incorporates a number of idealizations, including a clear life cycle and well-behaved connections between parents and offspring. Owing to these idealizations, the recipe may not accurately capture all cases of natural selection, and there are subtle differences among the informal recipe and the various mathematical representations of selection, such as the Price equation or replicator dynamics (Godfrey-Smith 2009; Frank and Fox, chap. 9). Yet, complexities aside, the recipe provides a familiar model that is sufficient to identify the differences between phenotypic and molecular signatures. The recipe also illustrates a general feature of natural selection: Darwinian processes can occur within populations defined at various levels of the hierarchy, from molecules to cells, or organisms to groups (Lewontin 1970; Goodnight, chap. 10). While the process of natural selection operates the same way at these different levels, it tends to leave different signatures, mainly as a result of the nature of the entities involved—organisms versus sequences—and the possible evolutionary trajectories available to those different entities.

Let me impose some further conceptual clarity. A *detection method* attempts to identify and evaluate *potential signatures* of selection in the data, where a potential signature is some predicted trace of a possible selection process. The goal of these methods is to confirm the selection hypothesis over rival hypotheses that invoke different evolutionary factors, such as demographic fluctuations or constraints. The primary epistemic risk is that signatures can be *artifactual*; a trace may cohere with a hypothesis that posits natural selection yet have a different cause. A detection method, when successful, should sort the authentic from the artifactual signatures.

The contrast I draw below compares canonical phenotypic and molecular methods. Although both aim to draw inferences about evolution by natural selection, phenotypic detection tends to focus on ecological causes, whereas molecular detection tends to focus on molecular variation. This contrast is pragmatic and involves comparing the general tendency for phenotypic versus molecular detection methods to focus on different sorts of information. It is a claim about practice, not about principle, and there will be instances of overlap. Ideally, molecular detection methods will connect sequence variation to functional variation at the phenotypic level and vice versa. A large part of evolutionary science involves the continuing development and refinement of detection methods to increase their scope and resolution. So the epistemic ground is shifting, and shifting fast, in ways that may blur the contrast I draw here. Nevertheless, the contrast is valuable, for it reflects a deeper divergence of philosophical perspectives on the nature of evidence presumed by the detections methods, and thinking through this contrast will help clarify our understanding of how evolutionary science works.

Phenotypic Signatures

Adaptations, the traits of organisms that are the products of direct natural selection, count as the most striking phenotypic signatures of selection. Constraints and other evolutionary factors leave their signatures on phenotypes as well, with vestigial organs as clear examples of nonadaptive phenotypic signatures. The phenotypic detection methods are well-developed tests in organismic evolutionary biology for testing adaptive hypotheses. Some of the core methods involve ecological or laboratory manipulations to test the evolutionary response of a lineage (e.g., Endler 1980); optimality models to determine how well suited target features of an organism are for postulated environmental conditions (Maynard Smith 1978b); long-term studies of both organisms and their environments to track microevolutionary changes (Grant and Grant 2008); and comparative studies to investigate evolutionary patterns across a clade (Harvey and Pagel 1991). Despite this wide diversity, these methods share a focal point: identifying the trait-environment interactions responsible for fitness differences.

While this statement may be a truism, it is worth exploring how some of the diverse methods converge. Endler (1986, 164), in his influential review on detecting natural selection at the phenotypic level, claimed: "The most successful studies are those in which it is possible to assign a direct cause-effect relationship among an environmental factor, the organism's biology and ecology, and the trait of interest." In addition to evidence that selection has occurred, Brandon (1990, 167) argues that an "ideally complete adaptation explanation" must include evidence for the ecological (causal) basis of selection. Wade and Kalisz (1990, 1953) argue that we need to appreciate the role of variable environments for "understanding the functional interaction of the phenotype and the environment that *causes fitness.*" Hitchcock (1995) argues that selection hypotheses, in contrast to drift hypotheses, posit a causal capacity responsible for differences in reproductive success (see also Frank and Fox, chap. 9). Selection hypotheses predict evolutionary changes as a result of these causal features operating in the local ecology of the species.

Detection methods attempt to provide evidence, either directly by engaging ecology or indirectly by quantifying downstream evolutionary changes, for these causal features. The ecological manipulations disturb the causal network to test for the predicted response. Optimality models directly represent the causal link as an optimization problem in an idealized environment. Long-term studies track evolutionary changes over time to infer the ecological cause posited by the selection hypothesis, such as tracking the availability of seed resources and observing how trait distributions (e.g., beak size in finches) change in response over numerous generations. Comparative studies contrast the traits and environments of different but closely related species to detect whether selection produced the evolutionary differences among the lineages. The importance of the causal link between a trait and reproductive success has even been raised as a concern for comparative methods: Leroi et al. (1994) argue that such comparisons can detect only *selection of* some trait, not *selection for* that trait (Sober 1984).

In short, phenotypic detection focuses on component ii of evolution by natural selection: differences in fitness. Although not the primary focus, studies of phenotypic evolution must also provide some evidence for component iii. The reliable inheritance of phenotypic traits is a substantive empirical commitment of selection hypotheses, a commitment shared by many constraint and drift hypotheses. It is worth noting that the patterns of descent can be complicated by phenotypic plasticity, and models of plasticity should ideally be incorporated into and evaluated by phenotypic detection methods (Scheiner, chap. 13).

The challenge for these detection methods is the lack of access to historical information, especially regarding variation (component i), and fine-grained environmental changes. One reason for this lack is that natural selection tends to "cover its tracks": when a new adaptive trait or trait value appears in a population, it changes the distribution of phenotypes and eliminates some variation; the preservation of ancestral trait values is rare. One dimension of the debate over the relative significance of evolutionary constraints emerges from this lack of information about historically available variation. Indeed, constraints are often understood as limitations on available variation (Maynard Smith et al. 1985). Some of this information can be extracted from phylogenetic comparisons with sister groups, but whether this information is sufficient depends on the details of the evolutionary process. In some cases, where the order of evolutionary changes or selective environments matters, and we have a suitably rich clade to evaluate this, phylogenetic comparison can provide enough information to get a clear picture of ancestral trait distributions. However, this need not be the case, and often we lack sufficient comparisons, or the evolutionary process does not show the necessary sensitivity to ancestral states (Beatty 2006). Also, the ecological causes of evolution by natural selection are often very fine-grained (Wade and Kalisz 1990). Changes in an ephemeral physical environment can be rapid and leave little or no trace of earlier environmental conditions. Sometimes the necessary historical information is available, and other times we can make inferences about past variation and environments based on extant comparative data, but the primary weakness of these methods is a lack of access to this information. And to be clear: this is not an argument against using these methods, only a diagnosis of their strengths and weaknesses, one that will help identify the contrast with molecular methods.

Molecular Signatures

Molecular detection methods rely on sequence data (protein, RNA, DNA). Identifying adaptations at the molecular level is much trickier than identifying them at the phenotypic level, for functional effects of a particular sequence of nucleotides or amino acids are mediated by the complexity of development (Love, chap. 8). As a result, these methods do not aim to detect the ecological bases of fitness differences, but focus on classifying sequence variation in an attempt to identify patterns of substitution that are signatures of selection (or other evolutionary factors). Innovations in sequencing techniques have helped generate massive amounts of molecular sequence data. Arranging and quantifying molecular sequence variation in a way relevant to evolutionary investigation requires some processing. Information about evolutionary history cannot be extracted from just any sequence—to start, proper sequence alignment is essential. Let me focus on DNA (nucleotide) sequence data, specifically coding (gene) regions (see reviews by Bamshad and Wooding [2003], Nielsen [2005], and Vitti et al. [2013]); for the sake of brevity I leave aside genome-wide comparative tests.

Owing to the degeneracy of the genetic code, comparing coding regions yields a mix of synonymous changes (those that do not affect the primary amino acid sequence) and nonsynonymous changes (those that do), and such regions provide target data for many attempts to detect selection at the molecular level. Obtaining data suitable for evolutionary analysis requires sampling the DNA sequences, often thousands of bases long, from the same part of the genome. Evolutionary analyses often focus on a single identifiable gene or pseudogene extracted from a number of individual genomes within and often across species. Once such sequences are obtained, they must be aligned, usually based on an assumed reference sequence. Each site in the gene sequence needs to align correctly with the corresponding sites from the other sequences.

Sequence alignment presents difficulties because insertion and deletion mutations can change the reading frame for the DNA sequence, thereby affecting the analysis. The reading frame is determined by the start codon in the genetic code, and since codons are three base pairs long, shifting the reading frame by a single base pair dramatically changes the corresponding amino acid sequence. Misaligned sequences artificially inflate the estimate of segregating sites by making conserved sites look variable. Sequence alignment is a challenge, and this introduces another source of uncertainty for molecular tests, but one I will not pursue here. Once we have a sample of aligned sequences, we can apply molecular tests for selection in an attempt to extract historical information about the evolutionary process.

Although different detection methods classify and use molecular variation in different ways, all treat each segregating site, each base pair difference detected between any two sequences in the sample, as a different allele. Many molecular tests for selection assume an *infinite sites* model of sequence evolution—infinite because it treats every mutation as occurring at a different site to avoid the confounding problem of the exact same site changing twice by mutation (multiple hits). Counting the number of segregating sites in the data provides an indicator of evolutionary rates. Analyzing these rates, and variation in the rates, is the basis for many detection methods. These methods often contrast different selection models with the neutral model of molecular evolution. The neutral model predicts a constant rate and assumes that mutations fix only by random genetic drift

(Kimura 1983). Thus, early molecular tests used acceleration or deceleration in evolutionary rates in comparison to the neutral rate as potential signatures of selection.

Core molecular detection methods include rates tests (dN/dS ratio), McDonald-Kreitman (MK) tests, hitchhiking tests (selective sweeps versus background selection), and frequency spectra tests. Rate testing involves sorting detected changes in the sequence data into synonymous and nonsynonymous ones. These tests assume that the synonymous sites evolve at a neutral rate. A significantly faster/slower rate of evolution at nonsynonymous sites is a signature of positive/purifying selection. MK tests use sequences sampled from across a set of closely related species. By making a set of comparisons between sequences sampled from within the species and those sampled across species, the MK test can detect whether evolutionary rates have accelerated or decelerated in various parts of the phylogenetic tree. For example, in the MK test, the selective signature of substitution is that the divergence in sequences between species is greater than the sequence polymorphism or diversity within species. The application of the MK test to *Drosophila* detected such adaptive evolution on the *Adh* gene (J. H. McDonald and Kreitman 1991). Hitchhiking tests look at how rates of recombination interact with selection. Given that recombination is never complete, adaptive mutations are often linked to a number of nearby neutral sites. When the mutation fixes in a population, these neutral sites hitchhike along, leaving a detectable trace in the amount of variation observed in a sample of gene sequences. Background (purifying) selection, by conserving an important protein sequence, eliminates neutral variation linked to deleterious mutations and thus affects the amount of variation detectable in the sequence data (B. Charlesworth et al. 1993). Finally, frequency spectra tests work with polymorphism (within species) data and look at the frequency of segregating sites in a sample of sequences. I take a closer look at the mechanics of frequency spectra in the next section.

All of these methods of molecular detection aim to identify signatures of selection by analyzing extant sequence variation both within and across species. Given agreement on the appropriate model of sequence evolution—a substantial assumption—these comparisons provide a rich source of historical information on the branching pattern of the gene trees and the rates of evolution. Regarding the recipe for natural selection, molecular detection primarily focuses on component i: variation. While some variation is lost owing to the fixation of specific sites by selection and neutral evolution, new variation is generated by mutation. Since molecular variation is generated at a substantial rate, there is always some genetic variation in the population to be analyzed. Different processes leave different patterns in the extant variation. In contrast to standard phenotypic methods, molecular detection requires no evidence to support component iii, for the reliable inheritance of sequence variation is a robust empirical fact. High-fidelity inheritance comes built into the sequence evolution models.

The weak point for molecular methods concerns component ii, the fitness effects of molecular variation. Beyond a minimal amount of functional information necessary for sequence identification and alignment, these methods do not attempt to identify the causal or ecological bases for fitness differences among variants. This lack of ecological information creates a problem. Many tests use sequence comparisons to provide different estimates of the *neutral parameter* (θ). This parameter is a product of the mutation rate (μ) and the effective population size (N_e) such that $\theta = 4N_e\mu$. Disentangling demographic factors that affect N_e, such as fluctuating population size, unequal sex ratios, or nonrandom mating, requires independent information. This entanglement of mutation and demography in the neutral parameter is a problem for detecting selection because the tests for selection involve some kind of comparison to baseline neutral molecular evolution. As a result, molecular methods face the significant challenge of distinguishing authentic signatures of selection from artifactual signatures generated by demographic fluctuations or changing mutation rates.

The Frequency Spectrum

Frequency spectra can be obtained for DNA polymorphism data (gene sequences sampled from a single species) and are the basis for sophisticated tests of selection versus neutrality, including the popular Tajima's D test (Tajima 1989; see also Yang 2006, 262–64). I first describe what frequency spectra are, then discuss the details of Tajima's D test.

The frequency spectrum for a sample of gene sequences is a set of counts of segregating sites at specific frequencies in the sample, often represented with a histogram. More precisely, for a sample of n sequences the frequency spectrum is a discrete distribution of the number of segregating sites at frequencies $x_i = i/n$ for $i = 1, 2, \ldots , n\text{-}1$, with a defined expectation under neutrality (Nielsen 2005; Yang 2006). On the neutral hypothesis, we expect many mutations detected at particular sites to occur in just one sequence of the set of sampled sequences, while the other sequences in the sample will have the ancestral nucleotide at those sites. We expect fewer detected mutations to occur in two of the sampled sequences, and so on.

Tajima's D test attempts to identify signatures of selection in the frequency spectrum. The test involves comparing two different estimates of the neutral parameter ($\theta = 4N_e\mu$) that are each sensitive to different features of the frequency spectrum. If gene sequences undergo only neutral evolution, and a number of background assumptions are met (Yang 2006, 262), then the two estimates should be equal. Hence $D \approx 0$ provides evidence for neutrality. If selection occurs, then the two estimates should differ in predictable ways. Purifying or negative selection (elimination of deleterious mutations) increases the prevalence of low-frequency segregating sites and decreases the prevalence of high-frequency sites. Positive selection (fixation of adaptive mutations) has the converse effect. Given these predictions, a test result of $D < 0$ provides evidence for purifying selection, and $D > 0$ for positive selection. Remarkably, using only extant DNA polymorphism data, Tajima's D test can detect whether selection has occurred. (The original development of the test assumes no recombination, so selection on nearby genes can confound the inference that selection is operating on the target gene.)

Tajima's D test exhibits the trade-off common to molecular detection methods: the test incorporates virtually no ecological information and so cannot disentangle the effects of selection from the effects of demographic factors. One demographic factor, population size fluctuation, can produce artifactual signatures indistinguishable from the signatures of selection. Tajima's D test assumes that the population has reached mutation-drift equilibrium, so recent changes in population size violate this assumption. A rapid population expansion will increase the prevalence of low-frequency segregating sites, creating an artifactual signature of purifying selection, and a recent population bottleneck will increase the prevalence of high-frequency sites, creating an artifactual signature of positive selection. These factors can be disentangled only if we have independent information on demographic fluctuations in the evolving population or on the fitness effects of the allelic variation.

Evaluating the Contrast

Both phenotypic and molecular detection methods aim to confront the challenge of testing holism and bring theory into contact with data from real biological systems. From a philosophical perspective, the contrast between phenotypic and molecular methods for detecting differing signatures of the same evolutionary process is striking. I see two reasons for why

the contrast exists. First, molecular methods confront holism with an interesting innovation; and second, there is a difference in the space of possible
evolutionary trajectories available to organisms versus sequences.

Models of Sequence Evolution

All molecular tests must presume probabilities of mutation and a model
of how background evolution of sequences tends to occur. The signature
of selection (or other evolutionary processes) is often detected by identifying departures from the background model of sequence evolution. Tajima's
D test, as it was originally proposed, adopts a simple idealized model of
sequence evolution: a nucleotide-based model that assumes infinite sites
and a single mutation rate that governs all possible changes for the entire
sequence. The trouble with this simple model is that sequence evolution is
much more complicated, often in ways that can confound molecular tests
for selection. To illustrate how sharply complexity can increase for models
of sequence evolution, consider the Jukes and Cantor (1969) model. This
model treats all possible nucleotide changes as equiprobable. This is an obvious idealization, for there is clear evidence that the probabilities of various nucleotide changes can vary considerably. Transitions (A↔G or C↔T)
are more likely than transversions (A, G ↔ C, T), owing to the biochemical nature of the nucleotides. A sequence evolution model requires two
parameters to be capable of representing this difference (Kimura 1980).
More complex nucleotide models can have up to eleven parameters, treating each possible nucleotide change as having a different probability. There
can also be rate heterogeneity across the sequence (mutational hotspots)
and across evolutionary time. Still, even these complex models are relatively tractable, since sequence space is bounded by four possible states (A,
G, C, T) at each site.

A major innovation in molecular evolution involved the development
of codon-based models of sequence evolution (Yang 2006). These models
abandon the four possible states of nucleotide-based models, expanding
sequence space to include the 61 possible codons (64 minus the 3 stop codons). Sequences are treated as sequences of codons rather than of nucleotides, and each codon can potentially change into any other. There are some
empirical constraints. Mutations still occur at the nucleotide level. Changes
to synonymous codons are more likely than changes to nonsynonymous
codons for coding regions. So in addition to a transition-transversion bias,
codon-based models need to represent synonymous-nonsynonymous bias.
Also, some codon transitions require two or more nucleotide substitutions.

These are extremely unlikely in short time steps and so are often treated as having a transition probability of zero. Even so, these models can have significantly more parameters than nucleotide-based models. Also, codon biases can be generated in different ways, and these biases can evolve (Hershberg and Petrov 2008). Codon-based sequence evolution models provide the basis for the more sophisticated detection methods in the literature.

Thus, molecular detection involves a complex and important model selection decision to identify a signature of selection in a target set of sequence data. The decision can be made on the basis of background theory. We can stipulate a highly idealized model of sequence evolution, then estimate one or two key parameters from the molecular data. This is the strategy behind the original Tajima's D test. Alternatively, the model selection decision can be based on molecular data using model selection statistics (Burnham and Anderson 2002; Posada 2012). We can rely on the data to select a model and estimate the necessary parameters. Using the data in this way raises an interesting philosophical concern about whether using the data twice is legitimate. I will resolve this below, but first let me follow up the conjecture about possible evolutionary trajectories.

Morphospace versus Sequence Space

The phenotypic signature of selection is written in ecology, in the trait-environment relationship that is the basis for fitness differences. The detection methods focus on isolating and testing for the causal links posited by selection hypotheses. When looking at phenotypic evolution, we often lack historical information on the variation within the lineage and fine-grained ecological features. The molecular signature of selection is written in sequence variation. Comparing many long sequences enables us to extract historical information about ancestry and rates of evolution. These detection methods focus on analyzing and classifying DNA and protein sequences to identify patterns of change indicative of selection. When looking at molecular evolution, we often lack sufficient functional information to determine the fitness effects of substitutions and disentangle the effects of selection from demography.

This observation about practice does not preclude the integration of molecular and phenotypic evidence for selection. Plumbing the biochemical and genetic bases of traits provides a more robust case for selection on the phenotypic level. Consider, for example, work that has identified the genetic and developmental basis of finch beak morphology (Abzhanov et al. 2004; Abzhanov et al. 2006; Lamichhaney et al. 2016b). This research fills in the details of the causal network in a way that can forge a link between phenotypic features that are relevant to finch fitness and molecular variation. Conversely, building the causal web up from molecular variation to fitness effects can enrich the support for selection at the molecular level. Recent work attempts to resurrect ancestral proteins from reconstructed gene sequences to analyze their functional effects (Thornton 2004; Harms and Thornton 2010). So there may soon be many ways to connect and integrate phenotypic and molecular detection methods. Yet it is still useful to analyze the differences between methods in order to guide such integrative efforts, and even if integration is successful, there is one salient difference in the signatures that I suspect will remain.

To help explain the striking difference between phenotypic and molecular signatures, let me offer a conjecture using a spatial metaphor for evolution. Phenotypic detection tracks one to a few trajectories through *morphospace*, where morphospace encompasses all the possible phenotypic changes available to a species lineage. The dimensionality of morphospace depends on the ancestral suite of phenotypic characteristics. Constraints can also play a role in structuring morphospace. The dimensionality of morphospace is historically contingent and varies from lineage to lineage, so that either there is no general morphospace that we can apply across the tree of life, or if there is, the dimensionality will be extremely high with myriad possible evolutionary trajectories. Comparisons of trajectories through different lineage-specific morphospaces will be difficult at best. In contrast, molecular detection tracks a multitude of trajectories through *sequence space*, where sequence space is bounded by the number of possible molecular states available to a lineage. Sequence space does generalize across the tree of life. Also, the number of evolutionary trajectories at the molecular level is vast, potentially as many as there are sites in the target sequence. The degree of recombination affects the number of independent evolutionary trajectories that we can track. With free recombination, each site evolves independently, whereas with intermediate or no recombination, an entire sequence (or even the whole genome) constitutes the unit of analysis. Regardless of the degree of recombination, there are many more trajectories to track and compare on the molecular level. Further, sequence space is well-defined by sequence evolution models. Nucleotide-based models allow 4 states; codon-based models allow 61 states. These sequence evolution models also specify the transition rules between these states (with differing degrees of complexity) to a degree of precision impossible for phenotypic models. Sequence space also generalizes across lineages. These differences, in the nature of the evolutionary space and the number of trajectories being tracked in the space, illuminate why molecular methods are better at extracting historical information. Molecular methods have many more comparisons to use in reconstructing evolutionary events—we are able to extract more information from sequences than from phenotypes, though there are limits (Sober and Steel 2011, 2014).

The relative advantage of phenotypic methods, their closer contact with ecology, has a different explanation based on causal distance. The effects of phenotypic variation on fitness are causally proximate, hence identifying causal links between a trait and fitness is a tractable task for the phenotypic detection of selection. The causal distance between molecules and the ecological interactions responsible for fitness differences is often much greater. Also, the causal web, including development and the long path from molecular variation to reproductive success, can be much more complex. This makes incorporating functional information in molecular detection methods difficult, often (but not always) intractably so.

Given this analysis, there are two obvious strategies for improving molecular detection methods: pursue functional information by exploring gene effects, protein expression, conserved noncoding regions, and even ecological consequences of biochemical variation; or gather more sequence data and develop more nuanced statistical tests. Pursuing the former option fills an obvious gap in the molecular tests, for without this functional information we cannot know the precise molecular target of selection. While some advocate the latter approach owing to the challenge of pro viding this functional information (e.g., Eyre-Walker and Keightley 2007), continuing research is beginning to reveal more of the causal network connecting molecular variation to the ecological bases of evolutionary fitness.

On the Nature of Evidence

Philosophers of science have developed sophisticated views on evidence and have connected them to evolutionary biology. For instance, Sober (2008) provides a formal account of evidence using probability theory and applies his account to testing selection versus drift and hypotheses about common ancestry. E. A. Lloyd (1988, 2005) takes a different approach, focusing on careful case studies to evaluate evidential standards. Rather than wade directly into this rich and varied debate, I approach the nature of evidence from a different angle. Let me start by describing a standard perspective on how testing works in science, then use the contrast between phenotypic and molecular tests for selection to sketch a novel way to think about testing.

Call the standard perspective *predict and test*. This perspective sees the detection methods as a direct confrontation between the prediction of some hypothesis and data, and it is this confrontation that produces evidence for or against the hypotheses under consideration. As part of so-called "background theory," the set of constitutive theories we use to articulate precise models to test in specific biological systems plays a role by identifying the hypotheses we need to test and helps generate the precise predictions. This sort of testing protocol fits with the account of science as proceeding through a cycle of bold conjectures and ruthless refutations (Popper 1963). This perspective is illustrated by the method of "strong inference" (Platt 1964). Strong inference proceeds by "(1) Devising alternative hypotheses; (2) Devising a crucial experiment (or several of them) . . . (3) Carrying out the experiment so as to get a clean result; (1') Recycling the procedure" (Platt 1964, 347). While Platt focused on experimentally driven sciences, the procedure of strong inference can be easily generalized by revising (2) and (3) to focus on generating predictions and rigorously testing those predictions by experiment or observation.

How does this perspective fit testing in evolutionary biology? The phenotypic tests instantiate *predict and test* quite well. Consider Darwin's finches and the hypothesis that finch beak size and shape are recent adaptations for utilizing locally available seed resources (Grant and Grant 2008). Framing a test for this hypothesis involves using background information to make the predictions precise (Forber 2011). This information can include a phylogeny, features of finch development, patterns of inheritance, description of foraging behavior, and so on. A test of the hypothesis that finch beak size and shape evolved by natural selection must make assumptions about finch phylogeny (e.g., ancestral states) and beak development (e.g., size and shape are not severely constrained). Framing the test also involves determining the predictions of the competing hypotheses. The selection hypothesis for finch beaks may predict a strong correlation between finch beak phenotypic features and the environmental variables that represent the availability and kinds of seeds in the local environment. In contrast, a drift hypothesis may predict no correlation, and a constraint hypothesis may predict no change from the ancestral state. Once the problem is framed, biologists can then look to the data to see if their predictions are borne out.

The molecular tests for selection do not instantiate *predict and test* nearly as well. As in phenotypic detection methods, specific tests must be framed using background information. For molecular methods this set of information often includes consensus phylogenies and the details necessary to align sequences for comparison. But the resemblance ends here. Molecular detection methods immediately confront a model selection problem: to identify the signatures of selection, molecular methods must select a sequence evolution model and estimate the relevant parameters for that model. While it may have once been feasible to make this model selection decision on the basis of background theory, the development of more sophisticated sequence evolution models has increased the range of possible models and number of parameters. Instead, the decision is based on a computational analysis of the target sequence data (e.g., Abascal et al. 2005; Tamura et al. 2011). Only once a sequence evolution model is in place can patterns of substitutions detected in the sequence data be identified as signatures of selection.

To capture what is distinctive about molecular detection methods, consider an alternative perspective: *extract the pattern*. This perspective sees detection methods as trying to extract and classify the particular patterns in the data, and the competing hypotheses provide the appropriate classification scheme by identifying certain relevant features that should (or should not) be present in the data. Background theory helps frame the test by determining the relevant competing evolutionary hypotheses, but it no longer plays exactly the same role in generating predictions. Instead, the target data plays a role in articulation by informing the choice of a sequence evolution model. *Extract the pattern* is based on the inferential procedure seen in models of statistical pattern recognition (Hastie et al. 2001; Bishop 2006). These models aim to classify data into categories based on either a prior theoretical framework or statistical information gleaned from the data. The models usually deploy a complex learning algorithm to "learn" the classification scheme, and once sufficiently "trained," these models demonstrate remarkable success at classification.

These pattern recognition models have an intriguing feature that helps contrast *extract the pattern* from the standard *predict and test*. The key feature is the way the data is used to test hypotheses: statistical information is distilled from the sequence data in order to bring the evolutionary hypotheses into contact with that data. More than just "background theory" is necessary for the test to generate evidence. The test is designed to be tailored to a target data set. A striking example of the *extract the pattern* perspective in action is the N. A. Rosenberg et al. (2002) study of the genetic structure of human populations. The identification of this genetic structure is crucial to evaluating whether selection operates on specific human genes; the analysis helps isolate the structural features that affect the baseline neutral rate of evolution that is essential to molecular detection methods.

More generally, in molecular detection there is a classification scheme that identifies the various signatures of selection in the sequence data. The scheme is determined in part by background theory, which provides a set of sequence evolution models, and in part by statistical information gleaned from the data, which instantiates the parameter values for the sequence evolution model and guides the model selection decision. This allows the molecular test to be deployed to recognize patterns in the data, providing evidence for hypotheses of natural selection or neutral evolution. The exact patterns can shift across different sets of sequence data, and patterns that count as signatures of selection can be realized in a number of different ways. Consider again Tajima's D test. Many different sets of evolutionary trajectories, generated by all sorts of different ecological selection pressures, can produce the same frequency spectrum. Furthermore, many different frequency spectra yield the same D statistic for Tajima's D test. Yet the test classifies these signatures of selection in the same way.

The *extract the pattern* perspective captures a sensitivity of molecular detection methods to the target data set, specifically, the ability of molecular tests to identify the baseline rate of neutral evolution, a rate that can vary across sets of sequence data. This perspective also helps capture the difference in scope between molecular and phenotypic detection methods. Rather than make predictions about the evolutionary trajectories of one or a few traits, molecular evolutionary hypotheses classify a broad collection of evolutionary trajectories across numerous segregating sites as indicative of different processes of natural selection or neutral evolution. Detecting evolution through sequence space versus morphospace requires different sorts of methods, methods that instantiate different responses to the challenge of testing holism.

Let me address one last concern about *extract the pattern* that emerged earlier. When molecular tests *extract the pattern*, they use the target sequence data twice: once to select and fit the sequence evolution model and again to detect the signatures of selection in an effort to provide evidence for or against selection hypotheses. Is this maneuver a legitimate response to the problem of testing holism? Or does reusing the data in this way compromise the test by insulating the selection hypothesis from risk? While there is a risk of merely accommodating data, model selection statistics are designed to assuage this worry (Hitchcock and Sober 2004; Epstein and Forber 2013). Further, molecular evolution sequence data contains rich and nuanced information about the evolutionary past that can be extracted with careful comparisons. The molecular detection methods may use the same data set for model selection and detection, but that data set is used in different ways. Fitting the sequence evolution model involves estimating mutation rates and various biases, whereas detecting selection involves comparing patterns of substitution. In Tajima's D test, two different estimates of the same parameter (θ) based on the same data are compared. This is legitimate because of what we know about the sensitivity of the estimation procedures; we know that the comparison will be informative because each estimate tracks a different statistical feature of the frequency spectrum. The *extract the pattern* perspective is a legitimate response to testing holism. However, there may be specific cases where the data used for the model selection decision unduly biases the detection method. Investigating whether these cases exist is part of what makes the development of molecular detection methods a theoretically interesting and challenging enterprise.

To conclude, let me respond to a skeptical challenge about this sort of inquiry: what is gained by philosophical reflection on the nature of evidence and the different ways we test hypotheses of natural selection? There is a gap, created by the problem of testing holism, between our constitutive theories (e.g., chaps. 9–15 below) and the biological systems we aim to explain and understand. How we bridge this gap depends on more than just the constitutive theories. In this instance, the nature of our source data has led to the development of a diversity of methods for detecting the signatures of selection in the adapted phenotypes of organisms versus the molecular sequences of their genomes. The same theories can be articulated in different ways. My analysis shows that we should not privilege a single approach. It also shows that diversity can help diagnose the strengths and compromises involved in our detection methods. For example, in this case, the causal connections uncovered by phenotypic methods complement and reinforce the signatures of selection detected by molecular tests. Reflecting on the role of theory, and on the fact that the theory does not dictate the strategies for making connections to real systems, is a useful enterprise for any evolutionary biologist. In short, we make progress by formalizing our constitutive theories as well as by reflecting on the variety and diversity of connections we can forge to explain and understand the biological world.

Acknowledgments

This project has benefited from the insight and valuable feedback of many colleagues, including the editors and reviewers. I owe my thanks to all of them, but especially to Michael Dietrich for setting me on the path. Partial support for the project was provided by the Mellon Foundation New Directions Fellowship.

FOUR

Modeling Evolutionary Theories

PATRICK C. PHILLIPS

Not only is algebraic reasoning exact; it imposes an exactness on the verbal postulates made before algebra can start which is usually lacking in the first verbal formulations of scientific principles.

—J. B. S. Haldane (1964)

The mistake is in thinking that through mathematical formulae, you can arrive at the truth. That's wrong. I used the naturalist's way of thinking . . . I just used the empirical evidence. I find that this invariably gives you better figures. The problem is the belief that mathematics is the royal road to truth.

—Ernst Mayr (Shermer and Sulloway 2000)

Ecology and evolutionary biology are often considered to be the most theory-rich fields within biology, at least by their practitioners. Although "theory" can be defined in a number of different ways, in the present context it refers to "a framework or system of concepts and propositions that provides causal explanations of phenomena within a particular domain" (Scheiner and Willig 2008, 21). While some form of conceptual framework is needed to make sense of nearly any observation within the life sciences, ecology and evolution have distinguished themselves by an insistence on the translation of general theoretical constructs to well-defined quantitative models. Indeed, within these fields "doing theory" is largely synonymous with exploring a question using a mathematical model. Within evolutionary biology, the coupling of evolutionary ideas to genetic mechanisms frequently provides the underpinnings of such models. Here, I address the question of why there has been such a strong emphasis on quantitative description within evolutionary biology and explore how the field as a whole has shifted from a focus on general conceptual models to more specific data-driven models, with a particular emphasis on population genetic models. The issues explored in this chapter are relevant to how all of the constitutive theories in section 2 get translated into models and are used for hypothesis testing.

Quantitative versus Qualitative Models

If you define scientific models only as certain mathematical structures, then science is the practice that involves the use and application of these kinds of models. Some bite this bullet but there are troubling consequences for doing so: Darwin's work on the foundations of evolutionary theory proceeded without the use of a single mathematical model. As a result, this work cannot count as scientific on this model-based definition of science. Those who think that it was not until the New Synthesis, with its mathematical machinery, that evolutionary biology became a science may be comfortable with this conclusion but perhaps not with the conclusion that much of molecular biology is not science either.

—Stephen Downes (2011)

Within science, the term "model" can refer either to the abstract representation of specific natural phenomena or to the formalized realization and implementation of a specific theory (Frigg and Hartmann 2012). These uses are not mutually exclusive. Within evolutionary biology most theoreticians usually try to base their models on underlying biological processes (e.g., mutation, Mendelian segregation) but also use various levels of abstraction within a theoretical context, with the degree of abstraction being dependent on the specific application and level of detail needed to examine a given question. One common view of scientific theories is that they consist of a coherent set of models (Downes 1992). This viewpoint suggests that the value of a theory can be determined by how closely its models map onto the real world (aka "target systems," Downes 2011). For the view of theory as a collection of models to hold true, the definition of "model" must be very broad, including verbal or semantic descriptions of natural processes (Downes 1992; Giere 2004). For instance, most "models" in molecular biology consist of diagrams representing biochemical interactions, the structures of proteins within a cell during signal transduction, or some similar (somewhat loosely described) putative causal associations between molecular elements. While there are certainly quantitative models in molecular biology, such as those surrounding metabolic theory

(Keightley 1989), they have tended to play a much smaller role in the development of the field than qualitative models of ordering relationships amongst molecular processes. In contrast, over the past century a "model" within evolutionary biology has come to indicate a precise mathematical construct consisting of a system of equations built upon an explicit set of assumptions, particularly within population genetics. As the quotation from Downes (2011) above clearly points out, using this as the sole definition of "model" excludes much of what would be classified as good science in the rest of the life sciences. Indeed, Darwin's original statements about the role of natural selection in generating evolutionary change, while logical in structure, are decidedly nonquantitative. There are many other instances of nonquantitative models that have had a substantial influence on evolutionary thought, and some evolutionary biologists have even argued that overly simplified models can actually prevent progress in understanding complex evolutionary processes (Mayr 1982; de Winter 1997).

Why then has "model" become equivalent to "a coherent set of mathematical equations" within evolutionary biology while it maintains a much more general meaning within the rest of biology? Darwin's case is a fine illustration of why the progression from verbal to mathematical models has been important to the field. Darwin was able to articulate the general structure of natural selection fairly easily but quickly ran into trouble when confronted with the question of how variation could be maintained under his model of blending inheritance, since in that case all individuals within a population would rapidly come to resemble one another, leaving no variation upon which natural selection could act (Bulmer 2004). Thus, while Darwin's general conceptual model was sufficient to help catalyze an entirely new branch of science, lack of mechanistic detail led to immediate problems. Indeed, this problem was not immediately solved even after the rediscovery of Mendel in 1900, as there was still a great deal of disagreement about how best to reconcile the patterns of discrete variation described by Mendel with the more continuous variation that was at the center of Darwin's thinking (Provine 1971). It was not until R. A. Fisher (1918) worked out a model of multilocus inheritance that the issue was finally resolved.

The formal framework developed by Fisher was critical for resolving nearly two decades of disagreement in the field, and an explicit quantitative specification of assumptions was important. One of the chief proponents of the anti-Mendelian biometrical school, Karl Pearson (1904), had concluded that continuous variation was incompatible with Mendelian inheritance because it predicted a resemblance between relatives that was lower than empirical observations. However, his models assumed complete dominance and too few loci. The underlying difference in the conclusions of Pearson and Fisher regarding one of the most fundamental concepts in genetics, the quantitative mapping of continuous variation onto Mendelian segregation, cannot be adequately evaluated except in light of the clarity of the underlying assumptions—something possible only in the context of an explicit mathematical model. This is a quantitative detail (the dominance coefficient under Mendelian inheritance) rather than a qualitative assertion. In the end, all models dealing with evolution depend on quantitative details, such as the relative strength of processes such as selection and mutation, and so must ultimately be quantitative in nature. In contrast, most research in molecular biology has tended to be largely qualitative, and so disputes about model applicability have in turn usually been successfully addressed using qualitative rather than quantitative predictions.

Another example of the uneasy balance between verbal and quantitative models in evolutionary biology is the controversy regarding the speciation mechanisms that emerged in the mid-twentieth century. Following arguments initiated by Ernst Mayr, a number of verbal models about the role of small populations and genetic interactions in "founder effect" speciation generated a great deal of debate that played out over the course of nearly four decades (e.g., Mayr 1954; Carson and Templeton 1984). Under Mayr's verbal model, speciation was most likely to occur in peripheral populations with a small number of founders because these populations were more likely to generate unique combinations of strongly interacting genes via random genetic drift. Mayr's (1959) initial concern was that existing single-locus population genetic models as exemplified by the work of Fisher (1922) and Haldane (1924) could not capture the complex influences of changes in population size, interactions between genes, and changes in the pattern of selection under environmental shifts. Take the standard equation for single locus change under natural selection:

(4.1) $$p' = p W_A / \overline{W},$$

where p' is the frequency of allele A in the next generation, with marginal fitness W_A and an overall mean fitness in the population of $\overline{W}$. In essence, Mayr was asserting that although this model is internally consistent, it is not entirely relevant to complex evolutionary scenarios because no organism exists as a single locus running around in nature, and thus the strength and pattern of natural selection depend on all of the other genes simultaneously present within the organism.

Mayr also relied on a second bit of quantitative modeling from Sewall Wright, but this time in support of his verbal model. Wright (1931) expanded on the deterministic model from eq. 4.1 to include the effects of migration and, especially, genetic drift to identify the expected equilibrium distribution of possible evolutionary outcomes under stochastic variation in allele frequencies:

$$(4.2) \qquad \Phi(q) \, \alpha e^{4Nsq_m} q^{4N(mq_m+v)-1} (1-q)^{4N[m(1-q_m)+u]-1},$$

where N is population size, s is the strength of selection, m is the migration rate, u and v are mutation rates, and q_m is the average allele frequency across all subpopulations experiencing migration. Wright (1932) built upon this specific analytical approach for one and two loci to create what he viewed as a more general model (his shifting balance theory) that posited that genetic drift could occasionally, albeit rarely, move a population to a genetic state in which novel gene combinations could rise sufficiently high in frequency so as to be favored by natural selection in a new environment even if they had been selected against in the original population because of an incompatible genetic background. Here the important point for Mayr was that unique evolutionary outcomes are possible when genetic drift is included in the general selectionist framework and that the influence of genetic drift will be enhanced in small peripheral isolate populations, which is where Mayr (1954) thought the initial steps of speciation occurred. For Mayr, Wright's movement from a simple quantitative model to an intuitive verbal model that encompassed the entire shifting-balance process was not a problem (Coyne et al. 1997; Whitlock and Phillips 2000).

Countering the arguments of Mayr and others (Carson and Templeton 1984), theoreticians who were bringing more rigorous models to the study of speciation genetics asserted that the details of those verbal models were too vague to formally evaluate, much less allow one theory to be clearly distinguished from another (Barton and Charlesworth 1984). In reality, speciation is likely to be complex and idiosyncratic, although this does not mean that many pieces of the process cannot be mathematically modeled (Coyne and Orr 2004; Poisot, chap. 12; Edwards et al., chap. 15). More recent quantitative models that have tried to more fully capture the intent of verbal models have concluded that the processes envisioned by Mayr can indeed occur under certain circumstances (Gavrilets and Hastings 1996; Gavrilets 2004).

The debate over mechanisms of speciation illustrates one of the chief values in moving from verbal to quantitative models: explicit enumeration of the biological conditions and assumptions underlying the process under study. Without this, many disagreements in evolutionary biology become highly semantic. To theoreticians the worst outcome is a "slippery balance theory" in which a debate emerges around the meaning of specific terms used in a verbal model and whether two researchers are really talking about the same thing (e.g., "species," Nathan and Cracraft, chap. 6). Better to be quantitatively precise in what you mean by writing it down as an equation, they would say. However, as exemplified by his statement in the epigraph to this chapter, Mayr would likely counter that quantitative models can give a false sense of certainty as well. Although assumptions can be clearly stated, whether or not they are appropriate from a biological standpoint is an empirical question. And saying that something is an empirical question does not necessarily mean that it is possible to resolve via a reasonable set of measurements in actual populations—some model parameters are difficult to measure and some processes are so deeply historical that critical characteristics that might be needed to properly parameterize a model may have been irrevocably lost in time (Jablonski, chap. 17).

So at no point in the debate on the role of founder effects in speciation was there a sense that the models would be used to actually analyze data of real species, other than to see if the baseline conditions of gene interactions and variable population sizes exist in nature. There was no attempt to specify a set of parameters that could be precisely measured within a given group of species in order to "test" the theory of founder effect speciation. Here both qualitative and quantitative models were primarily used to define the scope of the problem, including which evolutionary forces need to be investigated, and to explore how the relationship among those evolutionary forces might influence the outcome of the process. In this way, most evolutionary models have been used primarily to formalize the conceptual structure of the theory being tested and, at least during the first few decades of research in evolutionary genetics, were unlikely to be used to examine specific empirical examples.

During the initial creation of modern evolutionary theory, the primary problems addressed by modelers were the definition and characterization of the factors that can influence evolutionary change (e.g., mutation, population size, recombination, inbreeding, selection, migration), followed by an exploration of how these factors interact with one another (Provine 1971; Lewontin 1974). Although there were intermittent attempts to apply these models to explain evolution within natural populations (Millstein 2007), in reality there were simply not the tools available to generate the data necessary to relate the parameters specified in those models to what was actually going on within natural populations. Thus, while there was certainly a large number of empirical studies being conducted in the first half of the twentieth century, the role of models in supporting those studies was largely to guide the conceptual framework in which the data was interpreted, rather than to test specific hypotheses generated by these models. Most of the development of evolutionary models during the foundational period of modern evolutionary biology has been pragmatic in approach (Mindell and Scheiner, chap. 1), with a strong focus on underlying constitutive theories within a loose alignment of the fundamental principles of evolution itself. Part of what the preceding examples help to illustrate is the difficulty of translating general theoretical precepts, such as those outlined in this book, into specific actionable predictions that have sufficient quantitative detail so as to be useful in either testing specific evolutionary hypotheses or estimating underlying evolutionary parameters.

Overall then, the primary role of models during the beginnings of evolutionary theory was to resolve ambiguities in verbal arguments about evolutionary processes by making assumptions and quantitative outcomes explicit. In the process of doing so, those models helped to clearly define the set of evolutionary forces that should come to bear under any particular circumstance. Repeated conflicts due to a lack of specificity in qualitative models, as well as an ever increasing sophistication of quantitative models, have led to a strong shift in expectations within the field such that "model" now means "quantitative model with an explicit set of assumptions." In turn, these quantitative models have helped to specify what properties of biological systems should be measured and when one should measure them, although the transition from conceptual foundations to more precise empirical estimates is a relatively recent development.

Models versus Analysis

It can scarcely be denied that the supreme goal of all theory is to make the irreducible basic elements as simple and as few as possible without having to surrender the adequate representation of a single datum of experience.

Usually quoted as: Everything should be made as simple as possible, but no simpler.

—Albert Einstein (1934, *165*)

One of the interesting consequences of the strong reliance of evolutionary biology on explicit quantitative models is that "theory paper" has come to mean almost any study that uses mathematical approaches not involving the statistical analysis of data. Although some type of model is usually the starting point for this kind of analysis, the scientific output of the analytical process can be very different depending on how the model is analyzed. In particular, some approaches are more likely to yield general insights into a model, while others may be more useful for specific applications to actual data. In this regard, it is useful to make a distinction between approximation, calculation, simulation, and estimation.

Consider models of multiple interacting loci involving multiple evolutionary forces, which are some of the most challenging in evolutionary biology such that general solutions are not possible in most instances. It is not particularly difficult to write down the transition equations for each separate step in the evolutionary process (mutation, natural selection, genetic recombination, migration, genetic drift, and so forth), but pulling those processes together tends to create a complete mess. For instance, a two-locus model of mutation-selection balance with epistatic interactions between the loci results in an equation with more than eighty terms (Phillips and Johnson 1998). Nonetheless, this final (very long) equation is an exactly correct representation of the full set of assumptions that go into the model. It can therefore be used for calculation of the behavior of the model. Given a set of initial conditions and parameter values, the future trajectory of the population can be calculated exactly in an iterative fashion. While every analytical model uses some kind of computational approach, I use the term "calculation" here to indicate that the result is repeatable and exact, as in "the dynamics of this model can be precisely calculated."

Such calculations can be used to explore sets of parameter values, but because an exhaustive search of all parameter combinations is usually impossible, it is often difficult to feel confident that one is deriving general insights about system behavior from calculations alone. Theoreticians therefore often attempt to find an approximation that captures the essential quantitative and qualitative properties of the models in as few terms as possible (in Einstein's sense of "few"). For certain types of epistatic interactions in the two-locus model, it is possible to approximate the eighty-term equation with a single result involving only a few fundamental parameters, resulting in a much simplified equation that clearly illustrates how epistasis can structure mutation-selection balance (Phillips and Johnson 1998). The incomprehensible full equation reduces to a more manageable

$$(4.3) \qquad x = \sqrt{\ \mu/r + \sqrt{\ \mu/s}}$$

for the case of complete epistatic coverage of one locus with another, where x is the frequency of the gamete that carries the epistatic deleterious alleles, μ is the mutation rate, r is the recombination rate between the loci, and s is the strength of selection against the negative epistatic combination. The advantage of an approximation like this is that one can more readily see both the complex interchange between the genetic and evolutionary processes (mutation, recombination, natural selection) and a connection to the standard single-locus mutation-selection result for completely recessive, deleterious alleles: $q = \sqrt{\ /s}$. When linkage between loci is not too tight, the two-locus result is approximately the square root of the single-locus result.

The term "approximation" as used here is very context specific. In a sense, every model is an approximation of reality, so here approximation simply means that the essence of the model itself can be captured using a simplified representation (i.e., it is not an exact representation of the underlying model assumptions). Thus, a given result may be a good approximation of the model but still a poor approximation of nature.

In many cases the goal might be not to describe any particular evolutionary trajectory, but instead to set bounds on the behavior of a model (e.g., under what conditions will the model result in a stable equilibrium). Here approximations can be used in a very different manner to simplify the model sufficiently so that local stability analysis can be applied (Otto and Day 2007). Such boundary analyses have long been an important part of theoretical physics but only recently have been an increasing focus for evolutionary models (de Vladar and Barton 2011), particularly within the subfield of adaptive dynamics (Doebeli and Ispolatov 2014).

Because every aspect of the evolutionary process is at least partially stochastic (Mindell and Scheiner, chap. 1, principle 6, table 1.3), single outcomes are possible only for models that assume that these processes do not result in a probability distribution. The inclusion of random processes is usually achieved using two different approaches. The first is to assume (or calculate) the distribution of possible outcomes, as is done for instance through the use of a diffusion equation (Kimura 1983). The second is to simulate a process by using a random number generator to determine the outcome of any particular realization of that process. Such simulations are often used to explore (or verify) the consequences of the assumptions that underlie purely equation-based descriptions of a model.

There is something of a moral hierarchy within evolutionary theory as to how these different approaches work together. Following Einstein's dictum, the ideal is to have a simple approximation that captures the essential features of the model in a way that readily yields insights into the biological process being modeled. Calculations can be used to show that the approximation actually captures the model as formulated, while simulations can be used to see if the model as a whole adequately captures the finer details of actual biology. If approximations are not possible, then calculations can be used to explore the parameter space, even if it may be difficult to articulate general rules from the resulting output. At the lowest level, the entire process can be simulated and the myriad of outcomes explored.

There are three reasons why simulations are generally a weak way of evaluating a model. First, it may be difficult to adequately explore the parameter space and/or perform an adequate number of replicates to be sure that consequences of the model are properly understood. The problem here is that what is "adequate" is usually unknowable a priori. Second, it is usually difficult to create simple rules that describe simulation results, unlike approximations. This difficulty tends to restrict simulations studies to the specifics of the conditions being simulated. The final problem with simulations is more pernicious. Because simulations are often complex and implemented via computer programing, unknown biases ("bugs") may be introduced by the way that the simulation was implemented, even if they were not explicitly stated. Although in principle it should be possible to examine the computer code itself to detect such errors, in practice such problems are likely to remain hidden until the results are found to be incompatible with some other form of analysis. In the best of worlds, all three approaches—calculation, approximation, and simulation—yield similar results such that one is confident that a model is both general and valid.

A final category of analysis comes about when actual data is analyzed in the context of a model. Because of variation inherent in biological systems, using a model to describe a particular dataset and/or using data to test a given model depends on estimating model parameters. This entails layering statistical sampling on top of the other processes described by the model. The estimation process thus follows the same hierarchy described above. Sometimes it is possible to articulate an exact solution to an estimation problem, for instance the maximum likelihood estimator of an allele frequency (Excoffier and Slatkin 1995). Often, however, calculations can be quite complex, and it is increasingly common for simulations to be used to calculate the likelihood of observing the data, assuming a particular underlying evolutionary model (Bouckaert et al. 2014; Raj et al. 2014). Modern approaches to phylogenetics are great examples of how underlying evolutionary models can be mapped onto complex datasets using probabilistic approaches (Yang and Rannala 2012; Kearney, chap. 7). In general, evolutionary models developed primarily during the first half of the twentieth century have undergone serious restructuring at the beginning of the twenty-first century. Whereas the goal of the early models was to articulate the general processes that influence evolutionary change, the current goal, at least for models aimed at genetic data, is to use what is now known about those processes to make more specific inferences regarding their interaction and the magnitude of their parameters.

The Structure of Evolutionary Models

At their heart, models are idealized structures used to represent specific systems within the real world (Giere 1988). Why do some models end up having a very broad impact within a field while others do not? What features of evolutionary models have allowed them to retain their relevance for such a long period of time? I briefly explore this question as a means of illustrating some of the essential elements of models of evolutionary change.

The question of whether a model is generally useful is exemplified by that quote from Einstein: can the model be simple enough to be general, yet still capture sufficient quantitative detail to be useful (see also Levins 1966, 1993). Somewhat surprisingly, most meaningful models in evolutionary biology have been based on a fairly simple set of assumptions/ rules. Complexity usually arises from interpreting the full consequences of those rules, especially interactions between parameters (e.g., selection and genetic drift). Within ecology, there have been multiple periods during the development of formal theory in which debate has emerged as to whether it is actually possible to capture any realistic ecological circumstance using "simple" (as opposed to "systems") models (Odenbaugh 2006). In contrast, from the very beginning, models in evolutionary biology have tended to be as simple as possible. It is only recently that increasingly more complex models have been developed for application to more specific evolutionary scenarios and empirical datasets.

Although simplicity may facilitate applicability, the key to broad use of a model is that it be built upon general principles that can be applied across a wide variety of conditions actually found in nature. Broad application depends on whether the real-world elements upon which a model depends are shared across as wide a set of target objects as possible. So, for instance, the fact that every physical entity in the universe has mass and momentum and is made of atoms allows the laws of physics to be applied from stars to gecko toes. Similarly, one reason that evolutionary models have tended to be so successful is that they are built upon features that are broadly shared across organisms, with DNA being primary among these. Any system possessing the three basic tenets of evolution by natural selection—heritability, variability, and differential survival and reproduction—will necessarily experience evolutionary change (Frank and Fox, chap. 9). If every species had its own mode of inheritance and some unusual form of reproduction (imagine complex forms of budding rather than the usual production of gametes), they would all still evolve, but application of a small set of coherent evolutionary models that captured each special case would be much more difficult. The fairly limited forms of reproduction and genetic transmission observed in natural systems allow for a small set of evolutionary models to be applied quite broadly.

The chapters in section 2 explore the theoretical consequences of the coupling of genetics and evolutionary processes. Here, I briefly highlight one approach that was adopted early in the construction of evolutionary models—the analysis of relative frequencies—as an exemplar of how models have been structured to maximize generality.

Relative Measures of Competition between Types

Most models of evolutionary change via natural selection use the most trivial ecological model possible, exponential growth:

$$(4.4) \qquad N_i(t + 1) = \lambda_i N_i(t),$$

where $N_i(t)$ is the number of individuals of type i in the population at time t, and λ_i is the number of type i individuals produced within a given time interval (Fox and Scheiner, chap. 11, eq. 11.1). For asexual reproduction, the time interval can be taken as the generation time. For sexual reproduction, the time interval represents changes within a generation, since the transition to the next generation is strongly influenced by the complications of genetic transmission (Frank and Fox, chap. 9). This general definition of population growth can be readily expanded to include continuous reproduction, overlapping generations, and so forth, but the basic idea remains the same that natural selection is generated via direct competition through differences in reproductive rates (Tuljapurkar and Orzack 1980; B. Charlesworth 1994). At first glance using this approach seems ridiculous; we know that this model must in general be false because any species obeying exponential growth would either rapidly fill the universe with its numbers (if $\lambda_i > 1$) or be quickly consigned to oblivion (if $\lambda_i < 1$). However, a slight change in viewpoint allows this approach to be approximately correct, even under complex demographic conditions. First, if one divides the number of individuals of a given class by the total number of individuals within a defined population, then one can track the frequency of a type rather than its absolute numbers. Thus whether the population as a whole is growing or shrinking is immaterial, as it is the change in relative frequencies that determines the evolutionary outcome, as in eq. 4.1. This naturally leads to the frequency-change-based definition of evolution so abhorred by Mayr (Shermer and Sulloway 2000). The significance of focusing on relative frequencies, as in the earliest formulations of population genetics, cannot be overstated. In particular, it allows population genetic models to be correct to a first-order approximation even when ecological circumstances might be wreaking havoc with the actual demography of a given population.

Because moving to frequencies involves dividing by the total population size, the second consequence of this change in viewpoint is that differences in reproductive output among types also naturally convert to relative amounts (i.e., relative fitnesses), because it is only the ratio of fitnesses that becomes relevant when the ratio of population sizes is calculated (eq. 4.1). Thus it is only the increase or loss of fitness of a given type relative to the average fitness in the population that is responsible for determining frequency dynamics. This normalization allows frequency-based models to be approximately correct under a wide variety of circumstances because the change in frequency of any particular type tends to have a relatively small effect on the mean fitness of a population, even if differences in reproductive output are fairly large.

Naturally there are a number of complications that make this simple approach fail, most notably any frequency or density dependence that might arise from the interaction of different classes of individuals. Some models try to allow more explicit demography and ecology to be included (Coulson et al. 2010). One class of models, exemplified by the adaptive dynamics approach, are particularly geared toward the analysis of frequency-dependent events, such as those expected under more complex models of competition among individuals (Parvinen et al. 2003). Those models explicitly take density dependence into account in order to examine how diversity can arise as a consequence of competition (Dieckmann and Doebeli 1999). For the most part, however, those models ignore most of the troublesome complications caused by sex and genetic variation within populations (Orive, chap. 14), so they are currently poor general descriptors of evolutionary change (Waxman and Gavrilets 2005).

Overall, then, the vast majority of evolutionary models are built upon a relative frequency (and thereby relative fitness) approach, which allows evolutionary change to be predicted as a local perturbation of the current state of the population (Frank and Fox, chap. 9). That this approach works so well is perhaps the primary reason that simple evolutionary models have found such wide applicability.

Reintegration of Demography into Evolutionary Genetics

Part of the motivation for working in frequency space was to create a set of conceptual models that would be generally applicable across a wide set of circumstances, independent of demography. In finite populations not all frequency states are possible, since frequency must adhere to quantum changes in the single unit of inheritance (usually taken to be the "allele"). This leads to discrete frequency classes in units of $1/N$ (for haploids) or $1/2N$ (for diploids). Under these circumstances, Kimura (1983) showed that in many cases the relevant parameters are not the individual evolutionary forces (selection, mutation, migration) but the product of those parameters with population size. To disentangle population size from those parameters, recent models of molecular evolution are increasingly likely to include an explicit description of demography, including direct estimates of changes in actual population sizes (e.g., Pickrell and Reich 2014). Spurred by an explosion in the availability of DNA sequence data, it appears likely that evolutionary models will eventually move away from the general frequency approaches that have been the core of evolutionary modeling for a century toward more explicit ecological models as exemplified by eq. 4.4. Thus, while the core concepts of evolutionary theory have been largely illuminated using more simplified models, the application of those models to actual genetic data has become more and more dependent on their integration with models that more accurately reflect the population biology of the species in question.

Testing Evolutionary Models

While progress in molecular biology has consisted both of learning how things actually are in nature (the double-helical nature of DNA) and in finding methods for asking such questions (how to sequence DNA), the discoveries, both theoretical and experimental, that have had the most impact on population genetics are not those that revealed a particular fact about the world, but those that have provided methodologies for answering particular questions when desired. That is why theoretical investigations have played such a central role in population genetics as opposed to any other branch of biology.

—Richard Lewontin (2000, *20)*

Of course, [population genetics] also needs accurate numerical data, and these do not yet exist, except in a very few cases. . . . One of the important functions of beanbag genetics is to show what kind of numerical data are needed. Their collection will be expensive.

—J. B. S. Haldane (1964)

Applying evolutionary models to specific real-world circumstances can be challenging. During the development of formal evolutionary theory, many empirical studies were largely consistency arguments that helped to serve as exemplars for a given part of the overall conceptual framework. Part of the difficulty in bringing evolutionary models more directly to bear on empirical data has been the empirical (in)sufficiency of the data itself. This (in)sufficiency takes two forms. First, the data must actually measure the target system that is the subject of the model. In population genetics, this has always meant being able to measure the frequency of segregating alleles within a population, as well as the change in the frequency of those alleles over time. Even though a focus on allele frequencies has been the foundation of all models in population genetics since the 1920s, it was not until the late 1960s that data on allelic variation could be systematically collected at the level of proteins (Lewontin and Hubby 1966), and not until the late 1980s that this could be done with DNA (Kreitman 1983). Tellingly, nearly all of the core conceptual structure for the quantitative theory of evolutionary biology was developed before the material basis of inheritance was fully understood. The fact that the basic models in evolutionary genetics are largely independent of molecular details highlights that these models are used to understand the conceptual structure of the population genetic processes that lead to evolutionary change, rather than to describe the evolution of any specific system.

More recent population genetic models, particularly those examining molecular evolution, require much greater detail about the molecular processes that they are meant to represent (e.g., the difference in transition versus transversion mutation rates; McInerney, chap. 5), but in many cases some of this information is just beginning to come to light now that we are able to obtain DNA sequence in an unbiased fashion from a large number of individuals. For example, models in both population and quantitative genetics depend on the rate and distribution of effects of new mutations, yet the characteristics of new mutations are still only crudely known—certainly not at a level necessary for the precise parameterization of most of these models (Eyre-Walker and Keightley 2007; Keightley and Eyre-Walker 2010). Models in population genetics existed for nearly one hundred years before sufficient information about their target systems became widely available. Data needed for other models in evolutionary biology, for instance regarding details of the genotype-phenotype map (Hallgrimsson et al. 2014), still await accurate measurement.

The case of the estimation of mutational parameters helps to illustrate a second major issue with empirical sufficiency. Mutations are rare and therefore difficult to study without samples of very large size. In general, even when the information needed to help estimate model parameters is actually measurable, few evolutionary studies have been conducted at a size sufficient to critically evaluate a given evolutionary hypothesis. Evolutionary quantitative genetics is a fine example of this issue (Barton and Turelli 1989; Hersch and Phillips 2004; Phillips 2005). It has long been known that very large samples are needed to adequately estimate quantitative genetic parameters within natural populations (Klein 1974; M. Lynch 1988), yet most studies in evolutionary quantitative genetics have been conducted using experimental designs that are at least an order of magnitude too small to allow meaningful estimates. The reason for this is clear: an individual investigator can only do so much during the course of a given study. It is only now with the investment in human genetics involving hundreds of thousands to millions of samples that one begins to get a glimpse of how potentially difficult this problem can actually be (Visscher et al. 2012). Similarly, the new availability of high-throughput sequencing approaches is for the first time allowing sufficient information on DNA variation to be collected to test specific models of molecular evolution, such as those attempting to separate the effects of variable demography and migration from the effects of natural selection (Cutter and Payseur 2013). In phylogenetic studies, whole genome approaches are clearly showing that the evolutionary history of genes and species can be distinct and complex (Pease et al. 2016; Kearney, chap. 7). We are only now at the cusp of being able to rigorously evaluate evolutionary models using a scale of data that is sufficient to the task, and even here, this is only for a small subset of the many long-standing questions in evolutionary biology.

Limits to Model Inference

An essential feature of the question of empirical sufficiency is the historical nature of the processes being described in evolutionary models. Although all scientific fields rely on precise measurements to some extent, there is a long history in many fields of biology of being able to adequately rely on qualitative or categorical outcomes for causal inference, largely because the majority of these fields were built upon a paradigm of using experimental manipulations to test hypotheses (Platt 1964). Such manipulations are often selected so as to produce an effect of "sufficient" size that statistical analysis of quantitative variation is often beside the point. For instance, for most of the history of genetics a mutation that did not lead to an obvious phenotypic effect could not be detected, much less subjected to detailed functional analysis. The advent of genomics, reverse genetics, and systems biology is beginning to reverse this trend to some extent within molecular biology, but there is still an emphasis in this and related fields on limiting experimental tests to cases in which highly repeatable experimental manipulations can be conducted against a backdrop of very limited background variation. In contrast, because the objects of study in evolutionary biology are usually natural systems, effect sizes are precisely those supplied by natural variation, be they large or small. Most important, natural systems display inherent (irreducible) variability for many reasons, including genetic differences among individuals, macro- and microenvironmental variation, and the complex ecological circumstances in which these systems exist. This variability has led evolutionary biologists to become highly reliant on statistical analyses for most of their work. Indeed, regression analysis (Galton 1886), correlation (Pearson 1886), analysis of variance (Fisher 1918), and path analysis (Wright 1921) were all invented to address questions regarding natural variation or the analysis of quantitative genetic data.

Most conceptualizations of science are built upon the idea of testing a hypothesis, usually via experimentation. The central idea of an experiment is that one can change one or a small number of factors at a time while keeping all other things constant (*ceretis paribus*). This allows any change in the condition of the experimental observations to be ascribed to the factor that was varied. In the absence of an experimental effect, both treatment conditions are expected to be identical to one another, with this identity usually taken as the null hypothesis. When using observations from natural systems, there is no such thing as "keeping all other factors constant" or "varying only one element at a time." Causation can never be inferred from an observed difference; it can only be consistent with an expected outcome under certain assumptions. The null hypothesis of "no difference" is always false. For example, if one wants to ascribe the difference in genome structure between bacteria and mammals to differences in population size, it is possible to construct models that yield this pattern as an expected outcome, or perhaps more properly, the pattern is impossible to distinguish from an appropriate null model (M. Lynch and Conery 2003). However, it will always be formally impossible to demonstrate that the difference in population size was the cause.

To some extent, this overall statistical framework is the direction in which models in evolutionary theory have tended to migrate over the past forty years. Although the idea of using population genetic models to generate an adequate null hypothesis for natural observations has been around from the beginning (Wright 1931), it is the application of Kimura's elegant mathematical models to the analysis of emerging data about molecular variation that drove the null hypothesis approach forward (see also Lande [1977] and M. Lynch [1990] for similar approaches in quantitative genetics). Overall, this has been a very beneficial development for the application of evolutionary models to data. However, because they are used to analyze natural systems, there can be issues with naively applying these models as a null hypothesis. For example, if natural selection is in fact pervasive across the genome, then patterns of change generated by linkage among loci are likely to dominate most patterns of variation, whereas strictly neutral null models assume that loci are independent and that nonselective demographic factors apply equally at all sites across the genome (Hahn 2008; Kern and Hahn 2018).

This gap between model assumptions and nature is one of the essential problems with the statistical null approach. We know that any null hypothesis involving two populations, species, or other taxonomic units is false because the assumption of "all else being equal" must necessarily be incorrect; any two natural samples will differ in myriad ways unrelated to the hypothesis under test. One way of addressing this problem is to focus more on parameter estimation than hypothesis testing per se (Hahn 2008). Even here, estimating parameters based on false assumptions will yield biased estimates. The strength of quantitative models in evolutionary biology is that they make their underlying assumptions explicit. The peril of empirical data from natural populations is that they do not tell you which assumptions are valid in a given circumstance. Testing assumptions using additional modeling generates something of a Russian doll of nested assumptions. Ultimately, the validity of any given set of models is necessarily based on the consistency of their explanatory power across multiple natural systems because it is impossible to be certain that the model is perfectly applicable to any given case. Experimental evolution is an emerging paradigm that seeks to address this specific deficiency through the use of repeated, controlled experiments with well-defined initial conditions (Kawecki et al. 2012; Teotónio et al. 2017). This approach allows models—

and their associated hypotheses—to be tested using a more conventional framework. However, such experiments tend to be of limited spatial and temporal extent relative to evolutionary change within natural populations.

Evolutionary Models and the Development of Evolution as a Science

There are two possible explanations for why evolutionary theory has transitioned from general conceptual models during the foundational stages to a more recent emphasis on statistical models geared toward data analysis. Either this transition was driven by the need to initially enumerate and establish a general understanding of evolutionary forces before moving to more specific models applicable to actual data, or it was driven by the fact that evolutionary biologists simply did not have the appropriate data at their disposal in the first place. Conceptual models were all that were possible until such data became available. There is probably truth to both explanations, although it is impossible to read most theoretical arguments in evolutionary biology without the impression that they are formed within a vacuum generated by an absence of applicable data. It is the virtuous cycle of theory, data to evaluate theory, and new empirical discoveries leading to new theory that helps to generate meaningful progress within the field.

The transition from models that attempt to answer "what is possible" to models that attempt to infer "what has occurred" is therefore a sign of a healthy science. However, different subfields within evolutionary biology are at different points on this continuum (compare Orive, chap. 14, and Edwards et al., chap. 15, for instance). There is also the danger that the onslaught of genomic data that is finally providing the information needed to refine and test more sophisticated models within population genetics will serve as such a strong distraction that it will stunt the development of other areas of evolutionary theory or—worse still—allow us to forget the importance of other long-standing questions (Lewontin 1991). It is very unlikely that we have articulated all of the necessary components of a complete evolutionary theory (e.g., the properties of systems of complex interactions among functional components, Phillips 2008). Further, while genomic analyses will be an essential component of any complete framework of evolutionary models, genomic data by itself is insufficient to address the most important questions in evolutionary biology, which stem directly from the interaction of complex phenotypes with the environment.

Without question, over the last hundred years, quantitative models have developed an unassailable role within evolutionary theory. Indeed, they have come to define what "evolutionary theory" actually means in much the same way that quantitative models are now synonymous with the conceptual structure of modern physics (although see Mindell and Scheiner, chap. 1, table 1.2, for a more extensive treatment of the relationship between models and theories per se). The next hundred years are likely to witness increasing specialization and sophistication in the application of these models to specific empirical questions. The real mystery is whether or not there remain major conceptual breakthroughs in evolutionary theory that are as yet unrecognized and therefore remain unrepresented in existing models. The history of science in general suggests that this is likely to be the case.

Acknowledgments

I thank Henrique Teotónio, Sam Scheiner, and two anonymous reviewers for extremely valuable comments on an earlier draft of this chapter. This work was partially completed while the author was a Professeur Invité for the Laboratoires d'Excellence (Labex) MemoLife program at the École Normale Supérieure, Paris, and has been supported by funding from the National Institutes of Health (GM102511).

FIVE

Traits and Homology

JAMES O. MCINERNEY

In this chapter I give a brief historical overview of the concept of homology, why it has been a contentious idea, and why it has so long occupied evolutionary biologists. I present a set of ideas that have been developed to help understand homologous relationships and place these ideas in the context of other evolutionary biology theories.

The term homology is derived from two Greek roots: *homos* meaning similar and *logos* meaning discourse, investigation, concept, or doctrine. The concept of homology was first used in geometry (Spemann 1915). If two diagrams or figures had similar angles or corners, then those angles were considered homologous. Since then, the term has been used in many diverse disciplines including philosophy, mathematics, chemistry, and linguistics (Voigt 1973) and has different meanings in each field. In the biology literature there have been several uses of homology, both evolutionary and nonevolutionary, spanning areas of investigation as diverse as developmental biology and morphological analysis (G. P. Wagner 1989), molecular evolution (Reeck et al. 1987; Fitch 2000), and behavior (Hall 2013). In this chapter I detail the ways in which homology is interpreted today, starting with a brief account of the history of this idea, then distinguishing modern interpretations of homology in different contexts, and finally outlining some of the routes that can be taken to identify and interpret homologies.

Two hundred or so years ago, when organs or parts of organs in different organisms were found to be in the same position, they were not labeled homologous but (rather confusingly today) "analogous" (Hossfeld and Olsson 2005). The first documented use of "analogies" in the field of comparative anatomy is attributed to Camper (1784, 1785), who studied orangutan morphology. In 1825, there was another progression in the use of homology terms when Geoffroy St. Hilaire wrote: "when the development of organs is analogous, they are called homologous" (Hossfeld and Olsson 2005, 247). In 1848, Richard Owen, the anatomist, naturalist, and first director of the British Museum of Natural History, wrote the first major work on the notion of homology, attempting to clarify it and to show how this idea could be used to document differences in vertebrate skeletons. The title of the book—*On the Archetype and Homologies of the Vertebrate Skeleton*—provides an insight into the fact that Owen did not have an evolutionary concept for homology. At the time, the varieties of forms that were observed in nature were considered to be imperfect versions of an "archetype," an idealized organ or organism.

Owen hoped that there might one day be a "systematic Nomenclature of Anatomy," but until there was a degree of standardization, such a nomenclature would not be realized. Owen's overriding ambition, therefore, was to establish a formal system of identifying corresponding organs in humans and "lower animals." He argued that among scientists, there was a general sense of what kinds of organs could profitably be compared with one another: "No anatomist, for example troubles himself with the question of the amount of resemblance to a crow's or other bird's beak in the 'coracoid' bone of a reptile, or with the want of likeness of the kangaroo's coccyx to the beak of a cuckoo" (Owen 1848, 2–3). He concluded that it was desirable to establish for all animals "the relation in the part of the lower animal to its namesake or homologue in man" (3), defining "homologues" as "[t]he corresponding parts in different animals being thus made namesakes" (5).

In his book, Owen described three main kinds of homologies—general, special, and serial homologies—though only one of these, special homology ("The correspondency of a part or organ, determined by its relative position and connections, with a part or organ in a different animal," 7), is in common use today. Owen used the phrase "general homology" to mean a broader class of what he also called "homologs." For instance, when a particular bone is thought to be central to the overall structure of a vertebrate, then even though it might be a different bone in different animals, Owen called these general homologs. "Serial homologs" were repeated structures on an individual, such as hairs on the leg of an insect.

Owen's homology definitions were purely formal and did not refer to evolution as the causal agent for this similarity. In contrast, Haeckel (1866) set about defining the "phylum" and ascribing an evolutionary cause to the similarity between organs. This was transformational and moved homology away from the purely geometric "type" to the evolutionary "phylum."

The former was completely unconcerned with how homologs came to be, and indeed polyphyletic groups could be put together quite comfortably, while the latter was an evolutionary explanation and was tightly linked with monophyly (Hossfeld and Olsson 2005).

Some twenty-two years after Owen published his treatise, Lankester (1870) defined the word *analogy* to indicate morphological characters with superficial similarities. In the past hundred fifty years or so, there have been numerous efforts to succinctly characterize the ideas relating to homology, but it has not been without trouble. Voigt (1973) reported that by 1965, there were taxonomies of homology that had described 65 different kinds of homology and analogy. Wake (2003, 193) suggested that "the homology debate is the result of biologists attempting to save an ancient, vague concept." Tautz (1998, 17) has quoted John Maynard Smith as having said that homology has "become a word ripe for burning."

Homology as a concept needs clarification largely because there are different kinds of homologs and indeed clarification of homology concepts and models also requires discussion of analogy and homoplasy (common character states that were independently derived), as well as the routes that are taken to illuminate these ideas. The advent of developmental and molecular biology has brought to our attention new evolutionary processes that can give rise to complex kinds of evolutionary relationships, and consequently new types of homologies have been identified, interpreted, and described. Homology is fundamental for phylogenetic inference and the analysis of character evolution, speciation, and development. In particular, the accuracy of homolog identification underpins evolutionary interpretations. Unless we are interested in analogies and convergences, perhaps driven by selection, any attempted evolutionary comparison between non-homologous traits is meaningless. The first part of phylogenetic inference involves the identification of homologies; the second part involves understanding change or stasis in these homologies.

Homology as a Concept

In its current usage in evolutionary biology, homology is a concept that relates evolving entities to one another. However, the relationship of evolving entities to one another is a complex issue. On the one hand, there are evolving entities that are directly related to one another; these have been called replicators (G. P. Wagner 1989). On the other hand, there are evolving entities that are built anew with each generation. In the first case, we include genome sequences and also proteins that are transcribed and translated from these genomes directly. In the second case, we include characters that are built—often using the interactions of several genes and external stimuli—fresh in each generation and are therefore inherited indirectly, rather than directly. In the case of replicators, there is a direct line of inheritance through a molecule that is replicated by semiconservative replication, whereas in the second case, the homologous organs are constructed during development. In the first case (sometimes called "historical homology"), two evolving entities are said to be homologous if they are descended from a common ancestor. In the second case (usually called "biological homology"), two entities are said to be homologous if they "share a set of developmental constraints, caused by locally acting self-regulatory mechanisms of organ differentiation. These structures are developmentally individualized parts of the phenotype" (G. P. Wagner 1989, 62).

The biological homology concept is restricted to the description of developmental processes or morphological traits and was developed in order to include "iterative homology" (e.g., hairs on a primate's forearm, petals and sepals in flowering plants), while not excluding homologs that fit well into the historical homology concept. The impetus behind the development of the biological homology concept was to deal with issues such as the development of features that are clearly homologous but in different organisms; for example, they originate from different kinds of cells or are under the control of different genes. In addition, characters such as teeth, hair, petals, and so forth are serial homologs or iterative homologs that arise from very similar starting points. Clearly, such features would be problematic to interpret under the historical homology concept; hairs and teeth are not duplicates of one another, nor do they arise one from the other. Each individual originates from complex interactions of the same or very similar sets of genes in different places on the body, yet each tooth and each hair is quite clearly a separate organ.

An issue that is related to, but different from, the issue of very similar morphological characters newly built at each generation has been pointed out by Roth (1988). She showed that "genetic piracy" could take place, where genetic elements are coopted to control the development of a feature over time so that two homologous morphological or developmental traits would be controlled by entirely different sets of genetic material (Love, chap. 8).

In contrast to Richard Owen's original exposition, historical and biological homology concepts implicitly contain an evolutionary interpretation for the similarity of evolving entities. However, although from an ontological perspective there is agreement that homology should mean "descended from a common ancestor," from an epistemological perspective there is a surprising amount of discord, mostly due to the fact that evolutionary histories tend to be complex. For example, morphological characters can be interpreted under both the biological and the historical homology concepts, though again this connection is not always straightforward.

Characters and Character States

Patterson (1988) opined that homology should relate only to two objects that are recognizably "the same thing." In this interpretation, either characters are homologous or they are not. He gave the example of the cochlea of mammals, of which there are two kinds—spiral cochlea in therians and curved cochlea in monotremes. Patterson, opposed the idea that these two kinds of cochlea would be considered different "character states" of the same homologous character. Instead, he advocated a view that these were in fact two different homologous characters and felt that homology could be "framed at any level."

The opposite and arguably the more popular view is that spiral and curved cochlea are homologous and represent two different forms of the same character. This view advocates that each character, irrespective of whether it is a morphological trait or a nucleotide position in an alignment, can have one or more character states, while it is the organ or nucleotide position itself that is the homolog (Fitch 2000). In this view, evolving objects related through common ancestry are considered homologous characters and the different forms of this character are considered different character states. This perspective sees homology as a more inclusive idea and the challenge is first to identify homologs and then to accurately encode the different character states for analysis.

In the viewpoint advocated by Patterson, nucleotides in DNA sequences or amino acids in proteins would be homologous only if they were the same. This would mean that only the identical parts of sequences would be considered homologous or perhaps that two sequences would be considered homologous only if they were exactly identical. Under the more common viewpoint, sections of molecular sequences that can be traced back to a common ancestral sequence would be considered homologous, even if they are not very similar (Haggerty et al. 2014). Each viewpoint has its challenges, and disentangling the relationships between similarity and homology is more difficult than it might seem at first.

The Relationship between Similarity and Homology

All explanations for similarity between two evolving entities appeal to either homology or analogy. Homologous similarity is similarity that is due to common ancestry; analogous similarity is similarity due to convergence from a different evolutionary starting point. For the most part, evolutionary analyses tend to focus on homologous similarities, though, depending on the research question, analogies might be interesting as well (McInerney et al. 2011). The study of homology, therefore, is the study of what is conserved through evolution. However, while the nonrandom similarity of two structures or sequences might seem, from the last few sentences, like a prerequisite for inference of homology, there is not such a simple relationship. Consider, for instance, two DNA sequences or two protein structures that were derived from a common ancestor. If they evolved rapidly or for a long period of time, a comparison might show that their level of similarity is no greater than the random similarity of two unrelated sequences or structures. In this case, inference of homology based solely on nonrandom similarity would not be effective. The corollary of this situation is that two structures might display nonrandom similarity of sequence or structure but have arisen independently, i.e., they manifest analogous similarity. For example, convergences have been noted in the independent origins of fibroin-like protein motifs in several protein families (Gatesy et al. 2001) and in the evolution of protease active sites (Buller and Townsend 2013).

It is important to make the distinction between the convergence of traits that are completely unrelated and the convergence of the states of the same character, though both are different kinds of homoplastic event. In the first instance, there is homoplastic convergence from two nonhomologous starting points to two end points where the characters appear to have significant similarity. In the second instance, we speak about different states of the same homologous character and the independent evolution of the same character states in different lineages. This second situation is caused by repeated change of the states of the same character, otherwise known as superimposed character substitution. Analogy and homoplasy are nonsynonymous terms: analogy is the independent evolution of similar traits from nonhomologous structures or sequences, while homoplasy is a more general idea and covers both analogy and the independent evolution of character states.

Homoplastic convergences and parallelisms in morphological characters are often quite complicated. The study of closely related taxa and the underlying genetics involved in the homoplastic evolution of the same phenotype has often uncovered that the phenotype has evolved independently via different genetic mechanisms. A specific example is seen in the *Mc1r* gene, where a point mutation causes light coloration of the coat of American Gulf Coast beach mice, but the same phenotype is caused by a different mutation in Atlantic beach mice (Hoekstra et al. 2006). Clearly, the same phenotype can arise by different pathways. This suggests that for many morphological characters we might expect to see complex relationships between convergences and parallelisms. Indeed, it has been suggested that rather than distinguishing between convergence and parallelism, when it comes to morphological characters, it is sufficient simply to use the word "convergent" (Arendt and Reznick 2008).

Figure 5.1 illustrates the different kinds of evolutionary events that might be seen for a homologous molecular character. Figure 5.1A depicts a series of mutations where none are homoplastic. The ancestral character state for this character was an A nucleotide that in one lineage changed to a C nucleotide and in the other lineage changed to a T nucleotide. There are no other evolutionary changes, and the C character state for taxa 1, 2, and 3 is a synapomorphy for that group (a shared, derived state), while the T character state is a synapomorphy for taxa 4 and 5.

In contrast, all three of the other examples demonstrate homoplasy. Figure 5.1B provides an example of a convergence. In this case, the ancestral A nucleotide changed to a C nucleotide in one lineage, but a subsequent change in the lineage leading to taxon 1 converted the C to a T. On the other side of the tree, there was a change from an A nucleotide to a T nucleotide in the lineage leading to taxon 5. The net result is that taxa 1 and 5 now have the same character state, which has the effect of making it seem as if taxa 1 and 5 are more closely related to one another than they really are. What defines this kind of homoplastic change as a convergence is the fact that the two mutations that resulted in homoplastic similarity were different kinds of mutation. In one case, the mutation converted a C to a T nucleotide, whereas in the other the mutation converted an A to a T nucleotide. The hallmark of a convergent substitution is that change from two different starting points goes to the same end point.

Figure 5.1C depicts a parallelism where the same character state transformation occurs independently in two lineages. The ancestral A nucleotide changed into a T nucleotide in the lineage leading to taxon 1, and the exact same character-state transformation occurred in the branch leading to taxon 5. Because these are the only two contemporary taxa in the dataset with a T nucleotide, while all others possess an A nucleotide, again this

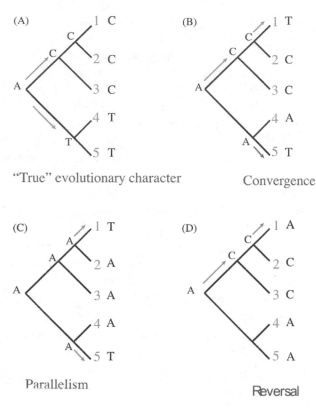

Figure 5.1. Evolutionary scenarios that demonstrate the different kinds of evolutionary transitions in molecular sequence data.

kind of homoplastic event is potentially misleading evidence of a close relationship between taxons 1 and 5 in the absence of other evidence.

Figure 5.1D depicts the last of the different kinds of homoplastic events, a reversal. In this case, the ancestral A nucleotide changed to a C nucleotide in one lineage and then subsequently in the lineage leading to taxon 1 reverted to its previous character-state. Again, this can provide misleading evidence of relationships, making taxon 1 appear more similar to taxa 4 and 5.

Homology Models

There are multiple kinds of homology relationships. First, we have homologs nested within homologs. Hands and paws of mammals, for instance, are homologs and they are all found on forelimbs, which are also homologous. For molecular sequences, we might consider, for instance, a set of ribosomal proteins that are found in several different species and, judging by their sequence similarity, are clearly homologous to one another at the overall sequence level. When we align these sequences, we consider the various nucleotides or amino acids embedded within them also to be homologs of one another. At another level, we might consider the structure of the ribosomes in these species and we might conclude that these ribosomes are also homologous. Uncertainty arises when particular characters are inserted in or deleted from DNA or protein sequences, or when organs or organelles have more or fewer components—as can be seen by the variation in protein content across the breadth of ribosomes (Wolfe et al. 1992).

Presence and absence of parts of a sequence or trait can result in conflicting views on homology. From the preceding example of ribosomal proteins, most would agree on the homology of the individual nucleotides, of the entire sequences, and of the entire ribosomes. However, some have sought to employ a different kind of model. Song et al. (2008, 3) put forward a homology model that sought to "distinguish multidomain [protein] homologs from unrelated pairs [of proteins] that share a domain." In their model, they consider two proteins that share a single domain not to be homologous at all but simply to "share a domain," whereas, other proteins that are homologous for most of their length, though perhaps with a domain inserted somewhere along the sequence, are indeed homologous. In their model, homology is both a concept and a quantity. Sequences are viewed as homologous if most of their sequence is homologous, but unrelated if they only share a domain. As an example, "chromosomal regions enriched with homologous gene pairs are likely to be homologous themselves. In contrast, enrichment with homologous domains does not support the inference that a pair of chromosomal regions is homologous." This model employs quantitative measures for homology and departs from the standard notion that homology is qualitative and should mean descended from a common ancestor and no more. Song et al. (2008) advocate that because some domains are promiscuous, it is acceptable to infer that the genes that contain those domains are not homologs. Genes are to be considered homologous to one another only if a sufficient quantity of the gene manifests homology.

For molecular sequence data, we additionally identify orthologs, epaktologs, ohnologs, synologs, and paralogs that are further divided into inparalogs and outparalogs. Short descriptions of these different kinds of homologs are given in table 5.1. Figure 5.2 illustrates the interplay of gene

Table 5.1. A Short Dictionary of Some of the Most Important Concepts in Homology

Homology: A relationship between molecular, morphological, or developmental characters based on common descent.

Historical homologs: Homologous characters due to common descent and inherited by replication (Fitch 1970).

Biological homologs: Biological characters whose development is regulated or constrained in the same way (G. P. Wagner 1989).

Orthologs: Genes that trace their most recent common ancestor to a speciation event (Reeck et al. 1987).

Paralogs: Genes that trace their most recent common ancestor to a gene duplication event (Reeck et al. 1987).

Inparalogs: Paralogs that reside in the same species (Sonnhammer and Koonin 2002).

Outparalogs: Paralogs that are found in different species (Sonnhammer and Koonin 2002; Nagy et al. 2011).

Epaktologs: Genes or proteins that have arisen by remodeling, such that two sequences appear to have the same homologous structures, but their histories are completely different and replete with domain swapping or other forms of remodeling (Nagy et al. 2011).

Synologs: Homologs that are in the same genome but may not have arisen by duplication or horizontal gene transfer (Lerat et al. 2005).

Ohnologs: Paralogs that have arisen as a consequence of a whole genome duplication event (a special case of homology) (Wolfe 2001).

Partial homologs: Two evolving entities (usually molecular sequences) that do not manifest homology along their entire length but instead share regions of homology (Hillis 1994).

duplication events and speciation events that can give rise to complex patterns of orthology and paralogy.

Homologs Are Not Always Natural Kinds

The *Stanford Encyclopedia of Philosophy* defines "natural kinds" thus: "To say that a kind is natural is to say that it corresponds to a grouping that reflects the structure of the natural world rather than the interests and actions of human beings" (http://plato.stanford.edu/entries/natural-kinds/; see also Nathan and Cracraft, chap. 6). Atomic elements, such as hydrogen or carbon, are considered to be natural kinds. If gene families or morphological characters were natural kinds, then the identification of homologs would be relatively simple and the partitioning of molecular sequences into gene or protein families would be uncontroversial. However, morpho-

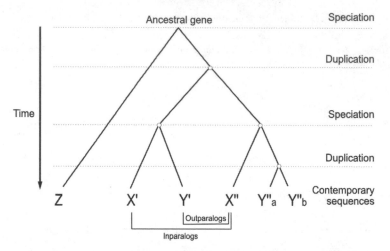

Figure 5.2. The evolution of a gene family. In this diagram, an ancestral gene has been duplicated into multiple copies through two rounds of speciation events and two gene duplication events. Duplication events create paralogous genes, and speciation events create orthologous genes. For instance, genes Y''_a and Y''_b are paralogs of one another because they can trace their most recent common ancestor to a gene duplication event, while both genes are orthologs of X'' because they can trace their common ancestor to a speciation event. X'' is paralogous to both X' and Y', though in the former it is an inparalog, because the paralogs are in the same genome, and in the latter it is an outparalog, because they are found in different genomes (Sonnhammer and Koonin 2002). Relationships between genes are reflexive (e.g., if X'' is a paralog of Y' then Y' is a paralog of X''). However, relationships are not transitive (e.g., Y''_a is a paralog of Y' and X'' is a paralog of Y', but Y''_a is not a paralog of X'').

logical characters can be constructed during development by overlapping sets of gene products and environmental stimuli, which makes the task of homologizing these characters quite difficult. If two proteins interact with each other in the same way in two different tissues, this is still not reason enough to consider those tissues to be homologous. Instead, the likelihood that similar gene expression or protein interaction networks indicate homology is related to how likely we expect such networks by chance. All cells have ribosomes, but this is not sufficient for us to identify morphological homologies as this is simply a feature of all cells (G. P. Wagner 2014).

On the basis of sequence similarity levels that are significantly higher than would be expected by chance, genes can be put into groups. However, problems arise if distant homologs are missed because they are too divergent, or conversely if in convergent evolution nonhomologous sequences are mistakenly identified as homologs. Additionally, placing genes into discrete gene families is not a simple task because of nonhomologous recombination, in which molecular sequence data tend to form large, connected sequence similarity networks (fig. 5.3). Promiscuous domains are those found in a large number of otherwise unrelated sequences (Basu et al. 2008), and these contribute disproportionately to cluster joining. Clusters in a graph are sets of nodes that are more highly connected to each other than expected (Enright et al. 2002). For many algorithms or approaches to making gene families, the task has been to identify where to "carve the network at its joints" (with apologies to Plato). The observed clustering in homology graphs has been used as the basis for partitioning evolving sequences into "gene families" (Tatusov et al. 2000; Enright et al. 2002). However, while this approach is popular and quite commonly

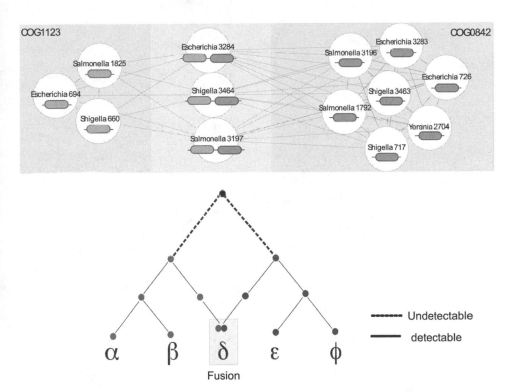

Figure 5.3. In the network in the upper panel, two unrelated gene families (COG1123 and COG0842) have undergone a merging event, and three sequences can be found that are composed of sequences that are homologous to these otherwise unrelated families. The lower panel illustrates how this kind of event might look on a network; technically, it is no longer a tree, because of the merging of branches at node δ. The dashed lines symbolize the fact that in many cases, it is not possible to see homologies, though we know that because of the diverging nature of mutations, such homologies exist.

used, it must be remembered that to cut this network into discrete groups is equivalent to denying the researcher the chance of analyzing weaker homologous relationships such as those involving only part of a gene or protein (Haggerty et al. 2014). Bearing this limitation in mind, a great deal can still be achieved by partitioning sequences into easily manageable and easily studied groups.

The most common approach to defining gene families involves the use of the TRIBE-MCL algorithm (Enright et al. 2002). This algorithm has the advantage of being fast in execution, and given the requirement for defined run-time parameters, the results are reproducible between research groups. The algorithm employs a graph-cutting approach that is not informed by biology. Instead, the user makes decisions about input parameters, and these have the effect of cutting the graph into large numbers of small clusters or smaller numbers of larger clusters (Enright et al. 2002). The choice of parameters can be somewhat arbitrary; however, the approach produces groups of genes that are usually quite amenable to further analysis. The approach identifies the between-cluster linkages and cuts them. This procedure, by necessity, removes statements of homology (i.e., evolutionary relationships), and these can no longer be seen in any downstream analysis.

It has been pointed out that this approach is problematic if genes have been extensively remodeled (Haggerty et al. 2014). In such cases, the clusters will share nodes that consist of fusion sequences (fig. 5.3). This presents us with a problem if we wish to say that a fusion gene belongs to only one family. In reality, fusion genes belong to both fusion-progenitor families and discretizing families. When we require a gene to be a member of only one family, we discard valid, useful evolutionary information. Similar problems have been noted for hybrid organisms, where reticulate evolution is ignored if only one evolutionary history is displayed (Mindell 1992).

To overcome the problems of analyzing genes or proteins with compound histories, it has been proposed that we should use "N-rooted fusion graphs" (Haggerty et al. 2014; Coleman et al. 2015), which are a kind of network with multiple root nodes and one or more special fusion nodes (fig. 5.4). Usually on a phylogenetic tree internal nodes have an in-degree of one (one ancestor) and an out-degree of two or more (two or more descendants) (Kearney, chap. 7). However, a fusion node has an in-degree of two (two ancestors) and an out-degree of one (a single merged sequence). The node with an in-degree of two reflects the flow of genetic sequence from two nonhomologous sequences into a single composite sequence. Recognizing composite sequences or taxa requires precise identification of homologies, particularly partial homologies (Haggerty et al. 2014).

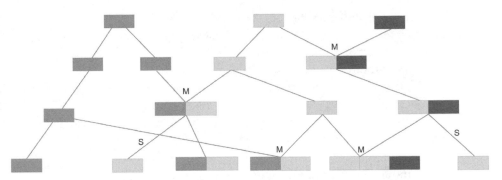

Figure 5.4. A hypothetical evolutionary history of a set of proteins that consists of several combinations of two fundamental domains. The letter "M" indicates a merger of evolving entities and the letter "S" indicates a splitting of the sequences. This evolutionary scenario is quite pervasive in all kinds of evolving entities.

Homology Networks

The analysis of sequence similarity networks has shown us that, contrary to a somewhat essentialist viewpoint (e.g., that homologs are natural kinds that form discrete groups), a network of homologies between genes and proteins underpins life on this planet. Sequence similarity networks are composed of nodes that represent genes or proteins, and edges connecting these nodes are simply statements of complete or partial homology. Tatusov et al. (2000) used network structures on a dataset of 17,967 proteins from seven genomes in order to define gene families. More recently network structures have been used in order to either analyze the nonrandom network of homologies (Enright and Ouzounis 2001), detect distant homologs (Bolten et al. 2001), or display relationships (Frickey and Lupas 2004; Halary et al. 2010). Halary et al. (2010) showed that the global structure homology relationships embedded in genomic data was highly connected, though with very clear breaks—which they termed "Genetic Worlds." Dagan and Martin (2007) carried out a cluster analysis of genome networks of prokaryotes and detected massive levels of horizontal gene transfer that were not restricted to particular kinds of gene or genome. Alvarez-Ponce et al. (2013) showed that the analysis of sequence similarity networks can be used to identify motifs that can falsify hypotheses of relationships between genomes. Bapteste et al. (2012) have formally described network motifs and what they might mean in evolutionary biology, and Dagan and Martin (2009) have described networks as more complete and accurate pictures of genome evolution. Most recently, sequence similarity networks have been used to show that some of the standard approaches in molecular evolutionary biology have shortcomings. In particular, standard analysis approaches tend not to address the merging of unrelated sequences into composite sequences, or the nonhomologous replacement of sections of chromosomes or genes (Haggerty et al. 2014).

This network viewpoint on the relationships of DNA and protein sequences challenges the idea of "related through common ancestry" and has promoted a view that a concept of "family resemblance" also should be adopted (Haggerty et al. 2014). The family resemblance idea was formulated by Wittgenstein (1922) and used to describe the similarities of games to one another. In the context of the shuffling of genes, protein domains, chromosomes, and even genomes (e.g., reassortment in viruses), we find that discretizing evolving objects and placing them into a single category or family often leads to denying their relationship to another grouping and leads to the problem of where to "carve nature at its joints."

Partial homology has sometimes been treated as an add-on to the concept of homology (Hillis 1994). However, partial homology is so commonplace that homology is properly viewed as a multilevel concept, with additional epistemic complications. We can, for instance, consider that two human chromosomes are homologs. However, in the event of replication slippage, or slipped-strand mispairing (Levinson and Gutman 1987), one chromosome has a section that is not homologous to any region of the other chromosome. This does not make those chromosomes nonhomologous, but it means that they are not homologous for their entire length.

The remodeling of genes results in partial homology relationships, and these relationships can often be quite extensive. When a sequence similarity network is generated, where the nodes represent discrete genes and the edges are statements of homology between these genes, we usually see in large datasets that a giant connected component emerges. A connected component is a portion of a graph where there is a path connecting every node in the connected component to every other node. A graph consists of one or more connected components. Giant connected components emerge because gene sequences are quite often partially homologous with other genes that are not themselves homologous to one another. Indeed, as the dataset increases in size, the graph tends to coalesce into a single connected component. Such a connected component (e.g., Halary et al. 2010, fig. 1) contains all of the sequences in the data that have a "homology path" linking them. Two nodes are said to be neighbors in the graph if they are directly connected to one another (i.e., they share some region of homology). If two nodes are indirectly connected through another node, that suggests sequence shuffling or remodeling, with the intermediate sequences manifesting partial homology. A geodesic path is the shortest walk that can be taken to connect two sequences. For large datasets, the length of this path is usually not very long, typically not more than six edges. However, this means that sequences can be connected via partial homology to hundreds of thousands of other sequences with which they share no direct homology, as long as the networks are accurate and not confounded by homoplasy. Precisely because of this remodeling, homology in molecular sequence data must be viewed as a multilevel phenomenon.

Character Identity Networks

A completely separate kind of network has been proposed as a general framework for identifying developmental and morphological homologies, the character identity network (G. P. Wagner 2007), which is similar to the model of the "kernel" proposed by Davidson and Erwin (2006). In this model, there are three tiers of developmental roles for genes. One tier specifies the positional information for the character; a middle tier that is the character identity network activates specific developmental programs; and finally the third tier involves the actual genes and protein products that make the morphological trait. The genes in this last tier—the "realizer" genes—are responsible for the different character states.

It is not sufficient to use gene interactions as indicators of homology. The interaction of two genes might be a shared, ancestral state (e.g., ribosomal proteins forming a ribosome). Additionally, gene duplications can result in interactions between duplicates. Therefore, the assignment of homology on the basis of gene interactions can be somewhat complex.

Homology Must Be Placed in Context

In some cases identifying homologs is trivial. For example, two 1 Kb nucleotide sequences that differ by one mutation can be easily said to be homologous. The hands of humans and chimpanzees are obviously homologous by any reasonable definition of homology. However, in a large number of cases, we run into problems and we must contextualize homology. We must specify what kind of homology we are dealing with.

An interesting situation has arisen from the study of transcription factors. The transcription factors *distal-less*, *engrailed*, and *orthodenticle* are all developmental regulatory genes, containing a homeodomain. In echinoderm evolution, the orthologs of these genes display very different patterns of expression, resulting in their involvement in several new roles in echinoderms and the loss of other roles that are conserved in arthropods and vertebrates (Lowe and Wray 1997). It is arguable that the genes are homologs, but their functions are not.

If we compare the flight of birds, bats, and butterflies, we can see that the organ for flight in butterflies is completely different from the organs used by birds and bats, and so this is a convergence in function and these organs are analogous. Comparing the forelimbs of bats and birds, we can see that similar suites of genes are responsible for the development of these organs and therefore they are homologous. However, birds and bats do not share a common ancestor that separates them from nonflying vertebrates (i.e., flight is not a monophyletic trait); therefore while the forelimbs are homologous, flight in these vertebrates is analogous.

Building phylogenetic trees from alignments of molecular sequences using standard approaches such as maximum likelihood or Bayesian inference requires a dataset of carefully aligned homologous sequences, usually derived from a cluster of sequences that are embedded in a sequence similarity network. These sequences would all be considered members of the same family. However, experience has told us that some of these sequences might have extra domains, and the normal way of treating these extra parts is to remove them, even though methods have been developed to analyze partial homologies (Song et al. 2008; Haggerty et al. 2014). Ignoring uncommon additional domains is perfectly acceptable, but it is an analysis of homologies in a particular context. Halary et al. (2010) chose not to carry out that particular kind of analysis and instead sought to broadly analyze all the homologies. At the opposite end of the scale is the analysis by Song et al. (2008), who consider genes to be homologs if they differ only slightly in domain content, but not homologous if they share only small promiscuous domains. Both kinds of study are analyses of homologies, but in different contexts.

Conclusion

Homology is an overarching concept in evolutionary biology, important for theories in developmental biology, population genetics, macroevolution, and phylogenetics. Phylogenetic relationships (Kearney, chap. 7), for instance, can be reconstructed only by the two-step process of first identifying homologies and then making inferences about changes in character states between different homologs. Homology is difficult to define with precision. Phrases such as "having an evolutionary link," "descended with modification," and even "shares a common history with" are all open to interpretation. The solution is to place the inferences of homology for any study into their appropriate context. In general, there are at least three things to consider when analyzing homologs. First, because genes, proteins, and even morphological and developmental characters are not always discrete entities forming unambiguous discrete relationships, each study needs clarity in what it considers to be homologous to what. Second, homology can be seen at many different levels—the individual nucleotide, the protein domain, the entire chromosome, the organ, or the trait. Therefore, the level at which homology is being assessed needs to be defined. Third, homology must have an evolutionary interpretation where entities are "related through common ancestry." If the traits or organs or sequences under consideration are not related through common ancestry in some way, they are not homologs.

Acknowledgments

I would like to thank Sam Scheiner and David Mindell for putting this book together. My work is funded by the Biotechnology and Biological Sciences Research Council (BBSRC) and The Templeton Foundation.

SIX

The Nature of Species in Evolution

MARCO J. NATHAN AND JOEL CRACRAFT

Much has been written about the "species problem"—the task of providing a functional species concept. Yet, to date, no consensus has been achieved on the individuation and definition of species, or whether a unique solution to the species problem exists. Some have even questioned whether contemporary biology requires a species concept at all. The goal of this chapter is to shed light on the sources of the disagreement. We begin by drawing attention to two distinct, and often incompatible, ways species figure in biology, namely, as units of classification and as units of evolution. Next, we introduce the species problem and discuss a variety of ontological issues that pertain to the nature and role of species in evolutionary theory. In the final sections, we explore the interface between philosophical reflection and biological practice.

Units of Classification or Units of Evolution?

One source of misunderstanding in species debates concerns whether species should be conceived as *units of classification* or as *units of evolution* (Dupré 1994). Both views are prominent within contemporary systematics. Historically, species as units of classification long preceded species as units of evolution. Linnaeus saw his system as a means of categorizing nature, and the categories (ranks) of his system (e.g., class, order, family, genus, species, subspecies) were intended as hierarchically nested abstract concepts. These "invented" devices (in the words of Winsor 2006) are not entities out there to be discovered but conceptual tools that purport to facilitate a scientific understanding of nature. Taxa, in contrast, are scientific hypotheses about the boundaries of evolutionarily related groups—species or groups of species—and thus are widely considered to be discoverable natural entities.

Understanding that taxa of a given rank are not anointed with some equivalence has empirical consequences. For example, many studies about biological diversity enumerate taxa ranked at the level of family or genus (e.g., Hominidae, Asteraceae). Although those groups may be discrete historical entities, they have no comparative equivalence merely because they are classified as families or genera. Each is simply a clade of species identified with a taxon name that happens to have a Linnaean rank. These taxa could just as easily have been ranked as a genus, tribe, or subfamily by another taxonomist. Categorical ranks have meaning only in relation to one another, hierarchically within a group. Although this distinction between taxa and ranks is widely recognized, it is frequently ignored.

Once the idea of natural classification, in which groupings represent things that exist in the world irrespective of the process that might have produced them, became more common in the nineteenth century, classification took on a larger intellectual role in debates about species (Wilkins and Ebach 2013). Darwin provided a rationale for seeing species and groups of species as the result of an evolutionary process, as historical entities (Kearney, chap. 7). Yet, the idea of using classification to impose order on our exponentially increasing knowledge of living and fossil diversity also set the stage for confusion over species themselves. Many taxonomic philosophies were being applied to ranking through the nineteenth and into the twentieth century, and this was amplified with the "modern synthesis" (Smocovitis, chap. 2). Arguably, evolutionary taxonomy fostered the breakdown of the recognition of "natural" groups, including species. The science of using taxonomy and classification to represent natural order became infused with the art of ranking that was designed to determine the Linnaean category into which the taxon should be placed. As G. G. Simpson (1961, 222–23) described it, "The eventual rank of the taxon thus initiated is usually proportional to the degree of distinction of the [adaptive] zone entered, hence the amount of basic divergence involved."

Looking at the literature in systematics, one might form the impression of chaos over what species are and how to classify them (Mayden 1997). Yet this apparent chaos seems due more to linguistic parsing than substantive differences. First, most known biodiversity was discovered and described prior to the past forty years. Contemporary arguments about species have had a marginal effect on current knowledge about global diversity, of which more than 95 percent includes arthropods and other highly speciose invertebrate groups. Second, much of that diversity was described using a basic operational idea: if a specimen is morphologically different from other described species taxa, it generally represents a new species. This emphasis on phenotype becomes less surprising when one realizes that most species are known from a relatively small number of specimens with restricted geographic distributions.

So why all the fuss? Contemporary debates concerning the nature of species often reflect different thinking about the speciation process, which influences how species are conceived or defined (Edwards et al., chap. 15). An example is the conflation of the process of reproductive isolation and speciation, on the one hand, and concepts of species, on the other hand, which is also entangled with how taxa are classified and ranked. Finally, views of species have been influenced by studies of particular kinds of organisms (animals versus plants versus bacteria) and investigators' conceptions of whether they are biologically different or how they evolved.

In sum, contemporary views of species and the process of speciation cannot be divorced from the historical imprint of classification. Different species concepts imply different methodological approaches to speciation and classification. Confusions over species-as-units-of-classification versus units-of-evolution are still with us, leading to the "species problem."

The Species Problem

Mayr (1996) argued that the "species problem" comprises two issues. The first issue is how to circumscribe species-taxa, i.e., how to group populations relative to other taxa. Many definitions of species, Mayr claimed, including the phylogenetic and cohesion species concepts, are "nothing but a recipe for the demarcation of species taxa" (Mayr 1996, 267). The second issue is how to rank taxa so that they fit the species category of the Linnaean hierarchy: "The species category is the class that contains all taxa of species rank. It articulates the concept of the biological species and is defined by the species definition. The principal use of the species definition is to facilitate a decision on the ranking of species level populations" (Mayr 1996, 267). Mayr's thinking does not address this problem because it focuses not on what species are, but on how to rank (classify) species given his own view of species.

In order to assess species concepts, one must determine what kind of entities fall under the concept(s). In short, the species problem has not two, but three components:

(i)   *Nature*: What is the ontological nature of species? What kind(s) of entities are they?

(ii) *Definition*: What is the conceptual framework for a definition of species? Which concept(s) is (are) consistent with that framework?

(iii) *Demarcation*: How do we individuate species-taxa that fit the definition(s)?

In principle, the nature of species is independent of definitions and their applications in practice. Still, many debates in both biology and philosophy have intermingled these problems. Conflation of ontological and epistemological questions is common, leading scholars to talk past one another. In an effort to parse out confusion and bring conceptual clarity, we provide a philosophical analysis of various approaches to species. One role for philosophy in the species debate—and within science generally—is to scrutinize the commitments that concepts implicitly presuppose. In what follows, we treat the expressions "species concept" and "species definition" as synonyms, a simplification consistent with most relevant literature.

Species in Evolutionary Theory: Philosophical Issues

Species are not the only ontologically controversial evolutionary entities. The nature of populations is also debated because under conventional speciation models, it is populations that become isolated and subsequently differentiate into taxa. Similarly, in population genetics, whether or not populations are treated as concrete individuals has consequences for the analysis of processes such as selection and drift (Millstein 2009; Frank and Fox, chap. 9). This section addresses issues that intersect ontological thinking about species, their origins, and their presumed participation in various evolutionary processes.

Individuals versus Natural Kinds

Scholars writing on the species problem are divided by whether they treat species as *individuals* or as *natural kinds*. What is at stake here is the "nature" component of the species problem: What kind of entities are species? Are they spatiotemporally bounded objects or are they categories of things? The crucial issue thus becomes: what makes a species an individual or a kind?

Authors seldom presuppose the same definitions, making the main claim difficult to assess. To illustrate, Ghiselin (1997, 37–49) proposes five necessary and jointly sufficient criteria for individuality: (i) It is possible for a class, but not for an individual, to have instances; (ii) Classes are spatiotemporally unrestricted, whereas individuals are spatiotemporally restricted; (iii) Individuals are concrete, as opposed to abstract; (iv) Individuals have no defining properties; (v) An individual is logically prior to being the member of any class, i.e., an individual is not an individual by virtue of its membership in some class. Contrast this with Rieppel's (2007, 376) characterization of individuality: "To consider species as individuals is to reject the idea that parts of species, i.e., the individual organisms that 'belong to' the species, share universal properties." Evidently, the individuality thesis bears very different burdens for these authors. For Ghiselin, individuality is a demanding ontological thesis that entails an outright denial that species have instances or defining properties, an explicit commitment to their physical concreteness and spatiotemporal restrictedness. In contrast, Rieppel's commitment to individuality basically amounts to a rejection of essentialism, the discredited thesis that all and only members of a species share a set of universal properties.

These considerations illustrate a general point. Authors who debate broad metaphysical positions, such as individuality, often have very different concepts, definitions, and commitments in mind. These are not nitpicky differences in terminology. Without some provisional agreement on these foundational matters, consensus or even meaningful debate is hard to obtain.

Concrete versus Abstract

Given these difficulties underlying individuality, the basic ontological question underlying the species problem might stem from a different, less theoretical issue, namely, whether species are concrete or abstract entities. This distinction might be orthogonal to the individuals-versus-kinds one (Ghiselin's multifaceted definition of individuality includes concreteness as a necessary condition). Yet concreteness is less theoretically loaded than individuality.

Most biologists, if asked whether species are concrete, would answer affirmatively. Yet, this does not necessarily imply any deep philosophical commitment. Still, some contemporary biologists take ontology seriously (Mayr 1996; Ghiselin 1997; Wheeler and Meier 2000; Hey et al. 2003). Here, the commitment to the concreteness of species and its metaphysical implications is not a by-product of technical jargon, but the result of conscious theoretical reflection: "The term 'species' refers to a concrete phenomenon of nature and this fact severely constrains the number and kinds of possible definitions" (Mayr 1996, 263).

The concreteness of species derives from the impression that what systematists count, organize, and classify are physical, natural entities. Sure, one cannot directly observe species like organisms or rocks. Still, the existence of species seems to many as clear as the existence of hydrogen atoms. This intuition seems directly opposed to the view that species can be conceptualized as abstract types, that is, objects that lack spatiotemporal dimensions, such as sets, numbers, properties, and other mathematical constructions.

Whether species are concrete or abstract is an ontological question that underlies much debate over the nature of species, evident, for instance, in the above discussion of individuals and kinds. There is, however, a widespread tendency, among both philosophers and biologists, to label the concreteness of species "realism" (Claridge 2010; Gourbière and Mallet 2010; Mishler 2010). This is a mistake. The reality and the concreteness of species are independent theses and should not be conflated. We suggest the introduction of a term of art, "concretism," to capture the very idea that species are physical objects, as opposed to being abstractions, while avoiding the quagmire of philosophical disputes between realism and antirealism.

In sum, *real*, *physical*, and *natural* are distinct predicates. There are myriad examples of entities, like tables and cars, that are perfectly real but not natural the way quartz crystals are. Likewise, fictional characters (Sherlock Holmes) and musical pieces (Beethoven's Ninth Symphony) are neither natural objects, *sensu* Mayr, nor physical. Hardly anything hinges on the reality of species, for virtually anything can be dubbed as real, in one sense or another. Hence, the claim that species are abstract entities should not be contrasted with "realism." It is "concretism" that underlies the species problem. Many biologists seem to agree (Hey 2001). Yet, we argue below, many entities designated to be species under some influential conceptions may be interpreted as being neither natural nor concrete.

Species Monism

Species monism is the thesis that there is a single correct species concept and that the job of systematists is to discover and apply it. This position is widespread. Advocates of influential species concepts, including the biological, phylogenetic, and evolutionary ones, typically believe that their very own definition provides the one true path to grouping organisms into species (Wiley and Mayden 2000b, 73; de Queiroz 2005, 6601).

At first glance, monism goes well together with concretism. Unsurprisingly, the two tenets are often conflated. If species are concrete, then shouldn't we be able to find a general way to characterize all of them unambiguously? Things are not that simple. The concreteness of an entity does not entail the existence of defining or diagnostic properties. For one, monism can also be reconciled with an abstract conception of species. Providing a set of conditions for organisms to belong to species, construed as kinds, one would thereby obtain a monistic definition of abstract species. Also, some pluralists have endorsed concretism (Ereshefsky 1992, 1999). Without entering into the details of the dispute, we can say that monism is independent of whether species are concrete or abstract. Consequently, these theses should not be conflated.

The monistic intuition is strong. However, a burning question remains: why are there so many concepts and why does no single concept cover them all (Wilkins 2003)? The large number of competing definitions, together with the failure of the biological community to reach consensus on which is better or more fundamental, has undermined faith in the existence of a single correct species concept in favor of an apparently less dogmatic pluralism.

Pluralism versus Heterogeneity

The puzzling nature of species is reflected in their multifarious definitions. Can any particular definition capture them all? Could there be different and equally legitimate kinds of species and, therefore, no way of encompassing all of them under a single overarching concept? Or, perhaps, what counts as a species cannot be captured independently of the particular inquiry or scientific goal at hand. This thinking leads to *pluralism*, the thesis that one needs multiple species concepts and definitions to individuate all kinds of taxa. Pluralism has thrived over the last few decades as an alternative to the monistic orthodoxy (Mishler and Donoghue 1982; Kitcher 1984, 1989; Dupré 1995; Boyd 1999; Mishler 1999; Hey et al. 2003; Rieppel 2007).

Biologists and philosophers alike have discussed species pluralism, often with different meanings and contexts. Hence, some caution is required in presenting the idea, which evolved along two related but distinct strands (Boyd 1999; Hey 2006). Species pluralism is often identified with the thesis that different kinds of species can be found in nature, and, consequently, different species concepts are required to account for this diversity. If birds evolve by different processes than bacteria, the argument goes, different concepts need to be adopted when studying these groups of organisms. We call this thesis "heterogeneity":

> *Heterogeneity*: Different speciation processes may produce different kinds of species, making species-taxa heterogeneous. The individuation of species thus requires different concepts to accommodate this diversity.

Biologists often conflate heterogeneity with pluralism. This is problematic because it obliterates the distinction between heterogeneity—the denial of species universalism (Hull 1997, 1999a)—and a stronger thesis that is independent of whether prokaryotes and eukaryotes can be clustered in the same kinds of species. This latter is the claim that assignments of species-level taxa are always relative to a particular scientific theory, aim, or classificatory purpose (Dupré 1981; Kitcher 1984; Ereshefsky 1992; Dupré 1995; Boyd 1999). We dub this "pluralism" to distinguish it from heterogeneity as defined above.

> *Pluralism*: There is no single correct species concept. Assignments of species-level taxa are always relative to the organisms and processes studied and the explanatory target at hand.

The general idea is the following. Suppose that species concept A individuates organisms x and y as conspecific, whereas concept B treats x and y as belonging to different species. Monists are committed to the claim that (at least) one of these incompatible hypotheses must be erroneous. Pluralists, however, can argue that both are equally correct since they are relativized to different concepts. If species are inquiry-dependent, they do not exist independently of the underlying theory and no concept is more fundamental than any other. From this standpoint, it makes no sense to ask whether the biological species concept or the evolutionary species concept is "better" *tout court*. The two concepts can be assessed only relative to, say, a given theory of origins. You specify your explanatory target, the pluralist claims, and I'll tell you which species concepts work better.

Unveiling the philosophical commitments of pluralism requires dispelling some common misunderstandings. First, from a methodological perspective, pluralism is more radical than heterogeneity. Whereas heterogeneity is consistent with the existence of a single correct way of clustering groups of organisms into species, pluralism overtly rejects this idea because, in principle, there is no single correct standard for uniting organisms or populations as members of a species. According to the pluralist, two incompatible groupings of organisms or populations into species can be equally correct, when relativized to different goals or theories.

Second, pluralism is sometimes criticized on the grounds that it fails to "settle" the species problem. Hey (2006, 448) laments, "Even if one commits to pluralism, there are still many ways of being philosophically pluralistic about species. . . . In short, species-concepts pluralism can be seductive, but it might not actually help to settle anything." Hey is correct that there is a "pluralism of pluralisms" in the literature. Yet this line of thought fuels a misunderstanding. Pluralism does not offer an easy answer to which species concept is better. Pluralism purports to offer a sketch of an explanation of why there cannot be, in principle, a single best definition of species. In short, monism and pluralism are not solutions to the species problem. They are frameworks within which to (we hope) find a solution. We should not expect either philosophical stance to settle anything.

Third, although some come close to endorsing the position that any grouping of species is acceptable or that clustering of populations into species is arbitrary (Dupré 1995), pluralism per se does not imply that anything goes. Relativizing species concepts and definitions to biological goals need not forgo objectivity. There might be independent reasons to prefer some theoretical goals over others, leading one to adopt the concepts posited in such frameworks (Kitcher 1989).

Finally, pluralism is commonly taken to be a form of philosophical antirealism. Hull (1999a), for instance, argues that whereas the realism-monism and the pluralism-antirealism combinations are rather natural, the association of monism with antirealism or of pluralism with realism would be weird. Nonetheless, a pluralistic conception that does not deny the reality of species has been articulated by various authors (e.g., Kitcher 1984; Dupré 1995; Boyd 1999; Dupré 1999; Wilkins 2003; M. H. Slater 2013). Specifically, if pluralism is intended as heterogeneity, then it is in tension with neither realism nor concretism. The claim that no single species concept applies to all organisms is compatible with both the reality and the spatiotemporality of species; at most, it amounts to a rejection of radical monism. In rejecting the realism-pluralism combination, we surmise, Hull refers to a stronger reading of pluralism, related to our reformulation above. Indeed, the simultaneous correctness of conflicting or incompatible species concepts is more problematic to reconcile with concretism (Cracraft 1983, but see Boyd 1999; Hull 1999a; R. A. Wilson 1999). According to the pluralist, species are inquiry-dependent; there is no such thing as a species independent of a theory that clarifies the intended meaning.

It should now be clear why we have insisted on keeping realism and concretism distinct: it helps avoid misunderstandings and reveals the likely source of disagreement. Hull sets up the wrong opposition. It is concretism, not realism, that is in tension with pluralism. Full-blooded pluralism (as opposed to heterogeneity) is a form of nominalism or conventionalism that goes well with the idea that species are sets or other abstract entities, but not with the idea that there are competing species concepts, all of which involve equally concrete entities. Some authors have understood and made explicit this antifoundational dimension according to which there are no bedrock entities that are theory-independent and thus constitute the foundation of a clustering of organisms into species taxa (Kitcher 1984; Dupré 1999). In contrast, other authors have attempted to promote a more ecumenical pluralism that retains the concreteness of species (Rieppel 2007). Such views, we maintain, have not (yet) solved the tension underlying the species problem. It is questionable whether a pluralistic view of species is compatible with their status as concrete natural objects.

In conclusion, both monism and pluralism are viable approaches to the species problem. But each comes at a cost. While monism can be straightforwardly reconciled with concretism, it faces the difficulty of explaining the plethora of competing species concepts and the sources of the disagreement. If there is a single adequate species concept, what's wrong with all the others? Pluralism accommodates this diversity. However, it contrasts with concretism, an assumption that many scholars—biologists in particular—do not find negotiable. The fundamental issue boils down to: if species are concrete evolutionary taxa, do we really need multiple concepts?

Species in Evolutionary Theory: Biological Issues

This section explores the interface between philosophical considerations and the biological world, including the theory, methodology, and data impinging on how species are conceived and used by biologists. This is not an idle exercise. As Ghiselin (1989, 65) pointed out, perhaps with a bit of overstatement, "The philosophy of biology . . . should be an effort to come to grips with, and solve, problems in both branches of knowledge." Our discussion mirrors that viewpoint, which is why we suggest that many differences among species concepts derive less from data or theory and more from different ways of seeing the nature of species entities and conceptually organizing those observations.

Individuality, Lineages, and the Boundaries of Species Taxa

How are we to interpret the alleged concreteness of species in relation to their role in evolutionary theory? For instance, how, exactly, should we conceive of spatiotemporal restriction as a species' boundary, which has clear implications for understanding and evaluating species concepts? Phylogenetic trees, as abstract representations of history, have branches (edges) that imply relative relationships among terminal concrete taxa, whether extant or fossil (Kearney, chap. 7). In a multidimensional tree-representation framework, Hennig (1966) depicted species branches as populational envelopes that have a temporal lower bound in their origin via speciation (a branching event) and then later in time, when they themselves are subdivided into taxa or became extinct. This depiction is arguably the most intuitive way to think about the temporal component, but not the only one. Many paleontologists, for example, have interpreted stratigraphic sequences of specimens as a species transforming in situ over time into another species without branching (speciation by anagenesis).

This way of representing the speciation process over time raises philosophical and scientific issues. For instance, it calls for an ontological position regarding lineages (Haber 2012), a term that has been applied to virtually anything having a history, from cells to people to stars, and is part of the general principles of the theory of evolution (Mindell and Scheiner, chap. 1, table 1.3). In the Hennigian view, lineages might be taken to be the populational envelopes (within branches), the species themselves (an individual branch), and also monophyletic groups of species. In some sense, these branches can be interpreted as abstract concepts. If we say that branches represent an individual, continuously evolving species, this might imply that all but terminal branches are a series of ancestral species. Yet specifying their ancestral status is nontrivial, both conceptually and methodologically. Given that these putative species end at a branching event whose subdivision is reconstructed, not directly observed, the "ancestral species" does not give rise to anything other than allopatric subpopulations. Those populations may or may not subsequently differentiate, and unless they do, there is no speciation, only population fragmentation. Today many, perhaps most, species are composed of populations that are fragmented across the landscape. Typically, there is interconnectivity among them over time, but a portion may be strongly isolated by a barrier that allows those populations to have an independent evolutionary trajectory. In a temporal context, therefore, the nature of a species would seem to be dependent on substantial theoretical preconceptions, and the notion of "ancestral species" is also ambiguous from an ontological perspective (Cracraft 1983).

The spatial component of spatiotemporality is, likewise, problematic. In the neontological world of extant species, distributions are fuzzy and indeterminate. Consequently, at best we might draw a boundary around all of the recorded observations, perhaps incorporating environmental information and modeling, and call that the distribution. As concrete entities, species exist in a place, but we cannot specify it in a precise manner. Distributions in the paleontological world entail if anything even more uncertainties.

Alternative species concepts rely on different criteria to individuate species boundaries. All concepts must and do have a notion of populational (intrapopulational or interpopulational) cohesion because none of them would place males and females, or life-stages, in separate species. This is a well-known observation. Still, some have argued against particular species concepts using the spurious claim that those concepts are not populational. Thus, interbreeding is universally seen as a cohesive process, which, in some sense, would seem to entail a boundary. Systematists have traditionally proposed shared phenotypes (e.g., diagnostic characters, behaviors) as indicative of boundaries, and in the fossil record the phenotype is all one has. Those shared phenotypes are often taken as evidence for the cohesive unity of populations into taxa, and differences among populations are interpreted as lack of cohesion across space.

One flip side of cohesion, of course, is reproductive isolation (Edwards et al., chap. 15). This is perhaps the most frequently invoked idea for a species boundary, as it is the cornerstone of the widely used biological species concept, where it has been invoked as the basis for individuating species (Mayr 1992). It would be a misreading of history to believe that the introduction of the biological species concept and reproductive isolation changed our worldview from one of seeing species as classes to one of seeing them as concrete entities. Moreover, reproductive isolation as a benchmark of species boundaries is fraught with difficulties and carries a steep burden as an arbiter of individuality or concreteness.

Species, Subspecies, and the Nature of Speciation

We now return to the influence of classificatory units on the evolutionary origin of species. In particular, we explore the ontological status of subspecies (or any infraspecific taxon) because they have long played a role in classification, evolutionary theory, and the species debate. Interestingly, subspecies taxa are used to characterize the taxonomy of only a small portion of Earth's biodiversity. Subspecies are most commonly applied to birds, mammals, some butterflies, and a few other groups in which attempts have been made to describe fine-grained geographic variation. Subspecies are used sparingly (when used at all) in most of the diverse groups of invertebrates. Formal infraspecific taxa, such as varieties, are relatively common in plant groups. But it is in birds and mammals that subspecies have played a large role in species concepts and evolutionary theory.

Subspecies have been said to give rise to species (Mayr 1942, 154–55; G. G. Simpson 1953b, 280–81), and like species, they could be judged to be concrete entities. Both Simpson and Mayr, like a host of systematic biologists after them, were conflicted over the ontological status of subspecies. Mayr (1969, 193), perhaps more than Simpson, saw subspecies as a concept of classification rather than of evolution. Yet both saw subspecies as things that often have something to do with the generation of species. Their ambiguity over subspecies characterizes much of the history of speciation analysis, namely, as a progression of differentiation from small differences (subspecies) to bigger differences (species) that are inferred to be sufficient to result in isolation. This view, and the conflicts it raises, is alive and well in various biological disciplines (Haig and Winker 2010).

These considerations raise important ontological issues. If one thinks of subspecies as units that can encompass any arbitrary set of organisms, or as part of a continuum, then how do we determine their boundaries? If subspecies are generated by a biological process of "subspeciation" (Phillimore 2010), or if subspecies can be actors in evolutionary processes, giving rise to species (Simpson, Mayr), they cannot be mere conveniences or artifacts of classification but must be concrete. But if subspecies are concrete entities with diagnostic morphological characters, what is the ontological difference between them and species? Are they different from species only in their degree of distinctness?

In deconstructing the process of "speciation," the main issue is perhaps not the species, in all its definitional guises, but the *taxon*. Indeed, ranking is getting in the way of thinking about the origin of taxonomic diversity (de Queiroz 2011). Would our ontology of the evolutionary process change if instead of speaking of the "origin of species" we were speaking about the "origin of evolutionary taxa"? As de Queiroz notes, Darwin himself was influenced by species-as-ranks, but by seeing branching as key, as did Hennig, he pointed to a different way of recasting the process: (1) conceptualizing a cohesive lineage of populations through time that becomes spatially isolated, and then (2) recognizing those isolates as having become differentiated to the point of being recognizable (diagnostic) as "evolutionary taxa." Effectively, recognizing two taxa that are each other's closest relatives with reference to a third is evidence that branching has occurred. Perhaps the concern about what rank they are becomes immaterial relative to envisioning them as markers of a branching history. Cartoons of the historical process, especially paleontological cartoons, tend to obscure and constrain our vision, which must be based on evidence provided by individual organisms and the similarities and differences among them. If one does not have an ontological commitment to taxa as concrete entities, then all bets are off.

In conclusion, a dissection of the ontological status of subspecies versus species creates problems for evolutionists who tie themselves to classification systems that treat subspecies taxa as if they were evolutionarily relevant rather than artifacts that systematists created at one time to deal with gradations in geographic variation. It is difficult to maintain that subspecies are not concrete and yet name them, or that they are discrete entities and yet equate them to species with respect to the processes that produce them (isolation and differentiation). This conceptual conflict has arisen from the long-term entanglement of classification and species biology, and it has served neither well.

A Hierarchy of Species Concepts

Extensive taxonomies of species concepts have been proposed (Mayden 1997; de Queiroz 1998). Yet minute differences among definitions and strategies for individuating species have inflated the rhetoric and obscured similarities while preventing discussion about their ontological implications. Our goal is to parse out some of the ontological underpinnings of four widely used groups of species concepts. Although this list is not exhaustive, most current definitions fall into one of these families. Advocates of these four concepts tend to see species as concrete units, as individuals out there in nature. The central differences lie in how they circumscribe the boundaries of those entities. Thus, an ontological understanding of alternative concepts stipulates whether they individuate the world differently— cut nature at the same or different joints. On this latter point, it is essential to understand that we are talking about the boundaries of taxa as distinct from populations. Systematists agree that a cluster of monophyletic species, a clade, is also a concrete historical unit. Species concepts must specify how those two concrete things are different, although some do not do so.

Various species concepts have been advanced as primary because they are theoretically based, as opposed to secondary ones, which are supposed to be more methodologically prescriptive, or diagnostic. This distinction is suspect as all species concepts entail some theoretical assumptions as well as other criteria noted above. Since the inception of contemporary biology, the fundamental function of species concepts has been to aid the discovery and recognition of diversity, not to convey a foundational, theoretical understanding of species, and this is still largely true today.

Evolutionary Species Concept (ESC)

The evolutionary species concept (ESC) was originally proposed by G. G. Simpson (1951, 1961), and further developed by Wiley (1978, 1981) and Wiley and Mayden (2000b, a):

> [A species is] an entity composed of organisms which maintains its identity from other such entities through time and over space, and which has its own independent evolutionary fate and historical tendencies. (Wiley and Mayden 2000b, 73)

It is avowedly a theoretical lineage concept, with "maintains its identity" signifying individuality and "independent evolutionary fate and historical tendencies" signifying divergence from other lineages as evidenced by character analysis (Wiley and Mayden 2000b, 75). Wiley and Mayden (2000b) argue for an ontological distinctness between species taxa and supraspecific taxa (two or more species-level taxa). Under their ontological distinction, parts of species (individuals, populations) have ancestor-descendant relationships, and, as lineages, species give rise to other species, but supraspecific taxa do none of these things. Despite its popularity and evolutionary context, the ESC raises some ontological quandaries. First, the boundaries of the entities specified by the ESC are ambiguous. The above definition does little to distinguish species from monophyletic clusters of taxa, which also have separate identities and histories as historical individuals. Moreover, independent evolutionary fates and tendencies, to the extent that they can be empirical descriptors, are not unique to species entities. For example, populations can maintain their identity over space and time, which is why they too can be considered individual entities (Millstein 2009). Populations can also split into subpopulations, suggesting that populations can have independence, leaving the distinction between population and taxon up in the air. Although applying the ESC must depend on character evidence to establish the population-taxon boundary, the definition itself entails a fuzzy ontology. Wiley and Mayden (2000b) endorse Hennig's (1966) idea of recognizing the "joints" of species at lineage branching events as clear theoretical boundaries for the beginning and end of a species taxon, but it must be made explicit whether a boundary splits populations or taxa.

General Metapopulation Lineage Concept (GMLC)

De Queiroz (1998) introduced the general lineage species concept as a simplifying solution to the species problem. He argued that entities identified or implied as species under the available species concepts, despite definitional differences over criteria for delimiting species taxa, are consistent with his lineage concept, which was subsequently renamed the "general metapopulation lineage concept" (GMLC, de Queiroz 2005). While a formal concept or definition has not been stated, the general idea is rather clear:

> I do not mean to say that there are no conceptual differences among the diverse contemporary species definitions but rather that the differences in question do not reflect differences in the general concept of what kind of entity is designated by the term species. All modern species definitions either explicitly or implicitly equate species with segments of population-level evolutionary lineages. I will hereafter refer to this widely accepted view as the *general lineage concept of species*. (de Queiroz 1998, 59–60)

Claiming all species concepts are consonant with a general lineage concept of species is little more than a tautology, given that species must necessarily be composed of population lineages, individual organisms, and so on. Moreover, the GMLC prompts the question: if two or more of those species concepts actually individuate species taxa differently, then why do those conflicting hypotheses of species suddenly become consonant under a general lineage concept? In short, the GMLC leaves us wondering how many species there are and, most important, what they are.

De Queiroz (1999, 63) proposes, "there is only one necessary property of species—being a segment of a population level lineage." Still, population histories can be expected to involve repeated splitting and coalescing over time, and the mere fact that they do sometimes split does not constitute a necessary and sufficient framework for species boundaries, theoretically or empirically. Under the GMLC, the boundary between an evolving population and an evolved taxon is not marked by anything other than a split, which makes it difficult to understand the boundary between population and species:

> [M]etapopulation lineages do not have to be phenetically distinguishable, or diagnosable, or monophyletic, or reproductively isolated, or ecologically divergent, to be species. They only have to be evolving separately from other such lineages. (de Queiroz 2005, 6005)

Phylogenetic Species Concept (PSC)

[A species is] the smallest diagnosable cluster of individual organisms within which there is a parental pattern of ancestry and descent. (Cracraft 1983, 170)

[A species is] the smallest aggregation of (sexual) populations or (asexual) lineages diagnosable by a unique combination of character states. (Wheeler and Platnick 2000, 58)

A species is the least inclusive taxon recognized in a formal phylogenetic classification. (Mishler and Theriot 2000, 46–47)

At first glance, these species concepts share similarities in acknowledging the theoretical distinction between populations (or population lineages) and taxa by stipulating the diagnosability of the latter. Thus, the concept of diagnosability (or detectable, character support) is critical for the notion of species units to be concrete. These definitions also set a boundary between a single taxon and a cluster of taxa by stipulating that species are the smallest taxonomic unit, which increases the objectivity of species in the sense that it eliminates infraspecific taxonomic units and any related ranking quandary. Thus, none of these concepts would admit subspecies to be concrete entities.

Over the years there have been substantial arguments among advocates of this species concept including whether phylogenetic species can and should be described as being monophyletic. The term "diagnostic" has taken on different meanings among investigators. None of these debates, however, seemingly creates a critical ontological divide over what species are.

Biological Species Concept (BSC)

Species is a dynamic rather than a static entity, and the essential feature of the process of species differentiation is the formation of discrete groups of individuals which are prevented from interbreeding with other similar groups by one or more isolating mechanisms.

—Theodosius Dobzhansky (1937a, *419)*

Species are groups of interbreeding natural populations that are reproductively isolated from other such groups.

—Ernst Mayr (1969, *26)*

The BSC is favored by many evolutionary biologists working on sexually reproducing organisms, but it has received substantial critical analysis from systematic biologists and philosophers (Kitcher 1989; Velasco 2008). Its theoretical lynchpin is the idea of reproductive isolation of populations, which is used to establish boundaries between taxonomic units (Edwards et al., chap. 15). However, it is well known that in many groups of organisms taxa that are phylogenetically and temporally deep can fail to develop reproductive isolation. This presents the first problem for the BSC, as it often cannot establish the appropriate boundaries among evolutionary units. The second problem is the ranking conundrum discussed earlier in which evolutionarily distinct taxa, sometimes not closely related, can be combined into a single biological species if it is known or suspected that their members might interbreed. Consequently, in many biological species, their boundaries belie the notion of the BSC as individuating a single concrete entity. And even when all the taxa combined in a single biological species are discrete evolutionary units, that biological species is acting as a higher taxon. There are many cases in which biological species are monotypic and thus can be taken as unitary concrete entities, but at the same time virtually all other concepts would also recognize them as such. Yet because the ontological status of biological species is ambiguous, its role in understanding speciation has been misunderstood. As noted, all species concepts include assumptions about reproductively cohesive populations.

Is There a Hierarchy of Species Concepts? Second-Order Pluralism

Some biologists, such as Mayden and de Queiroz, have attempted to reconcile the tension between concretism and pluralism by advancing a second-order pluralism with respect to species concepts. Roughly speaking, if the various competing species concepts are organized and visualized in a maplike structure, then they can all be equally correct, thus reducing all differences to an epistemological level, while maintaining both the monistic and the concretist intuition at the level of ontology. Let's assume that the dispute between species monism and pluralism is tied to the multiplicity of processes producing species-taxa. Thus, we might have agamospecies—a group of asexually reproducing organisms—on the one hand, and everything else on the other. One might argue in this case that there are two different kinds of species; hence the need for two definitions and a prima facie vindication of the heterogeneity thesis discussed above. But then, would instances of sympatric speciation or speciation through hybridization require different concepts? Even in a group dominated by allopatric speciation? But what if we expect that the evolutionary process, despite all the mechanisms that cause evolutionary change, always produces evolutionary taxa that have boundaries and can be distinguished from each other? Is that a (partial) solution to the species problem? Does it vindicate the existence of a single, general species concept?

One way or another, we need to be less prescriptive about how processes or products of evolution are imagined to be different across life. We cannot build a comparative framework for biology if nature is incompletely or erroneously partitioned into different kinds of species, for then regularities or patterns could be difficult to discern. So thinking and arguing about what is out there—not the semantics of a definition—will remain an important issue.

Regardless of the ecumenical effort to justify and give a role to several competing species concepts, both Mayden and de Queiroz turn out to be monists and, presumably, concretists with respect to the nature of species. Differences between competing species definitions are said to be mostly pragmatic or epistemological. Mayden and de Queiroz appear to claim that several species concepts can coexist at the level of individuation, but that is an attempt to reconcile monistic metaphysics, which treats species as concrete entities in nature, with a pluralistic idea that there is no single correct way of individuating them. However, this second-order pluralism does not do justice to pluralism because their frameworks do not agree with either of the two dimensions of pluralism discussed above.

Concluding Remarks

This chapter has attempted to uncover some philosophical issues that underlie the debate over the nature of species and, more generally, questions about the ontology of biological entities. Concretism and pluralism are both compelling assumptions, and yet they are hard to reconcile. The species problem thus confronts us with a dilemma. Concretism pins down the pretheoretical status of species but encounters the problem that no species definition may be comprehensive enough. Pluralism, in turn, acknowledges the difficulty of finding a single correct species concept and explains it by appealing to the fact that the notion of species is theory-dependent or inquiry-dependent. Recent attempts to reconcile the two intuitions such as those advanced by Mayden and de Queiroz do not resolve that tension.

The evidence for the existence of species seems to be independent of any particular theoretical framework in biology. Although different views about the process of speciation might imply different species definitions or concepts, that organisms are organized into species seems independent of any particular view of speciation. Indeed, the evidence led biologists to postulate the existence of species long before the post-Darwinian evolutionary framework. It could be argued that Aristotle had a notion of species that was not fundamentally distinct from some contemporary concepts, and deeply essentialist taxonomic systems such as that of Linnaeus were not radically modified by the evolutionary worldview. What profoundly changed with the work of Darwin and Wallace was our understanding of the origin and history of these entities, not the evidence that we have for believing that organisms in nature are clustered into species.

The idea that species-level entities are concrete and natural would seem to suggest that they are actors in one or more processes in nature. After all, if species were inert bystanders, how would we justify their prominence within biology? But what exactly do species do? Many evolutionary biologists think that species are active participants in multiple processes in nature. For example, species are said to speciate or go extinct, compete or predate, occupy niches or adaptive zones, interact with their environment and disperse. Species likely do none of these things (Holsinger 1984; Cracraft 1989; M. B. Williams 1989). Rather, these are all things that individuals or populations do. Species do little, or more likely, nothing. That is, they do not participate as actors in processes. Even assuming the lineage-through-time idea of "ancestral species" discussed earlier, individually isolated populations are the entities that differentiate and can be said to turn into species (i.e., speciate, in the context of allopatric speciation models). Therefore, even notions of "species as evolvers" (M. B. Williams 1989) might need to be revisited.

Our view of species-as-actors opens the door to thinking more mechanistically about many processes in nature and conceptualizing causation at a more appropriate hierarchical level, usually that of individual organisms, but there may be an argument for proposed selective processes of and among species within macroevolutionary theory (Jablonski, chap. 17). This view of species and their participation in causal processes should not be taken to mean that species are not crucially important for biology. They are the fundamental currency, along with individuals and populations, of evolutionary and environmental science. As we note, since the dawn of humanity there has been the clear idea that the living world around us is structured into different clusters of organisms with different characteristics. Knowing about this diversity has been the basis for human well-being, and over time knowledge about those "things" has become increasingly more sophisticated and fine-grained.

This chapter argues that debates about species entail deep ontological conceptions of the entities of nature that transcend conventional allegiance to a given species concept. Species entities are fundamentally important because they largely determine how we describe pattern in nature and think about causation, from describing and mapping diversity and building the tree of life (McInerney, chap. 5; Kearney, chap. 7; Edwards et al., chap. 15), to deciphering selection of natural units (Frank and Fox, chap. 9; Goodnight, chap. 10; Fox and Scheiner, chap. 11; Poisot, chap. 12; Scheiner, chap. 13; Orive, chap. 14), and describing geographic variation and speciation (Gillespie, chap. 16; Jablonski, chap. 17). Moreover, the structure of most environmental science is based on views about nature's entities, with species usually being the most important. With so many people, scientists and nonscientists alike, having such a long-standing stake in the intellectual debate about species, this is a game that is likely not to end anytime soon—and may very well never end.

Acknowledgments

We would like to express our gratitude to Patrick Forber, Roberta Millstein, Brent Mishler, and Matthew Slater for constructive comments on various versions of this chapter. They greatly improved our thinking and the presentation. We are especially grateful to the editors, Samuel Scheiner and David Mindell, for inviting us to contribute to this volume and providing feedback and support. JC acknowledges the National Science Foundation (awards 1241066 and 1146423) for research funding.

SEVEN

The Tree of Life and the Episodic Evolutionary Synthesis

MAUREEN KEARNEY

The Stars Are Indifferent to Astronomy

—Peter Caws (used as an album title by Nada Surf, *2012*)

The tree of life is the commonly accepted metaphorical representation used to depict the phylogenetic relationships of all organismal species lineages. It is grounded in phylogenetic theory—the theory that all species share a genealogical history as a result of the evolutionary process of descent with modification from common ancestors (table 1.3, proposition 3). And it is grounded in the theory of natural selection—the theory that evolutionary change over time is due to differential survival and reproduction of individuals and changes in variable traits (table 1.3, proposition 6). Phylogenetic theory and natural selection theory are closely related. Their interconnectedness was, in fact, foundational for the development of a general theory of evolution, as evidenced throughout *On the Origin of Species* (Darwin 1859).

Darwin (1859, 99) on phylogenetic theory: "The affinities of all the beings of the same class have sometimes been represented by a great tree. I believe this simile largely speaks the truth. The green and budding twigs may represent existing species; and those produced during former years may represent the long succession of extinct species. . . . As buds give rise by growth to fresh buds, and these, if vigorous, branch out and overtop on all sides many a feebler branch, so by generation I believe it has been with the great Tree of Life, which fills with its dead and broken branches the crust of the earth, and covers the surface with its ever-branching and beautiful ramifications."

Darwin (1859, 50) on natural selection theory: "As many more individuals of each species are born than can possibly survive; and as, consequently, there is a frequently recurring struggle for existence, it follows that any being, if it vary however slightly in any manner profitable to itself, under the complex and sometimes varying conditions of life, will have a better chance of surviving, and thus be naturally selected. From the strong principle of inheritance, any selected variety will tend to propagate its new and modified form. . . . I have called this principle, by which each slight variation, if useful, is preserved, by the term of Natural Selection."

A vast literature on Darwinism revolves primarily around the theory of evolution through natural selection, but Darwin's theory of phylogeny is just as prominent in the *Origin*. As Sober (2011, 9) notes: "The two big ideas in Darwin's theory are common ancestry and natural selection." The boundaries of these theories are fuzzy, yet logically distinct. If very different living species connect back to a common ancestor, then within-lineage changes must also have occurred in order for their distinctive features to have accumulated over time. Common ancestry of species, plus descent with modification occurring over time within species lineages, formed a holistic theory of evolution.

Darwin's theory of phylogeny was first visually captured in his iconic "I think" tree-like diagram in Notebook B, "The Transmutation of Species" (1837, fig. 58). Later, Darwin (1859, 411) wrote: "From the first dawn of life, all organic beings are found to resemble each other in descending degrees, so that they can be classed in groups under groups. This classification is evidently not arbitrary like the grouping of the stars in constellations." This is the first clear conceptualization of a theory of common descent of all species, which formed the basis for the nested hierarchy of species depicted in the tree of life. Darwin's "groups within groups" is an example of an inclusive hierarchy. Inclusive hierarchies contain nested entities, with each entity a subset of higher entities within the hierarchy. In Darwin's case, an inclusive hierarchy of all species resulting from common ancestry is most efficiently depicted by a nesting diagram such as a genealogical tree—a structure that intentionally, and importantly, represents taxa as having a part-whole relationship: "Thus, between A&B immense gap of relation, C&B the finest gradation, B&D rather greater distinction. Thus genera would be formed.—bearing relation to ancient types with several extinct forms" (Darwin 1859, 37).

Whereas inclusive hierarchies (e.g., species lineages, languages, cultures, families) are naturally represented by branching, tree-like structures or nested subset diagrams, exclusive hierarchies (e.g., military ranks, functions, or the Scala Natura) are typically represented by lists or ranks and the entities within such hierarchies do not exhibit a part-whole relationship.

Darwin was quite clear about his use of a tree diagram and what it could convey. His use of the term "simile" ("The affinities of all the beings of the same class have sometimes been represented by a great tree. I believe this simile largely speaks the truth" [Darwin 1859, 99]) and the broader context of his text illustrate that his tree of life was intended not as a literal portrayal of the evolutionary process leading to speciation and extinction but as a metaphor for the shared genealogy of species. This is an important point because recent challenges to the tree of life described later in this chapter are often based on observations of nondivergent genetic processes and do not make the distinction between a tree of life pattern of organismal-level genealogy and evolutionary processes.

At least as early as Darwin, then, we observe the mutualism between phylogenetic theory, natural selection theory, and a general theory of evolution. The modern evolutionary synthesis of the early twentieth century (Huxley 1942) profoundly expanded this evolutionary paradigm through the integrative work of diverse natural scientists. Building on Darwinism, the modern evolutionary synthesis unified theoretical and empirical results from previously disconnected disciplines into a working consensus about how evolution operates (Smocovitis, chap. 2). It also facilitated a new understanding of how much evolution explains about the natural world— the explanatory power of evolutionary theory was found to be expansive across the natural sciences. As Ernst Mayr (1963, 1) remarked: "The theory of evolution is quite rightly called the greatest unifying theory in biology." Indeed, the modern evolutionary synthesis gave birth to the new and general discipline of evolutionary biology, which integrated many subdisciplines. Evolutionary biology continues to grow and become refined as new organismal, genetic, ecological, systematic, and geological knowledge accrues. The discovery of DNA, plate tectonics, unique fossil forms, and ecological-evolutionary processes have contributed to an evolving evolutionary synthesis.

This growing evolutionary synthesis is not a smooth and linear process; instead it is characterized by periodic disruptions and reintegrations. Over time, new technologies or new fields of study emerge that are so intensely rich in data collection and/or research potential that synthesis is interrupted for a period of time while specialized studies rapidly advance. Recent examples of the latter include developmental biology and molecular biology. During the latter part of the twentieth century, as molecular biology and knowledge of the molecular basis of genetic change grew exponentially, a division between organismal biology and molecular biology emerged. Architects of the modern evolutionary synthesis such as Ernst

Mayr and G. G. Simpson remained prominent defenders of organismal and evolutionary biology and argued strongly against the reduction of evolutionary and organismal biology to genetics. Their arguments were rooted in the inherent complexity of organismal biology; evolutionary biology could not be reduced to genes, molecules, physics, or chemistry owing to emergent properties of complex biological systems. Mayr (1959) described population geneticists' view of evolution as "bean-bag genetics" and reiterated the importance of macroevolutionary issues such as transformation of species and of unified research approaches.

Today, a rich explosion of new data and process knowledge about organismal genomes is exponentially accumulating. Genomics is a powerful new tool that has contributed to major new discoveries about biodiversity, ecology, and the processes of genetic change. For example, the prevalence of lateral gene transfer—the movement of genetic material between organisms other than by parent-offspring vertical transmission—is remarkable in some taxonomic groups. This discovery has stimulated a dramatic contemporary debate that is challenging Darwin's tree of life itself. Some evolutionary biologists contend that a tree of life does not exist, citing genetic processes such as lateral gene transfer to claim that a tree-like pattern of species cannot be accurate (see "The Anti-Tree Argument" below). This debate has also attracted the attention of philosophers of biology since the subject matter intersects with many philosophical issues in evolutionary biology—the historical nature of organisms, species, lineages, and phylogeny, as well as more general issues regarding theoretical consilience versus pluralism. If the anti-tree arguments are correct, this would mark a profound turning point in the history of evolutionary biology because, as explained below, the tree of life has become utterly foundational in all of modern comparative biology. It would also require reconciliation with results from the continuing evolutionary synthesis of the last 150 years, which extends over many natural science disciplines (Smocovitis, chap. 2).

In this chapter, I review the growth of phylogenetics and the profound influence of the tree of life on all of biology. I summarize the current anti-tree-of-life argument, from both biological and philosophical perspectives, and suggest questions for further investigation in both those areas. I also consider the tree of life debate in terms of the future of the evolutionary synthesis. I argue that this recent challenge to the tree of life is a symptom of a reductive, gene-centric interpretation of organisms, species, and phylogeny. Further it is reflective of a current disconnection between several subdisciplines of evolutionary biology. Many contemporary evolutionary biologists are focused on the burgeoning power of genomics, but genomics has not yet fully integrated with other fields of biology such as paleontology, ecology, and morphology. I conclude that a resynthesis in evolutionary biology including new integrative approaches and conceptual work will lead to a more complete theoretical foundation. In the face of rapidly advancing knowledge gains and technology, there remains much exciting work to do toward a unified theory of the organism (from gene to form), species, and phylogeny across all forms of life.

The Growth and Influence of Phylogenetic Biology

The Development of Systematics

Aristotle (384–322 BCE) held an essentialist view of species as eternal and immutable and characterized features of organisms accordingly. This typological view of nature persisted for centuries and is reflected in early biological classifications that were based on logical divisions rather than inclusive hierarchies. Linnaeus's (1758) system of classification was fundamentally based on the Aristotelian tradition of logical dichotomization and became formalized under the familiar binomial system of taxonomic nomenclature, although that system is undergoing an evolutionary challenge today (de Queiroz 1988; de Queiroz and Gauthier 1994). It was Darwin's (1859) evolutionary theory that laid the groundwork for rejection of an essentialist notion of species and their classification, emphasizing the variability that must exist in order for natural selection to occur. Following on that, Ernst Mayr's (1942) seminal work on population variability versus typology became a cornerstone of the modern evolutionary synthesis (e.g., Dobzhansky 1937b; Mayr 1942; G. G. Simpson 1944b; Mayr 1964) and greatly influenced the field of systematics. The last half of the twentieth century witnessed several methodological revolutions in systematics (Kearney 2008), while at the same time the prominence of systematics within the broader field of biology grew steadily (O'Hara 1997; Baum and Smith 2013). Systematics has experienced some extraordinary paradigm shifts in its theoretical and methodological approaches. Looking back, we can see that each new approach in systematics contributed something important to the discipline as it is practiced today (see also Hull 1988).

Evolutionary taxonomy (e.g., G. G. Simpson 1961; Mayr 1969) made heavy use of Darwinian evolutionary theory and organismal expertise to classify organisms and assess the degree of evolutionary change within and between lineages. In contrast to pre-Darwinian taxonomic endeavors, evolutionary taxonomists built explicitly evolutionary trees rather than

Linnaean classifications—a significant conceptual leap in systematics practice. The variability required for natural selection to occur refutes typology, and therefore evolutionary taxonomists rejected prior typological classification methods.

Later, the numerical taxonomists advocated an approach to systematics that used computer-assisted, quantitative, explicit methods to analyze overall similarity of features of organisms (e.g., Sneath and Sokal 1973). To some extent, phenetics combined with the burgeoning application of computer science to biology may be viewed as a backlash against what were perceived as the subjective and unrepeatable methods of evolutionary taxonomy. But numerical taxonomy came under criticism for many reasons, including the fact that overall similarity is not a biologically meaningful basis for grouping organisms (Mayr 1964; Farris 1983). Further, its explicitly theory-free approach—described as the "look, see, code, cluster" method by Hull (1970, 31)—eventually contributed to its demise. Numerical taxonomy can be credited, however, for introducing quantitative approaches to phylogenetic analysis that still permeate the discipline today.

Hennig (1950) argued that taxonomy should reflect phylogeny, that genealogical relationships among species should be inferred from "special similarity" or shared derived characters, and that these relationships should be arranged in a hierarchical manner to reflect the theory of descent with modification. The critical distinction between monophyletic groups (groups of organisms consisting of a most recent common ancestor and all of its descendants, such as Mammalia), paraphyletic groups (groups of organisms consisting of a common ancestor and some, but not all, of its descendants, such as Reptilia), and polyphyletic groups (groups of organisms that do not include the common ancestor of all members of the group, such as Agnatha) is one of Hennig's most enduring legacies. Shortly after the translation of Hennig's book into English (Hennig 1966), systematics underwent another revolution with the development of cladistics (e.g., Kluge and Farris 1969; Eldredge and Cracraft 1980; Nelson and Platnick 1981). Expanding on Hennig's views, cladists argued against both evolutionary taxonomy and phenetics, the former considered not sufficiently objective and the latter considered flawed based on the use of overall similarity rather than homology to identify monophyletic groups. Cladists also introduced the philosophical principle of parsimony into systematics (Farris 1977; Wiley 1981) and later operationalized it into a parsimony method that minimizes hypotheses of homoplasy (McInerney, chap. 5) in phylogenetic analyses (e.g., Farris 1983). The use of parsimony was justified with appeals to explanatory power (most parsimonious phylogenetic hypotheses were said to explain as much of the available data as possible as homology, thereby avoiding ad hoc explanations of homoplasy [Farris 1983]), and later with appeals to Popperian falsificationism (Siddall and Kluge 1997; Kluge 2001). The latter justification was criticized by many (Sober 1994; Rieppel and Kearney 2002; de Queiroz and Poe 2003; Kearney and Rieppel 2006; Rieppel and Kearney 2007), but perhaps most effectively by Hull (1999b).

Felsenstein (1978) identified the conditions under which parsimony methods might be statistically inconsistent and thereby laid the foundation for the use of maximum-likelihood statistical methods in phylogenetics. The rise of those methods coincided with the increasing use of DNA sequence data in systematics and a concomitant interest in developing statistical models of nucleotide evolution for use in phylogenetic analyses (McInerney, chap. 5). Most recently, Bayesian inference methods have been applied to phylogenetics (Huelsenbeck et al. 2001) and continue to be the most prevalent method used in contemporary phylogenetics. Methodological debates regarding tree-building approaches in systematics have waned dramatically over the past two decades, while great progress in reconstructing phylogenies for many organismal groups has been realized. Over the same time period, understanding of the importance of phylogeny to all of comparative biology has grown immensely. "Tree-thinking" (Baum et al. 2005; Baum and Smith 2013) is now a common phrase in biology owing to the realization that we must account for the historical relatedness of species in any hypothesis-testing that is related to species or their parts.

Phylogeny Matters

Over a century after the general acceptance of Darwin's theory that all organisms and their lineages are linked through a history of common descent, biologists began to recognize that our studies of evolutionary and ecological processes all depend, to some extent, on species relatedness and clade placement in the tree of life. Phylogenetic legacies (both constraints and drivers) on genetic and phenotypic evolution, organismal development, assembly of ecological communities, patterns of biodiversity, ecosystem structure, and microevolutionary processes are now widely acknowledged. The classic work of Felsenstein (1985) greatly influenced this shift and arguably created a generation of phylogenetic comparative biologists. Felsenstein (1985) articulated that species, owing to their coexistence as part of the tree of life, are not independent entities and that this nonindependence must be accounted for in all comparative studies by taking phylogeny into consideration. One way to account for phylogenetic noninde-pendence is with the use of phylogenetic comparative methods (Felsenstein 1985; Harvey and Pagel 1991). Phylogenetic comparative biology contin-ues to pervade many areas of science. A special issue of the journal *Ecology* (July 2006) devoted to the influence of phylogeny on ecological processes demonstrated new approaches that recognize that the community ecology of species cannot be understood without consideration of species' history (Webb et al. 2006). Such a foundational phylogenetic perspective does not stop with ecology. Contemporary studies of molecular evolution, trait evo-lution, speciation and extinction rates, development, physiology, medicine, disease, environmental change, and the causes and consequences of pat-terns of biodiversity are commonly conducted now within a phylogenetic framework (Cavender-Bares et al. 2009; Losos et al. 2013). And phyloge-netic comparative methods themselves continue to evolve, an indicator of their high demand (Pennell and Harmon 2013).

"Tree-thinking" is not only common among biological researchers but also increasingly seen as a critical gap to be filled in science educa-tion (Baum et al. 2005; Baum and Offner 2008), and it is now included in biology textbooks. Undergraduate and graduate biology students often do not sufficiently grasp modern evolutionary concepts. Tree-thinking can uniquely correct several common misconceptions that students harbor about evolutionary patterns and processes, such as the errors regarding di-rectional evolution or the intricacies of ancestor-descendant relationships.

Trees in Demand and the Phylogenomics Era

With the growing reliance on tree-thinking and the increasing use of tree-based analyses in many fields, the need to quickly reconstruct bigger and more highly resolved phylogenies for all groups of organisms increased. Biologists urgently needed to "treeify" life in order to formulate and test hypotheses. White papers and major funding efforts encouraged efforts to-ward a complete tree of life for all species (e.g., Systematics Agenda 2000; the US National Science Foundation's Assembling the Tree of Life funding program). The demand for phylogenetic trees contributed to the already growing shift from traditional, time-consuming methods of coding mor-phological traits for phylogenetic analyses to rapid, large-scale, and more automated DNA-based phylogenetic analyses. While contributing to major phylogenetic progress for many groups, this shift also increased the exist-ing schism between DNA-based and phenotype-based phylogeneticists and between neontologists and paleontologists. Paleontologists also need to build and use phylogenetic trees for extinct groups in order to study eco-logical and evolutionary processes across the vast history of life (Jablonski, chap. 17). Given that over 99 percent of all species that ever lived are now extinct, paleontological systematics remains an enormous and critical en-deavor, but support and funding for paleontology- and morphology-based systematics decreased precipitously during this time period. A concomitant steep decline in the training of morphologists and whole-organism biolo-gists also occurred, and biologists are only now beginning to grapple with the magnitude of this legacy of declining expertise in organismal biology.

In spite of such disciplinary divisions, many authors continue to stress the importance of including fossil taxa and/or morphological data in phylo-genetic analyses for a number of reasons. Wiens (2004) noted that we need to include morphological data in order to realize a complete tree of life, to comprehensively characterize living taxa, and to identify areas in the tree of life where molecular results may be misleading. G. J. Slater et al. (2012) used simulations to show that including fossil data improves models of trait evolution in macroevolutionary studies. Quental and Marshall (2010) demonstrated that tree-based studies of evolutionary diversification are seri-ously inaccurate when extinct lineages are absent from phylogenies. These arguments point to a troubling vacuum of extinct biodiversity in modern approaches and the need to continue to include fossil taxa and organismal traits in phylogenetic studies. In order to do so, however, we will need to code morphological characters in addition to using DNA sequence data, and we will need to integrate these data sets using mixed analytical models. This will require un-siloing molecular and morphological systematists, re-viving support for the development of morphological and whole organism expertise, and working more on integrative phylogenetic models. Clades that contain significant numbers of both extinct and extant forms are partic-ularly challenging to analyze in today's phylogenetic paradigm owing to the lack of mixed models for molecular and morphological analyses, although some advances have begun (e.g., Zhang et al. 2016 and references therein).

It is also important to understand the evolution of gene histories in relation to the evolution of population and species histories. In a piv-otal paper, Maddison (1997) described the relationship between species trees and the gene trees they contain. This relationship continues to be the foundation of important investigations and methodological develop-ments in phylogenetics (S. V. Edwards 2009; Knowles and Kubatko 2011; Nakhleh 2013). The fact that different loci in the same organism can have gene histories that differ from the overall history of that organism some-times makes phylogenetic analyses based on sequence data challenging.

Gene-tree/species-tree discordance can be caused by a number of processes, such as incomplete lineage sorting, gene duplication, and reticulation. The latter may be caused by lateral gene transfer, introgression, or lineage fusion. Any single gene tree can mismatch with its containing species tree depending on which gene copies happen to have been sampled. And even with all gene copies sampled, there could be a mismatch between the species tree and the sum of all gene trees. Perhaps the most common assumption made when gene-tree/species-tree discordance is observed is the occurrence of lateral gene transfer—genes that moved horizontally across species lineages.

Knowledge of lateral gene transfer grew significantly through the work of prominent microbiologists. By the end of the 1970s, universally conserved nucleic acid sequences, especially from the small subunit rRNA gene, had become widely used for similarity-based phylogenetic analyses of prokaryotes (Woese and Fox 1977). These studies had enormous impacts, including the discovery of three domains of life: Bacteria, Archaea, and Eukaryotes. Equally important, ribosomal RNA trees demonstrated the huge and largely ignored diversity of prokaryotic lineages (e.g., Woese and Fox 1977; Woese et al. 1990; Pace 1997, 2004). While early systematics studies of prokaryotes were largely based on a single gene sequence, complete genome sequences soon became more numerous and available, resulting in huge gains in reconstructing the tree of life for many groups using phylogenomic methods. The availability of complete genomes also highlighted the prevalence of lateral gene transfer and caused some to question whether genes are transferred so rampantly between prokaryotic lineages as to render the concept of a tree of life for prokaryotic species meaningless. For example, Doolittle (1999, 2125) rejected the notion of a universal tree of life that includes microbes, "If instances of [lateral gene transfer] can no longer be dismissed as 'exceptions that prove the rule,' it must be admitted (i) that it is not logical to equate phylogeny and organismal phylogeny and (ii) that, unless organisms are construed as either less or more than the sum of their genes, there is no unique organismal phylogeny." This marks the genesis of the anti-tree-of-life argument.

The Anti-Tree Argument: Biological and Philosophical Perspectives

The recent challenges to the veracity of the tree of life attracted the attention of the scientific press and of philosophers of science, leading to some provocative and sensational headlines, e.g., "Why Darwin Was Wrong about the Tree of Life" (Lawton 2009), "Is It Time to Uproot the Tree of Life?" (Pennisi 1999), "Beyond the Tree of Life" (O'Malley and Boucher 2011). Such headlines were doubtless influenced by the sociopolitical incentives of critiquing something as high profile as Darwin and the tree of life. But at the same time, they disrupted conventional wisdom and fueled rich new discussions and investigations in both evolutionary biology and the philosophy of biology.

The Argument from Biology and Questions Raised

Following the discovery of widespread lateral gene transfer in prokaryotes, some microbiologists began to suggest that the tree of life is an erroneous metaphor for prokaryotic evolution and that the history of life cannot be conceptualized as a simple branching tree (Bapteste et al. 2004; Bapteste and Boucher 2008; Bapteste et al. 2009). As Doolittle (2000, 2125) summarizes: "Thus, there is a problem with the very conceptual basis of phylogenetic classification" and "If, however, different genes give different trees, and there is no fair way to suppress this disagreement, then a species (or phylum) can 'belong' to many genera (or kingdoms) at the same time: There really can be no universal phylogenetic trees of organisms based on such a reduction of genes." Going further, Morrison (2014, 635) made the case for using reticulate networks rather than trees as genealogical metaphors: "but I contend that if a set of gene trees is incompatible (i.e., the genes have different histories), then the associated species genealogy should be seen as a network, not a tree."

Morrison's comment demonstrates that the arguments against the tree of life actually apply to both the tree of life itself and the species that make up the tree of life. Several authors have argued that molecular mosaicism amongst prokaryotic lineages is so pervasive that distinguishable prokaryotic lineages simply do not exist (e.g., Zhaxybayeva et al. 2004). These rejections of prokaryotic species and prokaryotic phylogeny raise the interesting question of whether species and their genealogies are nothing more than the sum of their genes and the sum of all gene trees, respectively. Is there such a thing as an organism from this perspective? And should the tree of life be understood as the tree of lineages or the tree of (genetic) parts of those lineages? Certainly, Darwin's vision of the tree of life, which was conceived as a metaphor to depict the inclusive hierarchy of species lineages due to common ancestry relationships, differs significantly from the view of these authors.

Although it is common in the microbiology literature, not every microbiologist agrees with the anti-tree-of-life argument. While none seem to doubt the significance of genomic flux affecting prokaryotes, some argue for a core genome tree of life that aligns well with previous prokaryotic trees constructed with rRNA, thus protecting the possibility of a stable phylogenetic tree of life for all of life (Daubin et al. 2002; Ochman et al. 2005; House 2009; Andam and Gogarten 2013). These counterarguments rest on the premise that the vertical descent of genes is more common than the lateral transfer of genes and that the vertical signal is sufficient to overwhelm the noise of various lateral transfer events given sufficiently rigorous phylogenetic methods. That is a premise requiring further investigation, as are many other recent conjectures, some of which follow here.

First, from the perspective of phylogenetic effects, successful lateral gene transfer requires not just transferred gene(s) but incorporation of the transferred genes into the receiving genome and their subsequent persistence over evolutionary time. Further research is required to demonstrate whether gene transfer events between distantly related lineages are more likely to be successfully incorporated and passed on or more commonly lost via lineage sorting. A related question is whether transferred genes are never, sometimes, or always functionally beneficial to the receiving taxon. We also do not yet know how much of the total genome is potentially affected by lateral gene transfer. Operationally, if a transferred portion of DNA succeeds in becoming fixed in a new population, we do not know the likelihood that one affected gene tree will obscure the pattern of the dominant tree (the tree applying to most of the genome). When should we ascribe gene discordance to lateral gene transfer rather than to any number of other possible evolutionary processes (e.g., incomplete lineage sorting, gene duplication), or even simply to which gene copies have been sampled and included in the analysis? Importantly, if discordant gene trees are to be taken as evidence that there is no organismal tree of life and no reality of species, then what is the template against which we should contrast gene trees in order to conclude that they are discordant when using current species tree methods? In order for genes to be laterally transferred, they must be transferred between something and something. Most species tree methods rest on inferring the species phylogeny from multiple gene trees and then contrasting and reconciling the gene trees with the species tree in order to study the evolutionary processes occurring in the genome. Such an approach implicitly assumes the reality of organisms, species, and a tree of life. If the latter concepts are to be rejected, then these methods are intellectually inconsistent.

At its core, this debate rests significantly on species concepts, the most contentious issue in systematics throughout its history (Nathan and Cracraft, chap. 6). One might even argue that the central tenet of the lateral-gene-transfer-based argument against prokaryotic species is fundamentally a genetic species concept, and yet species concepts have not yet been addressed within the context of this debate. Whether lateral gene transfer is predominantly a population-level phenomenon—similar to introgression—or whether it occurs pervasively between evolutionarily stable lineages is still largely unknown. If the former, we can ask whether lateral gene transfer is really an issue for population genetics rather than species-level phylogeny reconstruction. In that case, we might also infer that reticulation events do not preclude the existence of a tree of life because these two phenomena are not mutually exclusive.

Finally, we must address the predictions raised by the assumption of a highly reticulate genomic tree of life: What does it predict we will find when we examine other lines of organismal evidence, such as geological or phenotypic data? Prokaryotic microfossils with stable and recognizable phenotypes and lineages persist in the fossil record for extraordinary timespans (House 2009). If those stable lineages have been rampantly exchanging genes with other prokaryotic lineages, it has evidently not prevented their long-term persistence at a higher level of organization. What does that mean about connections between the genome and the phenome in prokaryotes? If prokaryotic lineages that are very genomically unstable are capable of remaining quite phenotypically stable over long periods of time, that is a very interesting area to explore and again has not yet been addressed within the context of this debate.

Ensuing Philosophical Arguments and More Questions Raised

As is often the case with biological controversies, recent attacks on the tree of life have attracted the interest of philosophers of science. O'Malley and Dupré, for example, organized a group of microbiologists and philosophers of science in order to "question the Tree of Life," and a special issue of *Biology and Philosophy* was devoted to this topic (O'Malley et al. 2010). These activities began the work on the metaphysics of the anti-tree-of-life debate, and they raised many possibilities for future inquiry. O'Malley et al. (2010) begin by stating that evolution is now increasingly understood in terms of the descent of genes and that lateral gene transfer has forced evolutionary biologists to question the existence of a tree of life. As with the biological arguments, it is important to note that these philosophical arguments are, at least for the time being, largely gene-centric and prokaryotic-centric, and yet they draw wide-ranging conclusions about the reality of the tree of life. Also similar to the biological arguments about lateral gene transfer, the philosophical arguments sometimes blur population-level genetic processes with genealogical patterns of lineage diversification. For example, in Velasco's (2012) discussion of "nontreeness in systematics," he argues: "The empirical evidence is clear; there are a great many nontreelike processes that produce nontreelike genealogical patterns in nature. The realist is forced to become a realist about networks rather than trees" (628) and "On the smaller scale, it is obvious that forcing the history of Darwin's finches or human populations into a tree structure ignores information about migration and introgression between populations, which is essential to understanding their history" (632). However, migration and introgression are population-level phenomena, and these statements conflate population genetic processes with phylogenetic patterns of lineage history.

While caution is necessary when approaching these arguments owing to their gene- and prokaryote-centric limitations, it is clear that there is now fertile ground to explore related to the tree of life and the philosophy of biology. Some of the recent philosophical arguments regarding the nature of the tree of life closely mirror previous philosophical debates regarding the nature of species and historical individuality (Nathan and Cracraft, chap. 6; Hey 2001; Rieppel 2007; Ereshefsky 2010, 2011). For many years, we have debated the ontological status of species. Should they be defined by their evolutionary relationships (historical individuals resulting from speciation) or by the possession of certain traits or properties (kinds with membership criteria)? If species are identified by their genealogical history, as many contemporary biologists and philosophers of biology have argued (e.g., Ghiselin 1974b; Hull 1978; de Queiroz 1998, 1999, 2005, 2011), then no genotypic or phenotypic property can define a species. Instead, species are historical individuals bounded by time, space, and some type of homeostatic clustering process that allows a degree of coherence over some period of time (Boyd 1999). In other words, they exist as particular spatiotemporal lineages in the tree of life. A species is a species by virtue of its place in the single, unique history of life's genealogy.

In a post-tree-of-life worldview, are genes, organisms, and lineages still to be considered historical individuals? It would seem not, if the contention is that a certain percentage of lateral gene transfer is sufficient to eradicate lineages that were once historically individualized—but this question has not been addressed in the context of the current debate. If genes within an organism have different histories, does this signify that the organism has no history of its own? If so, what is the threshold number of relocated genes necessary to erase organismal history? And do we now require a new theory of the organism? These are far from novel philosophical questions at their core, but they are resurrected in a post-tree-of-life context. Throughout the history of philosophical thought, such questions of identity and sameness resurface in many different forms. For example, Theseus's Paradox asks whether something that has all its parts replaced over time remains the same thing. Does a ship that has all of its parts replaced sequentially over time in order to remain functional remain the same ship, even in the absence of any of its original components? Similarly, Heraclitus asked whether one can step in the same river twice given that all of the water in the river has changed by the time of the second step. The movement of genes across historically individuated lineage boundaries raises the question of whether a lineage retains its identity in the face of changing some of its parts. For decades, systematists have argued that at the level of evolutionary organismal lineages, there is only one history. Current tree of life discussions suggest that many philosophers may now disagree with this. It will be of great interest to observe this discussion expand as its focus broadens beyond the genome and to all taxonomic groups, including all of extinct life.

The debate has also exposed the inadequacy of our current theories of organism and form. The logical extension of the current argument predicts that an organism should be conceptualized solely as the sum of its genes, but no specific theory supporting organisms as simply the sum of their genomes has been articulated. Do organisms (even the most phenotypically simple microorganisms) have any emergent reality beyond their genomes? This question extends as well to the nature of species and phylogeny. Current anti-tree-of-life arguments implicitly or explicitly define phylogeny as nothing more than the summation of all the gene histories of all organisms that have ever lived. If we are to interpret a phylogeny of all species as simply the graph along which genetic information has been passed and nothing more, then a specific and new theory of phylogeny (the genetic theory of phylogeny?) must be articulated to support that.

But perhaps a unified theory of organisms, lineages, and phylogeny is simply not philosophically attainable for all of life. This is an interesting possibility and brings us full circle to very early debates regarding unity/disunity of the natural sciences (Dupré 1995). One idea is that we must adopt a pluralistic approach to species and to the tree of life in order to accommodate all of life. O'Malley et al. (2010, 442) suggest: "Foremost among the challenges of building a universal tree are the implications of [lateral gene transfer], a prominent and ineliminable feature of any adequate representation of prokaryote evolution." However, if we will require two approaches to phylogeny—one for prokaryotes and another for eukaryotes—then we will also require two theories of life's evolution. Here is an alternative conclusion: the essence of the problem is contained in a currently disconnected evolutionary biology that is itself in need of reunification.

The Tree of Life Reconsidered and the Need for Continuing Evolutionary Synthesis

For most of the last hundred fifty years, the tree of life has been understood to depict phylogenetic relatedness of species lineages. If one is attempting to depict phylogenetic relationships of species, and if our theory of phylogeny is one of common ancestry and descent with modification, then the logical way to depict those relationships is via an inclusive genealogical hierarchy illustrating the part-whole relationships of species lineages. To be sure, processes that ensure the coherence of lineages and their distinctness from other lineages vary amongst the vast array of life forms and are subject to myriad selective pressures. But fuzzy species boundaries are to be expected, predicted even, from an evolutionary perspective, as is the varying nature of that fuzziness across the entire tree of life. For example, most would agree that lineage boundaries are more permeable to genetic exchange in some organisms than in others. The species boundary may be thought of as a constraint on the evolution of its component parts owing to causal integration of those parts—genes, traits, development, organisms, and so forth. The degree of constraint will naturally vary across organisms along with varying degrees of organismal complexity and integration. We would predict species boundaries to be more indistinct when there are fewer constraints to change, e.g., in simple, unicellular organisms unburdened by constraints of complexly integrated development and morphology. Time is also a factor—evolutionary theory predicts that species boundaries will be fuzzier closer to the time of species divergence from common ancestors (Edwards et al., chap. 15).

Evolutionary theory also does not preclude that reconstructing the genealogy of life based on genomic data will be messy and challenging. In fact, given the tenets of evolutionary process theories (see chaps. 9–17), such a challenge is predictable using any data source. But the reality of the tree of life is not dependent on impermeable species boundaries or on simple divergent genetic processes or on imperfect methodological approaches. (The epigraph to this chapter suggests the question: are species and phylogeny indifferent to biology?) In fact, not only is the tree of life capable of conceptually accommodating underlying genetic processes such as lateral gene transfer, it is an indispensable prerequisite in revealing them. How can we infer that such processes occur across lineage boundaries if we do not accept that lineages exist? This points back to the critical importance of tree-thinking and the centrality of the tree of life in enabling all of comparative biology—including, but not limited to, the study of genetic processes such as lateral gene transfer.

A Holistic Phylogenetics

If lateral gene transfer and other processes cause investigators to question whether treeness exists for certain groups of organisms, it is now time to test this hypothesis with as much diverse data as possible. This will be challenging because evolutionary biologists are once again working in isolation from each other today. Neontologists and paleontologists are especially divided in today's systematics. However, only new synthetic work will enable these questions to be answered rigorously. For example, evidence of prokaryotic lineages persisting for billions of years in the fossil record (House 2009), and presumably living under the same conditions of lateral gene transfer as those affecting prokaryotic organisms today, must be explained—unless lateral gene transfer processes are assumed to have begun only recently. If well-preserved microfossils persist with stable morphologies for billions of years, we should be asking why organismal lineages persist despite potentially significant levels of genetic exchange. Such questions require interdisciplinary approaches.

Testing alternative phylogenetic hypotheses using different data sets is also critical, but more and more rare owing to fragmented disciplines in systematics. Genetic data is currently privileged in phylogenetic analyses, and it is increasingly uncommon to include data from morphology, fossil taxa, or any organismal data. In the rare cases when separate analyses of diverse data sets are pursued, instances of strongly conflicting tree topologies between DNA-only trees and those including morphology and fossils for the same taxa continue to be discovered (e.g., Gauthier et al. 2012; Wiens et al. 2012), and we have yet to fully explain them. Such extreme incongruence indicates either implausible degrees of morphological homoplasy or interesting problems with molecular or morphological evolution in certain groups (Losos et al. 2012). And conflicts between molecular phylogenies and those derived from morphological and fossil data are not restricted to tree topology, but extend to incongruent rates of diversification, the dating of speciation events, and other tree-based inferences (G. J. Slater et al. 2012). It is well-known that the inclusion of fossils improves estimates of phylogeny, estimates of trait evolution, and estimates of divergence dating, yet the use of fossils continues to diminish. More generally, the overall utility of phylogenetic comparative methods for macroevolutionary studies depends on the inclusion of fossil taxa and phenotypic traits (G. J. Slater and Harmon 2013). As Quental and Marshall (2010, 440) stated, "We need the fossil record if we are to understand all but the simplest features of the biodiversity dynamics that have led to the living biota." If biologists seek a unified theory of phylogeny, we will need to pursue better integrative models and use more complete data sets across the entirety of life.

Reunification and the Critical Role of Theory in Unifying

Periodic disunity and unity of the natural sciences is a recurring theme (Smocovitis 1992). Belief in a general unity of knowledge extends back at least as far as Plato, and a strong call for a unified biology emerged during the Enlightenment after long periods of time when disunity was the norm. Post-Enlightenment, the unity of science was a central principle of the Vienna Circle philosophers as they observed numerous "biological sciences" rather than a single unified biological science. The modern evolutionary synthesis and the unifying potential of evolutionary theory represented a tectonic shift. Biology itself was fundamentally realigned by the modern evolutionary synthesis, with evolution as its central principle. The social structure and culture of science does not necessarily incentivize integrative work, generally giving preference to siloed disciplines over interdisciplinary work. Important knowledge gains within different fields can remain disconnected, as they did prior to the modern evolutionary synthesis, until the need to integrate specialized results and develop unifying theories across disciplines becomes sufficiently stark that it stimulates the breakdown of disciplinary barriers. A new emphasis on the importance of convergence research is promising.

As theoretical frameworks continue to evolve (see chaps. 1, 9–17), they stimulate connections between subdisciplines because they help scientists identify links between their own work and other fields of science. As this pattern continues over time, errors of reductionism are mitigated. Theoretical work sometimes takes a subsidiary role during periods of intense technological advances, such as our new transformative advances in DNA sequencing, genomics, and computing. These areas are advancing so quickly in current biology that they are overtaking our ability to conceptually process the meaning of the resulting data. The current disruptive challenge to the tree of life signals that this may be a profitable time to engage in a new synthesis across systematics and other natural sciences. Such an extended synthesis will require new conceptual and theoretical work, not just big aggregated data sets.

Arguments between reductionist and systems approaches are a common occurrence in the history of most sciences. Both reductionist and systems approaches are profitable and, more important, they can be powerfully synergistic if deployed strategically within a theoretically unified field. Reductionism allows scientists to explain mechanisms and processes not applicable at higher levels (e.g., lateral gene transfer), but reductionism has serious limitations. In phylogenetics, genomics offers the possibility of massive new data sets for the study of organisms, species, and phylogeny, but such complex systems are not equivalent to the data we use in our efforts to apprehend them. Such a conflation is not just genome-reductionist, it is a classic error of operationalism—the error of defining a concept or entity by the operations used to determine it. An emblematic statement from Doolittle (1999, 2124) illustrates this problem: "in the words of Zuckerlandl and Pauling, 'the essence of the organism'—not only do genes reveal the phylogenetic pattern, they engender and embody it." This statement equates a complex system to its sampled constituent data parts (Sober 1994; Smolin 2006). Simply because we currently rely mainly on genetic or genomic data to identify species and lineages in contemporary systematics does not mean that species are ontologically nothing more than the sum of their genomes. In contrast, it is commonly thought that there exists a hierarchy of biological organization with increasing complexity (e.g., genes, genomes, transcriptomes, organisms, populations, species, clades). Species lineages contain gene histories, but they also contain much more (e.g., traits, complex interactions between traits, organisms, populations). Constant empirical and theoretical reintegration is required to move past these issues.

Conclusion

Charles Darwin's major contributions to biology include both a theory of natural selection and a theory of phylogeny. Both of these theories are central to an overall theory of evolution. Phylogeny is a fundamental conceptual framework for understanding biology because every organism has participated in the singular history of life. As a result, "tree thinking" (O'Hara 1988, 1997) has now transformed hypothesis-testing across many natural science disciplines. It has not yet pervaded all fields—for example, most theories and models that involve natural selection (e.g., Frank and Fox, chap. 9; Goodnight, chap. 10; and others in section 2) are ahistorical; tests of those theories and models can be enriched by consideration of phylogenetic history.

The mission of systematics, which has a very long and rich history, is to reconstruct and provide details on the tree of life so that comparative biological studies can proceed. The discovery of lateral gene transfer and reticulation has recently caused some to posit that the tree of life does not exist. This notion has been fueled by the investigation of rich new genomic data sets and major new discoveries regarding the behavior of genes. However, empirical research on lateral gene transfer is in its infancy, and the narrowness of the focal organisms and data presents a cautionary tale regarding wide-ranging conclusions. A reticulate network diagram may be a better representation of the relationship between genomes when genetic reticulation occurs, but it is not a better representation of the relationship between evolutionary lineages, unless lineages = genomes. In actuality, reticulate network and tree diagrams are both informative, but they should be specific and goal-dependent—it matters whether the goal is depicting genetic evolutionary processes, depicting the history of individual genes, or depicting the relationships of organismal lineages. Ideally, these processes and patterns will be used to inform each other, recognizing that the discovery of nontreelike genetic processes is an addition to Darwin's tree of life, not necessarily a negation of it.

The historically individual nature of species and phylogeny has long been a theoretical underpinning of systematics. If species are defined by their history, then no amount of lateral gene transfer and no degree of intact ancestral genetic similarity are required to make them a species. Gene flow between organisms is part of an organism's genetic history, part of its population history, and part of its species history. From a phylogenetic viewpoint, to refer to an organism as *Escherichia coli* is to say not that *E. coli* is a specific kind of organism, or that it has specific properties or traits, or that it has X percent of its original genome intact, but that the *E. coli* organism is part of a larger, historical entity (the lineage of *E. coli* plus all other subsuming lineages in the inclusive hierarchy of life). While the molecular revolution in biology brought about great progress, it also occasioned an eclipse of nongenetic organismal data and, consequently, a different way of seeing organisms and the tree of life. As a result, phylogenetic lineages are sometimes erroneously synonymized with genomic lineages. These two concepts will need to be decoupled if we wish species and the tree of life of which they are a part to be conceptually unified for all of life.

Over seventy years after the modern synthesis and the emergence of evolutionary biology as its own discipline, evolutionary biology requires major resynthesis. Recent debates about the validity of species and the tree of life in the face of genomic knowledge indicate that current subdisciplines within evolutionary biology are incompletely merged and indeed operating largely independently. This represents a natural cycle in the history of biology. We will always require continued integration across paleobiology, genetics, development, morphology, systematics, evolution, organismal biology, and emerging new fields if we expect a unified evolutionary biology to continue to grow. New synthetic theory development must also keep pace with new empirical discoveries, the accumulation of big data, and technological advances, because synthetic theories allow us to make sense of diverse kinds of data and their implications. Darwin himself did just that—synthesizing over morphology, development, biogeography, geology, and more—to bring us phylogenetic theory and the tree of life.

Acknowledgments

This chapter is based on work done while serving at the US National Science Foundation and the Smithsonian Institution. The views expressed in this chapter do not necessarily reflect those of the National Science Foundation, the Smithsonian Institution, or the United States Government.

EIGHT

Situating Evolutionary Developmental Biology in Evolutionary Theory

ALAN C. LOVE

Evolutionary developmental biology (evo-devo) is a complex array of research programs with two main axes: (a) the evolution of development, or inquiry into the pattern and processes of how ontogeny varies and changes over time; and (b) the developmental basis of evolution, or inquiry into the causal effect of ontogenetic processes on evolutionary trajectories, whether to constrain or to facilitate them (Love 2015c). Prominent among these research programs is the study of conserved genetic regulatory networks and signaling pathways. Others include studies of phenotypic plasticity, experimental embryology, and computational inquiry (e.g., simulations), as well as disciplinary contributions from systematics and paleontology. Instead of reviewing this heterogeneous landscape of empirical findings and theoretical commitments, the present chapter asks a different question: within the body of knowledge referred to as evolutionary theory, where do you put evo-devo? To answer this question, I limit my attention to investigations of the origin of evolutionary novelty (see also Jablonski, chap. 17).

The place of evolutionary developmental biology (evo-devo) within evolutionary theory is contested, especially with respect to the origin of novelty. How did feathers or flowers originate in ancestral lineages lacking these morphological structures? Evo-devo research aimed at explaining novelties has been a locus for criticizing population genetic frameworks: "It does not help much to say that there were one or two mutations that created eyespots and that these alleles were selected" (G. P. Wagner 2000, 97). This viewpoint feeds into claims that a standard outlook on evolutionary theory does not incorporate the critical explanatory role of development: "[evo-devo] emerged as a distinct field of research in the early 1980s to address the profound neglect of development in the standard modern synthesis framework of evolutionary theory, a deficiency that had caused difficulties in explaining the origins of organismal form" (Müller 2007, 943). The problem of explaining the origin of novelty from a developmental perspective represents a signature feature of evo-devo that distinguishes it from other strands of evolutionary research (G. P. Wagner et al. 2000; Moczek 2008; G. P. Wagner 2014). It appears regularly in descriptions of the nature and scope of evo-devo (Moczek et al. 2015; Lesoway 2016) and constitutes a recurring element in calls for an extended evolutionary synthesis (Pigliucci and Müller 2010b; Laland et al. 2015).

One interpretation of this line of criticism and the centrality of explaining novelty to evo-devo is that there are two distinct explanatory agendas within evolutionary biology. The first is associated with a perspective that emphasizes populations and function. Evolutionary change from one phenotype to another is explained via population processes such as natural selection, which sorts phenotypes, alters allele frequencies, and yields outcomes that are frequently though not exclusively adaptive (Hoekstra and Coyne 2007; M. Lynch 2007; see Frank and Fox, chap. 9). The second agenda is associated with organisms and structure (G. P. Wagner 2014). Evolutionary change from one ontogeny to another is explained by a variety of developmental genetic and epigenetic processes, which can be altered in different ways to produce novel morphologies in organisms (Amundson 2005; Calcott 2009). Although this interpretation bolsters the rationale for distinguishing evo-devo from evolutionary inquiry seen in population genetics or behavioral ecology, it does little for the task of situating evo-devo within evolutionary theory. For example, is the organism-structure explanatory agenda subsidiary to the population-function explanatory agenda, or a prerequisite for it?

One additional difficulty for situating evo-devo is that its boundaries are debated. Researchers from diverse disciplines that use an assortment of methods and approaches see themselves as working within evo-devo, sometimes to the exclusion of one another. Narrow depictions of evo-devo revolve around the comparative developmental genetics of metazoans (S. B. Carroll 2005a; De Robertis 2008), where the focus is on evolutionary adjustments to gene network regulation with an emphasis on *cis*-regulatory elements (Davidson 2006; S. B. Carroll 2008; Davidson and Peter 2015; Rebeiz et al. 2015), especially for the origin of novelty (Gompel et al. 2005; Davidson and Erwin 2006; Glassford et al. 2015). Broader depictions of evo-devo include this but also draw attention to comparative embryology and morphology, experimental investigations of epigenetic dynamics at different levels of organization, studies of phenotypic plasticity, and computational modeling or simulation-based inquiry (Hall 1999; Raff 2000;

G. P. Wagner et al. 2000; Müller 2007; Newman 2012; Love 2015b; Moczek et al. 2015; Lesoway 2016). These depictions stress the significance of multiple disciplinary contributors and methodological approaches, including the joint role of paleontology and systematics for supplying a necessary historical-phylogenetic dimension (Telford and Budd 2003; Jablonski, chap. 17; Raff 2007).

Given these divergent perspectives on the individuation of evo-devo, and given its persistent attention to explaining something that evolutionary inquiry anchored in populations and function purportedly does not (i.e., the origin of novelty), it is not surprising that questions about evo-devo's place coalesce around its relationship to established models from population biology and evolutionary genetics. For some evolutionary theorists, this is nonnegotiable: "The litmus test for any evolutionary hypothesis must be its consistency with fundamental population genetic principles . . . population genetics provides an essential framework for understanding how evolution occurs" (M. Lynch 2007, 8598). From this perspective, characterizing interspecific differences in developmental mechanisms is not identifying the mechanisms of evolution because it ignores the population-genetic processes "ultimately" responsible for evolutionary change. For others, the aim is to unify both evolutionary genetics and evo-devo by appeal to organizing mechanisms such as genetic regulatory networks (Laubichler 2009), concepts such as evolvability (Minelli 2010), or a small set of central principles: "Evolutionary theory is not just a collection of separately constructed models, but is a unified subject in which all of the major results are related to a few basic biological and mathematical principles" (Rice 2004, xiii). Part of the contested ground pertains to which kinds of "mechanisms" or "principles" are given precedence: should these derive from population genetics or evo-devo? For still others, constellations of disciplinary approaches in evo-devo are better characterized in terms of the complex explanatory projects they concentrate on (Love 2008; Brigandt 2010; Love 2010; Brigandt and Love 2012; Love 2013). This orientation suggests that no single theoretical framework based on a particular type of mechanism, small set of principles, or restricted suite of methods will be fundamental and serve to coordinate or organize all of the others within evolutionary theory, let alone evo-devo.

In light of these considerations, situating evo-devo within evolutionary theory is a complicated task. However, two shared premises can be derived from the above disagreements. First, evo-devo is distinguishable and different from what is typically associated with standard evolutionary theory (e.g., population biology models). Whether this distinction is based on an emphasis on an ontology of organisms rather than genes (G. P. Wagner 2014), on structure rather than function (Love 2011), or on the generation of variation rather than its differential preservation (Müller and Newman 2003), there is a robust basis for differentiating evo-devo. Second, there is a shared question about whether evo-devo should be hierarchically ordered in relation to other evolutionary models. Answers to this question vary. It would be helpful to have some notion of theory structure that facilitates situating evo-devo so that its distinctiveness and hierarchical ordering (or lack thereof) are explicated plainly. For example, a hierarchical ordering of evo-devo will be more useful when accompanied by a characterization of how or by what criteria that ordering operates.

The present chapter probes the value of a particular account of theory structure (Scheiner 2010; Zamer and Scheiner 2014; Mindell and Scheiner, chap. 1) for situating evo-devo within evolutionary theory. A necessary prerequisite to this probative endeavor is to distinguish two approaches for accomplishing the task: "abstract-structure first" and "concrete-practice first." After reviewing some findings from analyses employing the latter, I describe the framework of Scheiner and colleagues, which exemplifies the former. Using this framework, I situate evo-devo as an explanation of the evolution of morphology, especially the origin of novelty, and show how the framework respects evo-devo as a distinct area of inquiry, affirmatively places evo-devo in a hierarchical order within evolutionary theory, and (to some degree) supplies a characterization of how that ordering operates. However, the payoff for accomplishing these tasks is less clear than what might be expected. I derive two lessons from this unexpected result—one methodological, one substantive—and conclude with reflections on different strategies intended to advance scientific inquiry by explicating theoretical structure.

Approaches to Theory Structure

There are (at least) two distinguishable ways to approach the structure of evolutionary theory. The first is to identify or articulate an account of scientific theory structure abstractly and then ask how different areas of inquiry associated with evolutionary research fit within it. One potential advantage of this approach is the unification of different areas and an increased comprehension of the explanatory scope of theories. Another advantage could be isolating previously unnoticed similarities in unrelated models when comparing them within this abstract structure. A potential drawback is that this structure risks obscuring how theory guides investigation and explains phenomena *in vivo* because it is often formulated separately from the practices of research. Similarly, success in placing a theory within this abstract structure is no guarantee that there will be any payoff for ongoing inquiry.

The second strategy is to identify or articulate concrete theoretical practices in use within scientific communities to ascertain how different domains of evolutionary research are structured by these practices. One potential advantage is a clarification of why lines of investigation are organized in specific ways. Another advantage is making explicit the criteria of adequacy and related standards of evaluation; differences in these features can be a source of controversy within and across disciplines. A potential drawback is that this structure may apply to only one area of investigation and therefore cannot be leveraged to understand other areas of science. Similarly, detailing these practices may not be illuminating if there is a high degree of flux in the area under scrutiny, which makes it difficult to separate significant theoretical structure from idiosyncratic differences among individual researchers.

Neither of these strategies is perfect and each is better suited to achieving some goals than others. The first strategy ("abstract-structure first") seems apropos for epistemological endeavors that are one step removed from the daily use of knowledge in specific sciences, such as identifying relationships among disparate disciplines. The second strategy ("concrete-practice first") is germane when the aim is to understand why scientists do what they do, such as reasoning in a particular fashion or finding a type of explanation adequate. These two strategies are not mutually exclusive and can be complementary, especially with respect to their potential drawbacks. Importantly, they both focus on making theory explicit rather than leaving it implicit in particular sciences (Scheiner 2010; Zamer and Scheiner 2014). Abstract-structure-first strategies sometimes value distinctive epistemological properties (e.g., single definitions of terms with necessary and sufficient conditions) that appear beneficial at a distance but can act as obstacles to fecund lines of investigation operating without this constraint (M. Wilson 2006). Concrete-practice-first strategies sometimes value the specificity of local criteria of adequacy for research questions but lose sight of a bigger picture of knowledge relations within and among cognate sciences.

Results from Concrete-Practice-First Analyses of Theory Structure

In previous work, I adopted a concrete-practice-first strategy to analyze evo-devo theorizing and comprehend how it might fit into the broader context of evolutionary theory (Love 2008; Brigandt 2010; Love 2010, 2013,

2017b). Several key results emerged from these analyses. First, controversies over definitions of concepts, such as evolutionary novelty (Moczek 2008), are best understood in terms of divergent criteria of explanatory adequacy. Ernst Mayr defined novelty as "any newly acquired structure or property which permits the assumption of a new function," which fit within the framework of the modern synthesis: "The problem of the emergence of evolutionary novelties then consists in having to explain how a sufficient number of small gene mutations can be accumulated until the new structure has become sufficiently large to have selective value" (Mayr 1960, 357). In contrast, a different definition is found in most work in evo-devo: "A morphological novelty is a structure that is neither homologous to any structure in the ancestral species or [serially homologous] to any other structure in the same organism" (Müller and Wagner 1991, 243). Instead of functional properties ("selective value"), morphology in a phylogenetic context is the focus ("non-homologous structures"); instead of mutations in a gene pool, the developmental generation of qualitatively new variation is in view. The growth of cladistic methods for phylogenetic reconstruction and the emergence of molecular developmental genetics gave this structure-oriented definition operational traction (Love 2015a), especially in long-standing controversies like that of the origin of the chordate body axis (Gerhart 2000).

Although qualitative departures from an ancestral condition have been emphasized in evo-devo (e.g., West-Eberhard 2003), a continuum between qualitative difference and quantitative variant makes it difficult to distinguish novelty from nonnovelty in absolute terms. Biologists draw lines on this continuum differently (Brigandt and Love 2012; Palmer 2012), and there are typically precursors or homologous features at lower levels of organization for those structures deemed qualitatively novel (N. C. Shubin et al. 2009; Hall and Kerney 2012). For example, ventral cement glands in many larval anurans are a qualitative departure from the relevant ancestral condition (and thus novel), yet they share remarkable similarities with the dorsal casquettes in some larval teleosts as paired organs that secrete mucus, have mechanoreceptors, facilitate attachment, and arise via near-identical patterns of regulatory gene expression despite occurring in different bodily locations (Pottin et al. 2010). An appeal to deep homology could encourage seeing these traits as an exemplar of parallel evolution responding to similar ecological pressures in early life history and therefore not genuine novelties. In light of these issues, there is value in shifting attention away from identifying a single, correct definition of evolutionary novelty (i.e., delineating the entities a term classifies or "categorizes") and, instead, concentrating on characterizing the explanatory agenda associated with the concept (Brigandt and Love 2012). Different meanings of the term "novelty" indicate distinct explanatory expectations for the study of these morphological features.

Focusing on characterizing the explanatory agenda foregrounds questions about adequate explanations for derived body parts that usually lack homologous relations to structures in ancestral lineages and often possess the potential for new functionality. However, to understand explanatory expectations in different contexts, we need to ascertain how the concept is structuring the problem space. This is often indicated by allied concepts used with a specific definition of novelty. For example, discussions of novelty as nonhomology have stressed the importance of distinguishing character identity (e.g., the forewing of insects) from character states (e.g., wing blade versus protective cover) in order to clarify a particular sense of homology relevant to studying the developmental origin of new structures (G. P. Wagner 2014; McInerney, chap. 5). Designating the hierarchical level at which homology and novelty apply plays a role in dissecting how mechanisms of gene regulation have evolved to produce novel anatomical structures (N. C. Shubin et al. 2009). Linking novelty to evolvability accents abstract mapping relations between genotype and phenotype (Pavlicev and Widder 2015). Investigating the innovative ecological impact of a structural novelty makes natural selection germane to the explanation (Erwin 2012; Shirai et al. 2012).

A second result of the concrete-practice-first strategy is the identification of problematic assumptions about the structure of evolutionary theory that accompany attempts to situate evo-devo, especially in the context of an extended evolutionary synthesis (Pigliucci and Müller 2010b). One of these is an assumption that there is a single evolutionary theory under scrutiny. Spatial metaphors that presume a relatively cohesive unit are used to describe how the modern synthesis must be "expanded" to increase its content. This can be seen in the language used to contrast modern synthesis assumptions from those of an extended evolutionary synthesis: "Inheritance extends beyond genes to encompass (transgenerational) epigenetic inheritance, physiological inheritance, ecological inheritance, social (behavioural) transmission and cultural inheritance" (Laland et al. 2015, 2). It also can be observed in pictorial representations of this expansion and extension that use a Venn diagram structure to depict spatial increase within evolutionary theory. A related assumption, connected to the use of spatial metaphors, is that evolutionary theory has a relatively uncomplicated structure with a central "core" (Love 2013, 2017b). Often this core is equated with the apparatus of evolutionary genetics that undergirds an understanding of how genotypic and phenotypic changes occur in populations owing to natural selection, mutation, migration, and drift: "population genetics provides the fundamental theory of [evolution]" (Ridley 1993, vii).

Both of these assumptions are problematic in a variety of ways. First, if there are questions about how evo-devo is individuated, then these will not be discernable in a spatial container structure where evo-devo is assumed to be one cohesive part of evolutionary theory. Questions of individuation are relevant to how evo-devo is situated (Love 2017b). For example, it is plausible that different problem agendas rather than whole areas like evo-devo are pertinent parts of evolutionary theory (Love 2010). This yields distinct structures for evolutionary theory depending on how local criteria of adequacy are coordinated (Love 2008; Brigandt 2010). Second, biologists typically work with models or theory presentations rather than with an entire theory (Griesemer 1984), which means they idealize away from some elements to make others salient (Love 2013). Presumptions about what evolutionary theory is and how it is structured are guided by those elements salient to particular communities of research. Third, many researchers adhere to claims about the centrality of evolutionary genetics when discussing the structure of evolutionary theory: "If you can characterize formal population genetics, then you have characterized the 'guts' or 'core' of evolutionary theory" (E. A. Lloyd 1988, 8). This has led to a systematic neglect of the diversity of research that might be categorized within evolutionary theory (Love 2010).

The purpose of reviewing these results from earlier concrete-practice-first analyses is to establish a contrast class for the alternative strategy. In order to evaluate the virtues of an abstract-structure-first account, it is beneficial to see what has been learned by its complement. At the same time, since each strategy is suited to achieving some conceptual goals and ill equipped to address others, the contrast class also should make apparent the distinctive advantages that accrue under the aegis of a strategy prioritizing abstract structure.

An Abstract Account of Theory Structure

Scheiner and colleagues have advanced an abstract account of theory structure that is intended for all of biology and has been applied to ecology and other areas of the life sciences (Scheiner and Willig 2008; Scheiner 2010; Zamer and Scheiner 2014; Mindell and Scheiner, chap. 1). It instantiates an abstract-structure-first strategy and was motivated by the apparent lack of large-scale or overarching theories in biology.

> Theory is important because it clarifies thinking. It forces a modicum of formality onto data interpretation, thereby refereeing scientific disputes. It reveals assumptions hidden in specific models or experiments. It shows connections among disciplines, which is especially important in guiding interdisciplinary and transdisciplinary work. It defines risky or groundbreaking research. Finally, it clarifies the central questions being addressed by a scientific enterprise. Despite all of the benefits accrued by having explicit theories, biology appears to be bereft at the widest levels. (Scheiner 2010)

However, it is not clear that overarching theory is necessary (or sufficient) to achieve these various goals. For example, the clarification and organization of problem agendas and research questions within particular areas of scientific inquiry does not require this kind of theory or formal definitions (Love 2014, pace Zamer and Scheiner 2014). Making connections between approaches for interdisciplinary work can be done with other resources, such as structured problem domains with explicit criteria of adequacy (Love 2008; Brigandt 2010). Yet an abstract-structure-first strategy can be well suited to deriving wider epistemological pictures of sciences: "The purpose of this paper is to lay out a series of theories that encompass all of biology: an overarching theory" (Scheiner 2010, 294). Scheiner's overarching framework includes five general theories of ecology, evolution, organisms, genetics, and cells. Since our primary concern is situating evo-devo within evolutionary theory, I will focus only on evolution. (Notably, this idealizes away from questions about how evo-devo could be situated within general theories of genetics or organisms; see Zamer and Scheiner 2014.) We first need to describe the elements of abstract theory structure that will organize evolutionary theory.

Theories, on this view, are hierarchically structured and composed of three types: general theories, constitutive theories, and models (Scheiner 2010; Mindell and Scheiner, chap. 1, table 1.2). General theories are subdivided into three elements: background, general principles, and outputs. Constitutive theories are subdivided into three similar elements: background, propositions, and outputs. Models have four elements: two that are the same as general and constitutive theories (background, outputs) and two that are different (construction, tests). All three types share four components in their background (domain, assumptions, framework, and definitions), and models include a fifth: propositions. For general theories, general principles are subdivided into concepts and confirmed generalizations. The propositions of constitutive theories are separated into concepts, confirmed generalizations, and laws. General theories have constitutive theories as outputs, constitutive theories have models as outputs, and models have hypotheses as outputs. The two unique elements of models, construction and tests, have one corresponding component each: translation modes and facts. Each component of these types of representation has a succinct characterization (table 8.1). This theoretical ontology is rich and merits its own discussion, but our aim is to use it as a tool for situating evo-devo in evolutionary theory. If we take evolution to be the general theory, then evo-devo can be categorized provisionally as a constitutive theory. Before concentrating on how evo-devo might fit within this type of structure, it is necessary to describe the general theory of evolution within this framework.

In this framework, the domain of evolutionary theory is described as "the intergenerational patterns of the characteristics of organisms, including causes and consequences" (Scheiner 2010, table 1) or "patterns, causes, and consequences of change in the characteristics and diversity of life across generations and geologic eras" (Mindell and Scheiner, chap. 1). It has eight fundamental principles: (1) traits change across generations; (2) lineages diversify; (3) lineages are linked through common descent; (4) variation is a prerequisite for evolution; (5) variation arises from genetic properties; (6) evolution is due to deterministic and random processes; (7) properties of organisms condition evolution; and (8) rates of evolution are variable (table 1.3). This list is revised and augmented; Scheiner (2010) only has seven fundamental principles. "I recognize that others might disagree with the exact number and wording of these principles, but such disagreements over details do not alter the general form of my thesis" (Scheiner 2010, 294). While this may be true, disagreements may emerge about these principles from constituent theories that speak to their content.

Other subdivisions of this abstract theory structure are not discussed as extensively as general principles. For background, it is not entirely clear what counts as assumptions, framework, and definitions (e.g., what is necessary for a theory to work with clarity), or how they would be appropriately catalogued for a particular type of theory (e.g., whether there should be a relatively small set or whether one can have an extensive list). Within the general principles, various concepts and confirmed generalizations are noted in passing though not probed systematically. Some of the purported confirmed generalizations remain contentious or at least subject to several interpretations (e.g., how convergence fits with evolutionary change leading to lineage diversification, whether all variation arises from "genetics," and how we interpret "deterministic" and "random" processes

Table 8.1. Theory Components and Relationships (Mindell and Scheiner, chap. 1, tables 1.1 and 1.2). Abbreviations: GT = General Theory; CT = Constitutive Theory; M = Model

Component	Characterization	Location	Types
Assumptions	*Conditions or structures needed to build a theory or model*	*Constituent of background*	*GT, CT, M*
Concepts	Labeled regularities in phenomena	Constituent of fundamental principles or propositions	GT, CT
Confirmed generalizations	Condensations and abstractions from a body of facts that have been tested	Constituent of fundamental principles or propositions	GT, CT
Definitions	Conventions and prescriptions necessary for a theory or model to work with clarity	Constituent of background	GT, CT, M
Domain	The scope in space, time, and phenomena addressed by a theory or model	Constituent of background	GT, CT, M
Facts	Confirmable records of phenomena	Constituent of tests	M
Framework	Nested causal or logical structure of a theory or model	Constituent of background	GT, CT, M
Fundamental principle	Concept or confirmed generalization that is a component of a general theory	Constituent of GT	GT
Hypotheses	Tested statements derived from or representing various components of a theory or model	Constituent of outputs	M
Laws	Conditional statements of relationship or causation, or statements of process that hold within a domain of discourse	Constituent of propositions	CT
Model	Conceptual construct that represents or simplifies the natural world	n/a	n/a
Translation modes	Procedures and concepts needed to move from the abstractions of a theory to the specifics of model, application, or test	Constituent of construction	M

to understand the contingency of evolutionary change). The same holds for many concepts (e.g., gene, homology, or species). We lack an inventory of all relevant concepts or confirmed generalizations, which seems germane to discerning the output of constitutive theories (though also potentially tedious). Nevertheless, we can set this skeletal outline of evolutionary theory to one side in order to explore whether and how evo-devo can be understood as one of its constitutive theories.

Evo-Devo as a Constitutive Theory

A natural way to understand evo-devo as a constitutive theory is in relation to an earlier formulation of a general principle about variation (originally no. 5) from Scheiner (2010, table 1): "Variation among organisms within species in their genotype and phenotype is necessary for evolutionary change." This applies specifically to explaining the origin of novelties: "Recent debates about the origin of phenotypic novelties . . . as well as the emerging discipline of evo-devo . . . have focused on the meaning of and the mechanisms underlying fundamental principle 5" (Scheiner 2010, 295). However, Scheiner acknowledges that "this is an oversimplification, and often the debates were about several principles simultaneously" (295). The simplification is made explicit in the revised and augmented formulation of the principles (Mindell and Scheiner, chap. 1). While variation as a prerequisite for evolution remains (now no. 4 instead of no. 5), connections between evo-devo and no. 5 (variation arises from genetic properties) or no. 7 (properties of organisms condition evolution) stand out. Other fundamental principles quickly become salient, such as determining the significance of homoplasy due to shared developmental pathways (e.g., anuran cement glands and teleost casquettes). All of these are central in debates about the nature, scope, and significance of evo-devo (Moczek et al. 2015). This means we cannot simply locate evo-devo within a general theory of evolution but must scrutinize these structural relationships, in terms of both their existence and their interpretation. For the moment, let's treat the assumed relationship between general principles 4, 5, and 7 of this general theory of evolution and evo-devo *qua* constitutive theory as a helpful idealization (i.e., it represents things in a way that purposefully departs from other features known to be present). How does evo-devo fit into the abstract structure of a constitutive theory?

To help narrow our analytical task, we can focus on explaining the evolution of morphology, especially the origin of novelties, as a proxy for evodevo. This preserves an insight from the concrete-practice-first strategy that problem agendas are a significant unit of analysis in understanding how evolutionary theory is structured. Once the concept of evolutionary novelty is seen as representing an explanatory agenda, it facilitates making explicit the associated criteria of adequacy. These include the need to specify the relevant level of structural organization, the degree of generality that can be derived from studying particular novelties, and the disciplinary contributions necessary for an adequate explanation (Brigandt and Love 2012).

A prominent strategy of research in evo-devo involves investigating how developmental genetic changes contribute to the evolution of form and the generation of novel morphological structures.

> Evolutionary change in animal form cannot be explained except in terms of change in genetic regulatory networks architecture. (Davidson 2006)

> The evolution of development and form is due to changes within genetic regulatory networks. (S. B. Carroll 2008)

> Novelty requires the evolution of a new genetic regulatory network. (G. P. Wagner and Lynch 2010)

These changes can involve duplications followed by the differentiation of paralogous genes (Gompel et al. 2005), modifications of regulatory interactions (Shirai et al. 2012; Bloom et al. 2013), and the co-option of gene expression from one time or spatial location to another (True and Carroll 2002; Saenko et al. 2008). Thus, novel structures at one level of organization can arise from changes in homologous genetic regulatory networks (Shubin et al. 2009); their origins are explained by the recombination and redeployment of preexisting ancestral variation, rather than as a consequence of novel genes.

Sean Carroll, sharing some of the motivations emphasized by Scheiner and colleagues, has attempted to formulate these ideas into a theory for evo-devo that will articulate with an extended evolutionary synthesis.

> Do questions posed about evo-devo and evolutionary theory matter to anyone besides the specialists and a few future historians? I think the answers matter very much. . . . Without theories to organize and interpret facts, without the power of general explanations, we are left with just piles of case studies. Moreover, we are without the frameworks that enable us to make predictions. (S. B. Carroll 2008, 25)

For Carroll, theories are "structures of ideas that explain and interpret facts," and he offers eight principles (paraphrased here) that provide a basis for conceptualizing evo-devo as a constitutive theory of a general theory of evolution.

(1) *Mosaic Pleiotropy*: most proteins regulating development participate in multiple, independent processes that shape and pattern morphologically disparate body structures.

(2) *Ancestral Genetic Complexity*: morphologically disparate and long-diverged animal taxa share similar toolkits of genes that build and pattern morphology.

(3) *Functional Equivalence*: toolkit proteins typically exhibit functionally equivalent activities when substituted for one another; their biochemical properties and interactions have been largely conserved.

(4) *Deep Homology*: the formation and differentiation of many morphological structures are governed by similar sets of genes and conserved genetic regulatory networks.

(5) *Infrequent Toolkit Gene Duplication*: Duplications within toolkit gene families are not necessary for morphological novelty and have been rare because of negative effects on gene-dosage-sensitive developmental processes.

(6) *Heterotopy*: Changes in the spatial regulation of toolkit genes and the genes they regulate are associated with morphological divergence.

(7) *Modularity of Cis-regulatory Elements*: Large, complex, and modular *cis*-regulatory regions are a distinctive feature of pleiotropic toolkit loci.

(8) *Vast Regulatory Networks*: Individual regulatory proteins control many target gene *cis*-regulatory elements.

Notably, these are not exceptionless generalizations ("most," "rare," "typically"): functional equivalence does not hold for all *Hox* genes (e.g., Heffer et al. 2010; V. Lynch and Wagner 2010). Other pertinent principles are excluded, such as heterochrony.

According to S. B. Carroll (2008), there is a central "hypothesis" that can be distilled from the eight principles. It claims that *cis*-regulatory-element mutations are the primary source of morphological variation and therefore *cis*-regulatory-element evolution accounts for the bulk of morphological evolution, including the origin of novelties. Carroll claims that this hypothesis is empirically supported because: (a) *cis*-regulatory-element sequence changes are sufficient to account for the evolutionary divergence of traits and gene regulation among populations, species, and higher taxa; (b) *cis*-regulatory-element evolution is necessary for rewiring genetic regulatory networks; (c) gene duplication and coding changes alone are insufficient to rewire genetic regulatory networks; and, (d) *cis*-regulatory-element variation and divergence are detectable over shorter timescales and taxonomic distances than functional differences in transcription factors or toolkit gene duplications.

However, there are some gaps in this reasoning. First, the insufficiency of gene duplication and coding changes for rewiring genetic regulatory networks in (c) is consistent with their necessity for accomplishing it. Therefore, this alone would not privilege *cis*-regulatory-element variation over other molecular changes in accounting for morphological evolution. The purported sufficiency in (a) presumes a particular explanatory standard of what is involved in accounting for the evolutionary divergence of traits and gene regulation, which might be questioned. Additionally, (d) undergirds the primary role of *cis*-regulatory-element evolution only if other possibilities are exhausted by functional differences in transcription factors and toolkit gene duplications. Regardless, Carroll encapsulates his theory of morphological evolution in two statements: form evolves largely by altering the expression of functionally conserved proteins; and morphological changes occur largely through mutations in the *cis*-regulatory regions of regulatory genes exhibiting mosaic pleiotropy and target genes within the vast regulatory networks they control.

Recall that constitutive theories are subdivided into three elements (background, propositions, and outputs). Their background is composed of a domain, assumptions, framework, and definitions. Propositions of constitutive theories are separated into concepts, confirmed generalizations, and laws, whereas their outputs are models. The domain of evo-devo, understood as a constitutive theory, can be described as the genetic and morphological variation (or lack thereof) among organismal traits within and across species that is relevant to how ontogeny changes over time (the evolution of development) and how ontogenetic processes constrain or facilitate evolutionary change (the developmental basis of evolution). Although "species" refers to multicellular animals in Carroll's application of this theory, it is unclear whether this is the true extent of the domain. (For example, bringing plants into this perspective would not be that difficult.) Another possible limitation of scope concerns how "ontogeny" and "ontogenetic processes" are interpreted. Microbes do not have the same genetic architecture as plants and animals, but they do exhibit forms of growth, differentiation, and morphogenesis (Love and Travisano 2013). Therefore, evo-devo (as a constitutive theory) could facilitate a reorientation of evolutionary theory (the general theory) by supplying "a fundamental principle about the role of development in determining phenotype and the pathways of evolutionary change" (Scheiner 2010, 304). However, there is disagreement about whether "genuine" development is found only in multicellular organisms. If this is affirmed, then evolutionary theory plus a fundamental principle of developmental evolution would not be fully general with respect to all species. If it is denied, then evolutionary theory plus a fundamental principle of developmental evolution could be general with respect to the species in its domain. Thus, generality is not a single-dimensional property of scientific theories but rather should be keyed to different theory aspects (not just taxonomic scope). Generality with respect to developmental evolution can diverge from generality with respect to the evolution of genetic architecture, which in turn might diverge from generality with respect to the evolution of physiology (and so on).

Assumptions for the constitutive theory come in two forms: phylogenetic and developmental. Phylogenetic assumptions include a reliance on cladistics as the appropriate theory of classification that reconstructs evolutionary relationships among lineages. Developmental assumptions include a reliance on molecular developmental genetics, both its mechanistic models and its experimental procedures, although whether this can include more physico-chemical or mechanical mechanisms varies (Newman 2012; Love 2017a). Here we see a tension in how to relate evo-devo to general principle 5—variation arises from genetic properties—because this depends on whether variation arising from physico-chemical mechanisms is included within evo-devo. Significantly, assumptions from evolutionary genetics are not necessary to build Carroll's theory, though others have considered them germane (Rice 1998; Hansen 2006).

It is not clear what constitutes the framework ("nested causal or logical structure") for evo-devo as a constitutive theory apart from inferential relationships among Carroll's principles. That cis-regulatory elements are modular implies that mutations in these elements have the potential to change the spatial regulation of toolkit genes, thereby leading to morphological divergence via heterotopy. The claim that regulatory sequence evolution is the likely mode of genetic and morphological change uses mosaic pleiotropy, the modularity of cis-regulatory elements, and vast regulatory networks to infer two clauses: (i) a protein plays multiple roles in development, and mutations in its coding sequence are likely to have pleiotropic effects; and (ii) the relevant locus contains multiple cis-regulatory elements. These two clauses have generality by virtue of three other principles (ancestral genetic complexity, functional equivalence, and deep homology), are primary because of infrequent toolkit gene duplication, and take on evolutionary significance from heterotopy. Many operational definitions ("conventions and prescriptions necessary for a theory to work with clarity") of terms are required for this framework to hang together (e.g., *cis*-regulatory element, toolkit genes, homology, heterotopy, pleiotropy, modularity), as well as facts ("confirmable records of phenomena"), such as *Hoxa11* playing distinct roles in the formation of axial skeleton, appendicular skeleton, kidneys, and the reproductive tract (Zhao and Potter 2002).

Carroll's eight principles might be interpreted most perspicuously as propositions. Some correspond to concepts ("labeled regularities in phenomena"), such as mosaic pleiotropy, deep homology, heterotopy, and the modularity of *cis*-regulatory elements. Others correspond to confirmed generalizations ("condensations and abstractions from a body of facts that have been tested"), including ancestral genetic complexity, functional equivalence, infrequent toolkit gene duplication, and vast regulatory networks. It is possible to interpret some of them as laws ("conditional statements of relationship or causation, or statements of process that hold within a universe of discourse"). For example, we might reframe functional equivalence of distant orthologs and paralogs as a conditional statement of causation: if animal toolkit proteins are substituted for one another *in vivo*, then they exhibit functionally equivalent activities. Carroll's "hypothesis" is best interpreted as a confirmed generalization that yields a model since it does not relate to "tested statements derived from or representing various components of the theory." This also locates the claim in the constitutive theory, whereas hypotheses are outputs of models that represent the natural world at a lower hierarchical level in the typology of scientific theories (Scheiner 2010).

More could be said about evo-devo as a constitutive theory. For example, we have neglected translation modes ("procedures and concepts needed to move from the abstractions of a theory to the specifics of a model, application, or test"). One way this might occur is by taking a general characterization of evolutionary novelty and narrowing it to an operational definition in order to test a hypothesis about mechanisms underlying the origin of a particular morphological feature (e.g., feathers). Another question is how a constitutive theory of evo-devo formulated in terms of Carroll's eight principles compares with constitutive theories meant to apply to similar domains. A preliminary sketch of a constitutive theory of phenotypic novelty has some propositions overlapping with Carroll's and others that differ (Zamer and Scheiner 2014). Still another question is whether this constitutive theory of evo-devo could be constructed by combining elements of a general theory of evolution and a general theory of organisms or genetics

(Zamer and Scheiner 2014; Mindell and Scheiner, chap. 1). However, an assumption contained in the question, and presumed in the title of this chapter ("situating evo-devo in evolutionary theory"), suggests a more fruitful line of inquiry.

To what degree could we think about situating evolutionary theory (or at least parts of it) within evo-devo? (More broadly, should we think about situating a theory of organisms or genetics—or least parts thereof—within evo-devo?) Its diversity of questions and methodologies and its facility in exploring neglected questions about evolution argue in favor of its integrative potential (e.g., Moczek et al. 2015). An immediate objection relates to the question of taxonomic scope: evo-devo lacks the requisite generality because many species (seemingly) do not exhibit development (Zamer and Scheiner 2014). But if generality is not a single-dimensional property of scientific theories, then there may be routes to probing whether a general principle of developmental evolution could be general with respect to particular features of species within the domain of evolutionary theory. Although we typically talk about viruses as being assembled rather than developing, conceptualizing the process of assembly and its material resources in terms of an expanded conception of development could be theoretically fruitful (Griesemer 2014). Additionally, physiological innovations, such as the ability to exploit citrate as a carbon source (Blount et al. 2008), could be conceptualized similarly with attendant benefits and therefore mitigate concerns that evo-devo's theoretical perspective about the origin of novelties applies only to morphology.

We are now in a position to affirm that this abstract-structure-first strategy achieves several of the aims identified earlier. Assuming this architecture for a general theory of evolution, it situates evo-devo within evolutionary theory to explain the evolution of morphology (or traits), especially the origin of novelty, and recovers at least some of the rationale for why evo-devo is a distinct area of inquiry. It affirms the hierarchical ordering of evo-devo within evolutionary theory, displaying how it can be understood as fleshing out several principles of the general theory (variation is a prerequisite for evolution; variation arises from genetic properties; properties of organisms condition evolution). In particular, the principle that properties of organisms condition evolution links up directly with the developmental basis of evolution (how ontogenetic processes constrain and facilitate evolutionary trajectories). However, there were tensions (e.g., variation arising from nongenetic properties of developing organisms), and it is not clear that we have done much to characterize how the ordering operates. We assumed a particular structural view of the general theory of evolution and adopted a narrow interpretation of evo-devo (Carroll's eight principles) as a constituent theory. And the abstract-structure-first strategy does not foreground particular examples of novel traits (unlike the concrete-practice-first strategy), which makes it more difficult to see how this strategy benefits investigators of citrate metabolism innovation or the origin of anuran cement glands. Additional scrutiny of this hierarchical ordering and its associated assumptions seems warranted.

Situating Evo-Devo: To What End?

The value of an abstract-structure-first strategy for situating evo-devo in evolutionary theory cannot be discerned from the above analysis alone. Notice that several of this strategy's potential benefits require comparison and contrast with other constitutive theories, including the unification of different domains or the identification of previously unnoticed similarities in divergent areas of science. A natural question to ask is whether any other value of this strategy is discernable. No hidden assumptions have come to light, and guidance for interdisciplinary work appears available already in concrete-practice-first results, as does the clarification of central research problems. To some degree the abstract-structure-first strategy may have clarified thinking, but it does not help in refereeing scientific disputes by formalizing the significance of data. In fact, some of the potential drawbacks of the strategy have surfaced because the details of how theory both guides investigation and explains pertinent phenomena are opaque apart from what was gleaned from concrete-practice-first analyses. As cautioned above, success in placing a theory within an abstract structure is no guarantee that there will be payoff for ongoing inquiry.

Perhaps this is too pessimistic a conclusion. Although we often focus on successes, in both the sciences and philosophy, failures can be highly informative (Firestein 2016; Redish et al. 2018). For example, heuristic reasoning breaks down in systematic ways and allows scientists to metabolize their errors productively (Wimsatt 2007). The manner in which the abstract-structure-first strategy fails (or falls short) may harbor lessons that contribute to clarifying our thinking about evolutionary theory and evo-devo. Two broad lessons—one methodological and one substantive— can be gleaned from our exploration of this approach to scientific theory structure.

First, the presumed hierarchical ordering of evolutionary theory and evo-devo involved a narrow interpretation of evo-devo. This interpretation does not include what many researchers think is critical for explaining the origin of novelty, such as experimental investigations of epigenetic dynamics at different levels of organization or phenotypic plasticity. In particular, this puts pressure on the fifth principle from the general theory of evolution: "variation arises from the genetic properties of organisms." Do we need to reformulate this principle? Would this encourage revisions of other principles or require new principles, or even suggest a different conception of hierarchical ordering? This is a methodological lesson: simultaneously explicate the structure of a general theory and its purported constituent.

Following this methodological lesson makes possible a reciprocal feedback between general theory structure and constituent theory structure. Some of this feedback is conceptual—we may need to revisit the theoretical ontology and scrutinize whether it adequately captures the structural features of interest or whether elements need revision. Much of the feedback will revolve around the diversity of theoretical content and how it is interpreted. Diverse understandings of evo-devo and divergent views on its location or ordering within evolutionary theory are two places where this methodological lesson is applicable. This is observable when Müller (2007) claims that evo-devo's "results take evolutionary theory beyond the boundaries of the Modern Synthesis," but does not mean cis-regulatory element variation and evolution, of which he is relatively skeptical: "Further experimental proof will be necessary to determine the extent to which gene regulatory change has a causal role in evolution" (Müller 2007, 944). However implausible this may appear (Rebeiz et al. 2015), it highlights that researchers structure evo-devo differently owing to a divergence in explanatory standards. Müller thinks that comparative experimental epigenetics is crucial to explaining the evolution of form and origin of novelty; Sean Carroll does not. Why? For Müller, explanation "necessarily includes many more factors than the evolution of gene regulation alone, notably the dynamics of epigenetic interactions, the chemicophysical properties of growing cell and tissue masses, and the influences of environmental parameters" (Müller 2007, 944). The introduction of epigenetic interactions and physico-chemical properties as relevant factors (Newman and Bhat 2008; Newman 2012), as well as the possibility that novel traits may begin as conditional structures due to developmental plasticity (West-Eberhard 2003; Moczek et al. 2011; Palmer 2012; Scheiner, chap. 13), forces a revision of the fifth general principle in the theory of evolution because variation arises from more than the genetic properties of organisms. This methodological lesson is also applicable when we consider evo-devo as a constituent theory of a general theory of organisms (Zamer and Scheiner 2014). And since evo-devo continues to grow and transform both conceptually and empirically (e.g., through the incorporation of microbiome-related considerations), the identification of these kinds of differences facilitates ongoing inquiry when the structure of a general theory and its constituents are treated simultaneously.

There are few biologists who would deny evo-devo any place in understanding evolutionary change, but there are diverse views on its location. S. B. Carroll (2005b, 1164) expresses this in frustration: "I am not convinced that what we have learned about the evolution of form is being adequately considered in comparative genomics and population genetics, where the potential role of regulatory sequence evolution appears to be a secondary consideration, or ignored altogether." He goes further and urges a reversal of orientation about what should be central: "the evolution of form is the main drama of life's story, both as found in the fossil record and in the diversity of living species" (S. B. Carroll 2005a, 294). This reversal has been analyzed in terms of standards that deem "causal-mechanistic" explanations from evo-devo superior to explanations citing different "forces" operating on populations. Evolutionary theory, from this perspective, must be restructured so that developmental considerations take a more foundational position (Laubichler 2010). In opposition to this radical maneuver, other biologists have recommended supplementing existing models of evolutionary change with the conceptual content of evo-devo (Minelli 2010) or subordinating evo-devo entirely (M. Lynch 2007). A *via media* is the interpretation of two distinct explanatory agendas (as described earlier) or the recognition that multiple perspectives are required (cf. Scheiner 2010):

> Theodosius Dobzhansky was famous for his challenging dictum that, "Nothing in biology, makes sense except in the light of evolution." He meant that everything about an organism could be explained as an adaptation to selective conditions. . . . I began to wonder whether the opposite was true: nothing in evolution makes sense except in the light of cell, molecular, and developmental biology. Today it is clear that each of these ways of looking at biology is necessary; it is the properties of the cellular processes as much as the selection and segregation-modification of DNA that explain. (Kirschner 2015, 203).

If we engage in explicating the structure of a general theory at the same time that we explicate the structure of a constituent theory, then it becomes possible to dissect where differences in criteria of adequacy lead to divergent evaluations of the place of evo-devo in evolutionary theory. Shuttling back and forth between these conceptual tasks helps elucidate nodes of disagreement, which may lie in how structure is understood for one or the other (or both).

The second broad lesson that emerges from our exploration of abstract scientific theory structure is substantive: the value of explicating theory structure is always tied to the goals of the endeavor. A primary motivation for Scheiner and colleagues is to achieve an overarching conceptual framework for biology (Scheiner 2010). The value of unification looms large in this picture. Others have advocated a different value—integration—as a way to understand how connections can be drawn between different theories (Odenbaugh 2011). This integration happens in a more piecemeal fashion as different theories and their models intersect to generate more robust perspectives for whichever domain is in view (Wimsatt 2007). One place to observe this in explanations of the origin of novelty is with respect to the question of whether natural selection has explanatory relevance.

From one vantage point, the structure of evo-devo as a constituent theory is motivated by a concern that natural selection has been given too prominent a place in evolutionary explanations. One of the goals in articulating a structure for evo-devo, evolutionary theory, and an extended evolutionary synthesis is to foreground this so that the value of developmental considerations in evolutionary explanations is unambiguous, especially with respect to biases in the kind and distribution of phenotypic variation, and so that form or morphology is not subjugated to function: "the diversity of organismal form is only partly a consequence of natural selection— the particular evolutionary trajectories taken also depend on features of development" (Laland et al. 2015, 3). This perspective has been seen from many different angles throughout our discussion.

However, from a different vantage point, the aim could be to comprehend how natural selection helps to explain the origin of novelty, such as by increasing the frequency of particular kinds of variants through altering the genetic background (Godfrey-Smith 2014). Thinking in terms of function, some have argued that there are two distinct possibilities for the origin of a novelty: exaptation or developmental capacitance (Moczek 2008). Exaptations are traits either originally selected for another purpose or by-products of a different trait's formation. When exposed to a novel selective environment, they present opportunities for new adaptive functions. Developmental capacitance represents processes that buffer against the manifestation of phenotypic variation until a threshold is reached. For both possibilities, the emphasis is on what conditions make it possible for natural selection to act and thereby solidify a new trait via its functionality. Discerning the variational properties of phenotypes can inform how selection contributes to the stability of a trait and how new structures become individuated evolutionarily (Pavlicev and Widder 2015).

The value of explicating evo-devo's theoretical structure for the origin of novelty and relating it to a general theory of evolution is in clarifying what kinds of explanatory projects biologists are engaged in. Different sets of goals generate different kinds of structure, including what counts as general for an aspect of theory, and this should be captured in our account of theoretical structure. This implies that taxonomic scope is not the primary arbiter of generality; distinct structures are relevant for evolutionary theory and evo-devo because they facilitate diverse kinds of inquiry with different types of generality (Love 2010, 2013, 2017b).

If we apply the methodological and substantive lessons in the context of other constitutive theories (chaps. 9–17), then similar payoffs can be secured. For example, if we simultaneously explicate the structure of a general theory of evolution and the constituent theory of macroevolution (Jablonski, chap. 17), then we might observe that a particular form of hierarchical ordering also encourages reciprocal feedback that forces a revision in general principles, such as rates of evolution are variable. The nature of this heterogeneity requires integrating paleontological and neontological analyses to show how origination and extinction rates calibrate our understanding of rates of evolutionary diversification (Hunt and Slater 2016). Similarly, a recognition that the value of explicating theory structure is linked to the goals of the endeavor might mean that in some instances there is a focus on unification across constituent theories and in others an awareness that specific constituent theories lead to a different picture of evolutionary change, such as the predominance of random walks as a mode of trait evolution in the fossil record (Hunt et al. 2015). As with the origin of novelty in evo-devo, different sets of goals can generate different kinds of theoretical structure when relating a general theory of evolution to constituent theories.

Conclusion

We began with two preliminary conclusions about evo-devo in the context of evolutionary theory: (a) evo-devo is different in nature, scope, and style when compared with population biology, evolutionary genetics, and allied inquiry, and (b) the question of whether evo-devo is hierarchically related to other elements of evolutionary theory must be addressed. Using an abstract-structure-first strategy, Scheiner and colleagues provide an account of how a general theory of evolution and a constitutive theory of evo-devo can be structured. This account demonstrates the distinctness of evo-devo and argues that it is subsidiary to the general theory of evolution. Unfortunately, it did not fulfill some of its stated goals, such as facilitating the resolution of scientific disputes, identifying hidden assumptions, or connecting disciplines in illuminating ways, and was problematic because it assumed a narrow depiction of evo-devo. The latter revealed a tension in how to understand the fifth principle of a general theory of evolution (variation arises from genetic properties), which many evo-devo proponents dispute. These issues could be identified in part because we contrasted the results of this strategy with what already had been achieved by adopting a concrete-practice-first strategy.

And yet our examination generated two lessons: we should simultaneously explicate the structure of a general theory and its purported constituents because this facilitates reciprocal interactions between them, and the value of explicating scientific theory structure is always tied to the goals of the endeavor. The second substantial lesson points in the direction of seeking different kinds of structure to accomplish different kinds of goals. There is more than one way to structure evo-devo, and this plurality is a source of strength; it is what permits a wide range of inquiry based on diverse investigatory and explanatory aims. Thus, the goal of the present volume—to establish an overall conceptual framework—is one among many that we might have in attempting to situate evo-devo within evolutionary theory. Scheiner and colleagues proceed from a commitment to unification, even in the midst of discerning this plurality: "I recognize that theories can take many forms, and I do not mean to imply that the structure presented here is the only possible one. However, this structure appears capable of embodying a wide variety of theories within biology" (Scheiner 2010, 296). Unification is a significant goal, but it is not preeminent. No single goal governs endeavors focused on explicating and characterizing structure for evolutionary theory. This is because the suitability of theoretical structure for some goals does not transfer to other goals. And, importantly, this is pertinent to pedagogy. Although any kind of theoretical structuring is preferable to a presentation mode reliant on lists of facts, some kinds of structuring are better suited than others to distinct teaching tasks (Love 2010).

On a final note, Scheiner holds that his account of theory structure can help to pick out risky or bold research: "A bolder study addresses a broader domain or a more general theory" (Scheiner 2010, 304). We have already noted that what counts as "broader" or "more general" will be qualified with respect to different theory aspects. However, it is significant that the methodological lesson of jointly articulating structure for a general and constitutive theory demonstrates that conceptual endeavors themselves count as "bolder studies." They are likely to have implications for general theories (evolutionary and otherwise), which is exactly what arose in problematizing the narrow reading of evo-devo as a constitutive theory and worrying about the fifth general principle. If "the riskiest and most iconoclastic research is that aimed at disconfirming an established principle" (Scheiner 2010, 305), then research aimed at questioning an established principle of a general theory from the perspective of a constitutive theory is both risky and iconoclastic. That it would so heavily involve conceptual analysis rather than empirical investigation should encourage new and continuing collaborations among biologists and philosophers to explicate the structures of evolutionary theory and its constituents for the benefit of future research and teaching.

Acknowledgments

I am grateful to Sam Scheiner for the invitation to contribute this chapter and for an inordinate and wholly undeserved amount of patience in waiting for the completion of various drafts of the manuscript. Sam and six anonymous referees provided helpful input and spirited criticism on earlier drafts of this material. The conclusions and any remaining errors are mine alone. The research and writing of this chapter were supported in part by a grant from the John Templeton Foundation ("Integrating Generic and Genetic Explanations of Biological Phenomena"; ID 46919).

PART 2

Constitutive Theories

NINE

The Inductive Theory of Natural Selection

STEVEN A. FRANK AND GORDON A. FOX

Darwin (1859) got essentially everything right about natural selection, adaptation, and biological design. But he was wrong about the processes that determine inheritance (Smocovitis, chap. 2). How could Darwin be wrong about heredity and genetics, but be right about everything else? Because the essence of natural selection is trial and error learning. Try some different approaches for a problem. Dump the ones that fail and favor the ones that work best. Add some new approaches. Run another test. Keep doing that. The solutions will improve over time. Almost everything that Darwin wanted to know about adaptation and biological design depended only on understanding, in a general way, how the traits of individuals evolve by trial and error to fit more closely to the physical and social challenges of reproduction.

Certainly, understanding the basis of heredity is important. Darwin missed key problems, such as genomic conflict. And he was not right about every detail of adaptation. But he did go from the absence of understanding to a nearly complete explanation for biological design. What he missed or got wrong requires only minor adjustments to his framework. That is a lot to accomplish in one step.

How could Darwin achieve so much? His single greatest insight was that a simple explanation could tie everything together. His explanation was natural selection in the context of descent with modification. Of course, not every detail of life can be explained by those simple principles. But Darwin took the stance that when major patterns of nature could not be explained by selection and descent with modification, it was a failure on his part to see clearly, and he had to work harder. No one else in Darwin's time dared to think that all of the great complexity of life could arise from such simple natural processes. Not even Wallace.

Now, more than a hundred fifty years after *The Origin of Species,* we still struggle to understand the varied complexity of natural selection. What is the best way to study natural selection: detailed genetic models or simple phenotypic models? Are there general truths about natural selection that apply universally? What is the role of natural selection relative to other evolutionary processes?

Despite the apparent simplicity of natural selection, controversy remains intense. Controversy almost always reflects the different kinds of questions that various people ask and the different kinds of answers that various people accept as explanations. Natural selection itself remains as simple as Darwin understood it to be.

Deductive and Inductive Theory

There are two different ways we can think about natural selection. In the deductive way, we use our understanding of natural selection to make predictions about what we expect to find when we observe nature. For example, we might be interested in how a mother's resources influence her tendency to make daughters versus sons. A female wasp that lays her eggs on a caterpillar will sometimes have a large host and sometimes a small host.

We can make a model to predict what sex offspring the wasp will produce when faced with a large versus a small host. We make that deductive prediction by calculating the number of grandchildren that we think the mother can expect based on the size of the host and the sex of the offspring. A simple interpretation of natural selection is that the process favors a mother that behaves in a way that gives her the highest number of grandchildren—the highest fitness—within the limits of what she can reasonably do given the biology of the situation.

Perhaps the simplest deductive theory of natural selection concerns the change in gene frequency. A gene associated with a higher fitness than average tends to increase in frequency, a simple mathematical deduction. We can deduce exactly how fast a gene will increase in frequency given its fitness relative to the average. Although the mathematical deduction is very simple, in practice it is difficult to know in advance what fitness is associated with a particular gene. That fitness will depend on the gene itself and what it does inside cells, and also on the interaction of that gene with other genes and with the environment (Scheiner, chap. 13).

The value of deductive theory is, of course, that we can compare our predictions to what we actually observe. When reality differs from what we observe, then we know that some aspect of our initial understanding is incomplete or wrong. Much of the theory of natural selection develops deductive predictions, which can then be tested against observation.

Inductive analysis turns things around and begins with observations of what actually happened in nature. Suppose, for example, that we know the frequency of a gene at two points in time. From the observed change in frequency, we can infer the fitness of the gene by inductive reasoning: the strength of natural section that would be required to cause the observed change. Although we can induce the power of the unseen cause of natural selection, we cannot rule out other processes that might have caused the change in frequency. For example, it might be that the frequency changed by random sampling of alternative genes rather than by differences in fitness caused by effects of those genes. Inductive studies often seek to determine which of the various possible causes is most likely given the observations.

In our wasp example, we might have begun with the observation that mothers gain greater fitness when laying daughters rather than sons on large hosts. We could inductively estimate the strength of natural selection when comparing the production of daughters versus sons on a given host size. To the extent that we identify natural selection as a primary causal force, we would be estimating the strength of that cause in shaping the decision behavior of mothers faced with hosts of different sizes.

Natural selection itself may be thought of as an inductive process. With each step in time, gene frequencies change. Characters become more prevalent when they are correlated with genes that increase in frequency. Roughly speaking, natural selection inductively assigns the likely causes of improved fitness to those characters that are correlated with reproductive success.

When thinking about natural selection, we must always be clear about which of the different points of view we wish to emphasize. We may have a deductive prediction to test against observation. Or we may have observed data that we can use to induce the likelihood of alternative underlying causes. Or we may think about how natural selection itself works as an inductive process that associates actual changes in gene frequencies, or other informational units, with underlying factors that can potentially act as causes.

In this chapter, we focus on inductive perspectives of natural selection in relation to underlying causes of fitness. The notion of cause here is subtle. The population geneticist C. C. Li observed that there are many formal definitions of causation, but it is often not necessary to adopt any one of them. "We shall simply use the words 'cause' and 'effect' as statistical terms similar to independent and dependent variables, or [predictor variables and response variables]" (Li 1975, 3).

Partitioning Causes of Change

We follow Li's suggestion to learn what we can about causation by study-ing the possible relations between potential causal factors. The structure of those relations expresses hypotheses about cause. Alternative structural relations may fit the data more or less well. Those alternatives may also suggest testable predictions that can differentiate between the relative like-lihood of the different causal hypotheses (Crespi 1990; Frank 1997, 1998; Scheiner et al. 2000).

In evolutionary studies, one typically tries to explain how environ-mental and biological factors influence characters (Mindell and Scheiner, chap. 1). Causal analysis separates into two steps. How do alternative char-acter values influence fitness? What fraction of the character values is trans-mitted to following generations? These two steps are roughly the causes of selection and the causes of transmission.

Domain of the Theory

In a broad sense, the domain of the theory is evolutionary change in re-sponse to natural selection. This domain of natural selection is not the whole of evolution. For example, smoky pollution might darken the color of trees in a nearby forest, causing a change over time in the average color-ation of the population. In this case, tree color did not change by selection. Instead, the change was simply a consequence of a changed environment. However, natural selection remains the only force that could potentially explain a consistent tendency toward adaptation—the match between an organism's characters and the environmental and social challenges faced by that organism.

Following our distinction between deductive and inductive perspec-tives, the theory of natural selection has two complementary subdomains. Deductively, we may begin with known or assumed characters and with known or assumed fitnesses, and then work out how selection will change the characters over time. Although that sounds simple, it can be challeng-ing to work out how various characters and various selection processes in-teract over time to cause evolutionary change. In this case, it is often useful to partition selection into different causes (e.g., how a character changes the fitness of a neighboring sibling, and how the same character changes the fitness of the individual bearing the character).

Inductively, we may begin with observed characters and observed fit-nesses, and then work out how the various characters and their interactions caused the values of fitness that we observed. Again, this is not easy to do in practice. For example, we might see that bacteria secrete a digestive enzyme that causes some external food source to break down into components that are more easily taken up by cells. We might have measurements on how much enzyme is secreted by different kinds of cells and on the fitnesses of the various cell types. For a given fitness, can we infer how much of that fitness value is caused by the amount of enzyme secreted and how much of that fitness is caused by the amount of digested food taken up? The amount secreted is a direct cause of the fitness of the associated cell type, but the amount of food taken up depends on the amount of enzyme secreted by all neighbors—a partition of causes between direct and social aspects of selection.

Basic Models

Prelude

Improvement by trial and error is a very simple concept. But applying that simple concept to real problems can be surprisingly difficult. Mathematics can help but can also hinder. One must be clear about what one wants from the mathematics and the limitations of what mathematics can do. Useful mathematical modeling involves some subtlety. The output of mathematics reflects only what one puts in. If different mathematical approaches lead to different conclusions, the approaches have made different assumptions. There is a natural tendency to develop complicated models, because we know that nature is complicated. However, false or apparently meaningless assumptions often provide a better description of the empirical structure of the world than precise and apparently true assumptions.

The immense power of mathematical insight from false or apparently meaningless assumptions shapes nearly every aspect of our modern lives. The problem with the intuitively attractive precise and realistic assumptions is that they typically provide exactness about a reality that does not exist. One never has a full set of true assumptions, and we generally cannot estimate large numbers of parameters accurately. Worse yet, model error may grow multiplicatively with many parameters, so that even if we can estimate the parameters, the resulting predictions are often so broad as to be useless (Walters 1986; Hilborn and Mangel 1997). By contrast, false or apparently meaningless assumptions, properly chosen, can provide profound insight into the logical structure of nature. Experience has supported this truth over and over again.

Table 9.1.   The Theory of Evolution by Natural Selection

Domain: Evolutionary change in response to natural selection.

Propositions:

1. Evolutionary change can be partitioned into natural selection and transmission.
2. Fitness describes the evolutionary change caused by natural selection.
3. Information can be lost during transmission of characters from ancestors to descendants.
4. The balance between information gain by selection and information loss by transmission can be used to explain the relative roles of different evolutionary forces.
5. Fitness can be partitioned into distinct causes, such as the amount of change caused by different characters.
6. Characters can be partitioned into distinct causes, such as different genetic, social, or environmental components.

Six propositions, shown in table 9.1, provide the logical structure of the theory of evolution by natural selection. Proposition 1 says that we can account for evolutionary change in populations by ascribing it to two basic causes: selection and transmission of information between generations. This first proposition requires a definition of fitness, such as the evolutionary change caused by natural selection (proposition 2). We now consider how these concepts are used in models of fitness and frequency change in populations, which include both deductive and inductive approaches.

Frequency Change and Selection

In a basic model of fitness and frequency change, there are n different types of individuals. The frequency of each type is q_i. Each type has R_i offspring. The average reproductive success is $\overline{R} = \sum q_i R_i$, summing over all of the different types indexed by i. Fitness is $w_i = R_i / \overline{R}$, used here as a measure of relative success. The frequency of each type after selection is:

$$(9.1) \qquad\qquad q_i' = q_i w_i.$$

To obtain useful equations of selection, we must consider change. Subtracting q_i from both sides of eq. 9.1 yields:

$$(9.2) \qquad\qquad \Delta q_i = q_i(w_i - 1),$$

in which $\Delta q_i = q_i' - q_i$ is the change in the frequency of each type. Rearranging shows that $q_i'/q_i = w_i$, a mathematical expression of proposition 2.

We often want to know about the change caused by selection in the value of a character. Suppose that each type, i, has an associated character value, z_i. The character z can be a quantitative trait or an allele frequency from the classical equations of population genetics. The average character value in the initial population is $\bar{z} = \sum_i q_i z_i$. The average character value in the descendant population is $\bar{z}' = \sum_i q_i' z_i'$. For now, assume that descendants have the same average character value as their ancestors, $z_i' = z_i$. Then $\bar{z}' = \sum_i q_i' z_i$, and the change in the average value of the character caused by selection is:

$$\bar{z}' - \bar{z} = \Delta_s \bar{z} = \bullet\ q_i' z_i \rightarrow\ q_i z_i = \bullet\ (q_i' - q_i) z_i,$$

where Δ_s means the change caused by selection when ignoring all other evolutionary forces (G. R. Price 1972b; Ewens 1989; Frank and Slatkin 1992). Using $\Delta q_i = q_i' - q_i$ for frequency changes yields:

(9.3) $$\Delta_s \bar{z} = \bullet\ \Delta q_i z_i.$$

This equation expresses the fundamental concept of selection (Frank 2012a). As defined in proposition 2, frequencies change according to differences in fitness (eq. 9.2). Thus, selection is the change in character value caused by differences in fitness, holding constant other evolutionary forces that may alter the character values, z_i.

Frequency Change during Transmission

We may consider the other forces that alter characters as the change during transmission. Define $\Delta z_i = z_i' - z_i$ as the difference between the average value among descendants derived from ancestral type i and the average value of ancestors of type i (proposition 3). Then $\bullet\ q_i' \Delta z_i$ is the change during transmission when measured in the context of the descendant population. Here, q_i' is the fraction of the descendant population derived from ancestors of type i.

Thus, the total change, $\Delta \bar{z} = \bar{z}' - \bar{z}$, is exactly the sum of the change caused by selection (proposition 2) and the change during transmission (propositions 3 and 4):

(9.4) $$\Delta \bar{z} = \sum \Delta q_i z_i + \sum q_i' \Delta z_i,$$

a form of the Price equation (G. R. Price 1972a; Frank 2012a). We abbreviate the two components of total change as:

(9.5) $$\Delta \bar{z} = \Delta_s \bar{z} - \Delta_c \bar{z},$$

which partitions total change into a part ascribed to natural selection, Δ_s, and a part ascribed to changes in characters during transmission, Δ_c (proposition 4). The change in transmission subsumes all evolutionary forces beyond selection.

Characters and Covariance

We can express the fundamental equation of selection in terms of the covariance between fitness and character value. Combining eqs. 9.2 and 9.3 leads to:

$$(9.6) \qquad \Delta_s \bar{z} = \sum \Delta q_i z_i = \sum q_i (w_i - 1) z_i.$$

The right-hand side matches the definition for the covariance between fitness, w, and character value, z, so we can write:

$$(9.7) \qquad \Delta_s \bar{z} = Cov(w, z),$$

which we can rewrite as a product of a regression coefficient and a variance term:

$$(9.8) \qquad \Delta_s \bar{z} = Cov(w, z) = \beta_{zw} V_w,$$

in which V_w is the variance in fitness and

$$\beta_{zw} = \frac{Cov(w, z)}{V_w}$$

is the classic statistical definition of the regression of phenotype, z, on fitness, w. The statistical covariance, regression, and variance functions commonly arise in the literature on selection (Robertson 1966; G. R. Price 1970; Lande and Arnold 1983; Falconer and Mackay 1996).

Gene Frequencies

Treating the character z as an allele frequency, we can relate the above equations with the classical equations of population genetics. Assume that each individual carries one allele. For the ith individual, $z_i = 0$ when the individual carries the normal allelic type, and $z_i = 1$ when the individual carries a variant allele. Then the frequency of the variant allele in the ith individual is $p_i = z_i$, the allele frequency in the population is $p = z$, and the initial frequency of each of the N individuals is $q_i = 1/N$. From eq. 9.6, the change in the allele frequency is

$$(9.9) \qquad \Delta_s \bar{p} = \frac{1}{N} \sum (w_i - 1) p_i.$$

From the prior section, we can write the population genetics form in terms of statistical functions:

(9.10) $$\Delta_s \bar{p} = Cov(w, p) = \beta_{pw} V_w.$$

For analyzing allele frequency change, the population genetics form in eq. 9.9 is often easier to understand than eq. 9.10, which is given in terms of statistical functions. This advantage for the population genetics expression in the study of allele frequencies emphasizes the value of using specialized tools to fit particular problems.

By contrast, the more abstract statistical form in eq. 9.10 has advantages when studying the conceptual structure of natural selection and when trying to partition the causes of selection into components (proposition 5). Suppose, for example, that one wishes to know only whether the allele frequency is increasing or decreasing. Then eq. 9.10 shows that it is sufficient to know whether β_{pw} is positive or negative, because V_w is always positive. That sufficient condition is difficult to see in eq. 9.9, but is immediately obvious in eq. 9.10. One use of the kind of theory discussed in this volume is to understand the fundamental relationships between models that appear initially to be quite different from one another (see also Phillips, chap. 4).

Generalizing from the Basic Models: Scale, Distance, and Invariance

This volume focuses on the structure of evolutionary theory. To consider the fundamental role of natural selection within that broad theory, this section discusses a few key conceptual issues. In the first subsection below, we show that selection can be described in several equivalent forms: as a variance, as a distance, or as a gain of statistical information. Each of those descriptions is especially useful in a different context. The second subsection considers the common alternatives for analysis of evolutionary change: the change in phenotypic characters or the change in fitness. We show that these two alternatives for the analysis of evolutionary change are just alternative coordinate systems (like Cartesian and polar coordinates) that can readily be related to one another. Each alternative is especially useful under particular circumstances. These first two subsections provide three different ways to describe selection and two different coordinate systems for evolutionary change. Those combinations show the connections between various approaches and give us some freedom in developing evolutionary models.

The third subsection examines what we need to know in order to address particular questions about natural selection. For example, if we want to study questions about the detailed dynamics of evolutionary change for a particular trait, we typically need to know a lot about the trait's genetic architecture. On the other hand, if we want to ask whether selection is likely to account for some part of the difference between two populations, the answer will not generally depend on such details. Understanding what information is sufficient to answer a question provides crucial guidelines for the development of useful models.

Variance, Distance, or Information

The variance in fitness, V_w, arises in one form or another in every expression of selection. Why is the variance a universal metric of selection? Clearly, variation matters because selection favors some types over others only when the alternatives differ. But why does selection depend exactly on the variance rather than on some other measure of variation? We will show (proposition 2) that natural selection moves the population a certain distance. That distance is equivalent to the variance in fitness. Thus, we may think about the change caused by selection equivalently in terms of variance or distance.

We begin by noting from eq. 9.2 that $\Delta q_i / q_i = w_i - 1$. Then, the variance in fitness is:

$$(9.11) \qquad V_w = \sum q_i (w_i - 1)^2 = \sum q_i \left(\frac{\Delta q_i}{q_i}\right)^2 = \sum \frac{(\Delta q_i)^2}{q_i}.$$

The squared distance in Euclidean geometry is the sum of the squared changes in each dimension. On the right is the sum of the squares for the change in frequency. Each dimension of squared distance is divided by the original frequency. That normalization makes sense, because a small change relative to a large initial frequency means less than a small change relative to a small initial frequency. The variance in fitness measures the squared distance between the ancestral and descendant population in terms of the frequencies of the types, as proposition 2 and eq. 9.2 imply (Ewens 1992; Frank 2012b, a).

When the frequency changes are small, the expression on the right equals the Fisher information measure (Frank 2009). A slightly different measure of information arises in selection equations when the frequency changes are not small (Frank 2012b), but the idea is the same. Selection acquires information about environmental challenges through changes in frequency. Although this point may seem abstract, it may be a more accurate description of the process of adaptation than to say that phenotypes have fitnesses and populations climb fitness peaks, because individuals and populations respond to the environment rather than possess fitnesses.

Thus, we may think of selection in terms of variance, distance, or information. Selection moves the population frequencies a distance that equals the variance in fitness. That distance is equivalent to the gain in information by the population caused by selection.

Characters and Coordinates

We can think of fitness and characters as alternative coordinates in which to measure the changes caused by natural selection in frequency, distance, and information. Using eq. 9.2, we can rewrite the variance in fitness from eq. 9.11 as:

$$V_w = \sum q_i (w_i - 1)^2 = \sum \Delta q_i w_i.$$

Compare that expression with eq. 9.3 for the change in the character value caused by selection. If we start with the right side of the expression for the variance in fitness and then replace w_i by z_i, we obtain the change in character value caused by selection. We can think of that replacement as altering the coordinates on which we measure change, from the frequency changes described by fitness, $w_i = q_i'/q_i$, to the character values described by z_i.

Although this description in terms of coordinates may seem a bit abstract, it is essential for thinking about evolutionary change in relation to selection. Selection changes frequencies. The consequences of frequency for the change in characters depend on the coordinates that describe the translation between frequency change and characters (Frank 2012b, 2013a).

Eq. 9.4 provides an exact expression that includes four aspects of evolutionary change. First, the change in frequencies, Δq_i, causes evolutionary change. Second, the amount of change depends on the coordinates of characters, z_i. Third, the change in the coordinates of characters during transmission, Δz_i, causes evolutionary change (proposition 3). Fourth, the changed coordinates have their consequences in the context of the frequencies in the descendant population, q_i' (proposition 4).

In models of selection, one often encounters the variance in characters, V_z, rather than the variance in fitness, V_w. The variance in characters is simply a change in scale with respect to the variance in fitness—another way in which to describe the translation between the coordinates for frequency change and the coordinates for characters. In particular,

(9.12) $$\Delta_s \bar{z} = Cov(w,z) = \beta_{zw} V_w = \beta_{wz} V_z,$$

thus

$$V_z = \frac{\beta_{zw}}{\beta_{wz}} \ V_w = \gamma \ V_w.$$

Here, γ is based on the regression coefficients. The value of γ describes the rescaling between the variance in characters and the variance in fitness. Thus, when V_z arises in selection equations, it can be thought of as the rescaling of V_w in a given context (Frank 2013a).

Sufficiency and Invariance

Having seen that there are alternative ways to model evolution and adaptation (Phillips, chap. 4), and that they are all related to one another, it seems appropriate to ask: What do we need to know to analyze natural selection? The notion of *sufficiency* is useful here. Informally, a statistic is sufficient for estimating a quantity if no other statistic can be calculated from a sample that provides more information. A familiar example is that of the normal distribution. If we know the variance, σ^2, then the mean in a sample, x, is a sufficient statistic for the true mean, μ, because we cannot calculate any other statistic that provides more information about the true mean.

We compare two alternative modeling approaches. One provides full information about how the population evolves over time. The other considers only how natural selection alters average character values at any instant in time.

A full analysis begins with the change in frequency given in eq. 9.2. For each type in the population, we must know the initial frequency, q_i, and the fitness, w_i. From those values, each new frequency can be calculated. Then new values of fitnesses would be needed to calculate the next round of updated frequencies. Fitnesses can change with frequencies and with extrinsic conditions. That calculation provides a full description of the evolutionary dynamics over time. The detailed output concerning dynamics reflects the detailed input about all of the initial frequencies and all of the fitnesses over time.

A more limited analysis arises from the part of total evolutionary change caused by selection. If we focus on the change by selection in the average value of a character at any point in time, we have

$$\Delta_s \bar{z} = \sum \Delta q_i z_i = Cov(w,z) = \beta_{zw} V_w$$

from eqs. 9.6 and 9.8.

To calculate the change in average value caused by selection, it is sufficient to know the covariance between the fitnesses and character values over the population. We do not need to know the individual frequencies or the individual fitnesses. It is sufficient to know a single summary statistic over the population, the covariance. Put another way, a single assumed input (the covariance) corresponds to a single output (the change in average value caused by selection). We could, of course, make more complicated assumptions about inputs and get more complicated outputs.

Invariance provides another way to describe sufficiency and the causal effect of selection in populations. The mean change in the character value caused by selection does not depend on (is invariant to) any aspect of variability except the covariance. Many alternative populations with different character values and fitnesses have the same covariance and thus the same change in the character value caused by selection. The reason is that the variance in fitness, V_w, describes the distance the population moves with regard to frequencies, and the regression β_{zw} rescales the distance along coordinates of frequency into distance along coordinates of the character. Thus, simple invariances sometimes can provide great insight into otherwise complex problems (Frank 2013b).

For example, Fisher's fundamental theorem of natural selection is a simple invariance (Frank 2012c). The theorem states that at any instant in time, the change in average fitness caused by selection is equal to the genetic variance in fitness (discussed below). Fisher's theorem shows that the change in mean fitness by selection is invariant to all details of variability in the population except the genetic variance.

Causal Models

We now turn to emphasize inductive approaches. Eq. 9.12 describes associations between characters and fitness. In that equation, we know only that a character, z, and fitness, w, are correlated, as expressed by $\text{Cov}(w,z)$. We do not know anything about the causes of correlation and variance. But we may have a model about how variation in characters causes variation in fitness. To study that causal model, we must analyze how the hypothesized causal structure predicts correlations between characters, fitness, and evolutionary change. Alternative causal models provide alternative hypotheses and predictions that can be compared with observation (Crespi 1990; Frank 1997, 1998; Scheiner et al. 2000).

Regression equations provide a simple way in which to express hypothesized causes (Li 1975). For example, we may have a hypothesis that the character z is a primary cause of fitness, w, expressed as a directional path diagram $z \rightarrow w$. That path diagram, in which z is a cause of w, is mathematically equivalent to the regression equation

(9.13) $$w_i = \varphi + \beta_{wz} z_i + \varepsilon_i,$$

in which φ is a constant and ε_i is the difference between the actual value of z_i and the value predicted by the model: $\varphi + \beta_{wz} z_i$.

Multiple Characters

Proposition 5 asserts that we can partition fitness into the amounts of change caused by different characters. Here we show how this can be done. To analyze causal models, we focus on the general relations between variables rather than on the values of particular individuals or genotypes. Thus, we can drop the i subscripts in eq. 9.13 to simplify the expression, as in the following expanded regression equation

(9.14) $$w_i = \varphi + \beta_{wy \cdot z} z + \beta_{wy \cdot z} y + \varepsilon.$$

Here, fitness w depends on the two characters z and y (Lande and Arnold 1983). The partial regression coefficient $\beta_{wz \cdot y}$ is the average effect of z on w holding y constant, and $\beta_{wy \cdot z}$ is the average effect of y on w holding z constant. Regression coefficients minimize the total distance (sum of squares) between the actual and predicted values. Minimizing the residual distance maximizes the use of the information contained in the predictors about the actual values.

This regression equation is exact, in the sense that it is an equality under all circumstances. No assumptions are needed about additivity or linearity of z and y or about normal distributions for variation. Those assumptions arise in statistical tests of significance when comparing the regression coefficients with hypothesized values or when predicting how the values of the regression coefficients change with context.

Note that the regression coefficients (β) often change as the values of w or z or y change, or if we add another predictor variable. The exact equation is a description of the relations between the variables as they are given. The structure of the relations between the variables forms a causal hypothesis that leads to predictions (Li 1975).

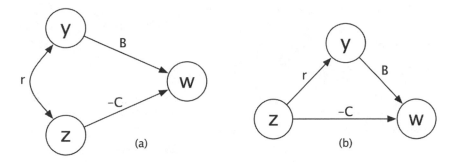

Figure 9.1. Path diagrams for the effects of phenotype z and secondary predictor y on fitness w. (a) An unknown cause associates y and z. The arrow connecting those factors points both ways, indicating no particular directionality in the hypothesized causal scheme. (b) The phenotype, z, directly affects the other predictor, y, which in turn affects fitness. The arrow pointing from z to y indicates the hypothesized direction of causality. The choice of notation matches kin selection theory, in which z is an altruistic behavior that reduces the fitness of an actor by the cost C and aids the fitness of a recipient by the benefit B, and r measures the association between the behaviors of the actor and recipient. Although that notation comes from kin selection theory, the general causal scheme applies to any pair of correlated characters that influences fitness (Lande and Arnold 1983; Queller 1992). From Frank (2013a).

Partitions of Fitness

We can interpret eq. 9.14 as a hypothesis that partitions fitness into two causes (proposition 5). Suppose, for example, that we are interested in the direct effect of the character z on fitness. To isolate the direct effect of z, it is useful to consider how a second character, y, also influences fitness (fig. 9.1).

The condition for z to increase by selection can be evaluated with eq. 9.12. That equation simply states that z increases when it is positively associated with fitness. However, we now have the complication shown in eq. 9.14 that fitness also depends on another character, y. If we expand $\mathrm{Cov}(w,z)$ in eq. 9.12 with the full expression for fitness in eq. 9.14, we obtain

$$(9.15) \qquad \Delta_s \bar{z} = \beta_{wz} V_z = (\beta_{wz \cdot y} + \beta_{wy \cdot z} \beta_{yz}) V_z.$$

Following Queller (1992), we abbreviate the three regression terms. The term $\beta_{yz} = r$ describes the association between the phenotype, z, and the other predictor of fitness, y. An increase in z by the amount Δz corresponds to an average increase of y by the amount $\Delta y = r \Delta z$. The term $\beta_{wy \cdot z} = B$ describes the direct effect of the other predictor, y, on fitness, holding constant the focal phenotype, z. The term $\beta_{wy \cdot z} = -C$ describes the direct effect of the phenotype, z, on fitness, w, holding constant the effect of the other predictor, y.

The condition for the increase of z by selection is $\Delta_s \bar{z} > 0$. The same condition using the terms on the right side of eq. 9.15 and the abbreviated notation of the previous paragraph is:

$$(9.16) \qquad\qquad rB - C > 0$$

This is Hamilton's famous equation for kin selection (Hamilton 1964a, 1970). We use this equation here to emphasize the fact that Hamilton's model is a particular case of a much more general relationship. The condition in eq. 9.16 applies whether the association between z and y arises from some unknown extrinsic cause (fig. 9.1a) or by the direct relation of z to y (fig. 9.1b).

This expression describes the condition for selection to increase character z when ignoring any changes in the character that arise during transmission. Thus, when one wants to know whether selection acting by this particular causal scheme would increase a character, it is sufficient to know if this simple condition holds.

Testing Causal Hypotheses

If selection favors an increase in character z, then the condition in eq. 9.16 will always be true. That condition simply expresses the fact that the slope of fitness on character value, β_{wz}, must be positive when selection favors an increase in z. The expression $\beta_{wz} = rB - C$ is one way in which to partition β_{wz} into components. However, the fact that $rB - C > 0$ does not mean that the decomposition into those three components always provides a good causal explanation for how selection acts on the character z.

There are many alternative ways in which to partition the total effect of selection into components. Other characters may be important. Environmental or other extrinsic factors may dominate. How can we tell if a particular causal scheme is a good explanation?

If we can manipulate the effects r, B, or C directly, we can run an experiment. If we can find natural comparisons in which those terms vary, we can test comparative hypotheses. If we add other potential causes to our model, and the original terms hold their values in the context of the changed model, that stability of effects under different conditions increases the likelihood that the effects are true.

Three points emerge. First, a partition such as $rB - C$ is sufficient to describe the direction of change, because a partition simply splits the total change into parts. Second, a partition does not necessarily describe causal relations in an accurate or useful way. Third, various methods can be used to test whether a causal hypothesis is a good explanation.

Partitions of Characters

We have been studying the partition of fitness into separate causes, including the role of individual characters. It is sometimes useful to partition individual characters into separate causes as well, such as contributions of different alleles, or of other factors (proposition 5). Each character may itself be influenced by various causes. We can describe the cause of a character by a regression equation:

$$z = \varphi + \beta_{zg}g + \delta$$

in which φ is a constant traditionally set to zero, g is a predictor of phenotype, the regression coefficient β_{zg} is the average effect of g on phenotype z, and $\delta = z - \beta_{zg}g$ is the residual between the actual and predicted values. For predictors g, we could use temperature, neighbors' behavior, another phenotype, epistatic interactions given as the product of allelic values, symbiont characters, or an individual's own genes.

Fisher (1918) first presented this regression for phenotype in terms of alleles as the predictors. Suppose

(9.17)
$$g = \sum_j b_j x_j \, ,$$

in which x_j is the presence or absence of an allelic type. Then each b_j is the partial regression of an allele on phenotype, which describes the average contribution to phenotype for adding or subtracting the associated allele. The coefficient b_j is called the average allelic effect, and g is called the breeding value (Fisher 1930; Crow and Kimura 1970; Falconer and Mackay 1996). When g is defined as the sum of the average effects of the underlying predictors, then $\beta_{zg} = 1$, and

(9.18)
$$z = g + \delta$$

where $\delta = z - g$ is the difference between the actual value and the predicted value.

Transmission

Now we turn to include transmission in our models. To do so it is useful to note some facts. If we take the average of both sides of eq. 9.18, we get $\bar{z} = \bar{g}$, because $\bar{\delta} = 0$ by the theory of regression. If we take the variance of both sides, we obtain $V_z = V_g + V_\delta$, noting that, by the theory of regression, g and δ are uncorrelated.

Heritability and the Response to Selection

To study selection, we first need an explicit form for the relation between character value and fitness, which we write here as

$$w = \varphi + \beta_{wz}z + \varepsilon.$$

Substituting that expression into the covariance expression of selection in eq. 9.12 yields:

(9.19) $$\Delta_s \bar{z} = \text{Cov}(w,z) = \beta_{wz}V_z = sV_z,$$

because φ is a constant and ε is uncorrelated with z, causing those terms to drop out of the covariance. Here, the coefficient $s = \beta_{wz}$ is the effect of the character on fitness. Expanding sV_z by the partition of the character variance (proposition 5) given in the previous section leads to:

(9.20) $$\Delta_s \bar{z} = sV_z = sV_g + sV_\delta = \Delta_g \bar{z} + \Delta_n \bar{z}.$$

We can think of g as the average effect of the predictors of phenotype that we have included in our causal model of character values. Then $sV_g = \Delta_g \bar{z}$ is the component of total selective change associated with our predictors, and

(9.21) $$\Delta_g \bar{z} = \Delta_s \bar{z} - \Delta_n \bar{z}$$

shows that the component of selection transmitted to descendants through the predictors included in our model, Δ_g, is the change caused by selection, Δ_s (proposition 1), minus the part of the selective change that is not transmitted through the predictors, Δ_n (proposition 4). Although it is traditional to use alleles as predictors, we can use any hypothesized causal scheme. For example, one of the predictors could be the presence or absence of a particular bacterial species in the gut. When one adds gut bacteria as predictors, or new alleles not previously accounted for, the expanded causal model typically assigns greater cause to the totality of predictors, Δ_g, and less cause to the remaining component of change, Δ_n. Thus, the separation between transmitted and nontransmitted components of selection depends on the hypothesis for the causes of the phenotype.

If we choose the predictors for g to be the individual alleles that influence the phenotype, then V_g is the traditional measure of genetic variance, and sV_g is that component of selective change that is transmitted from parent to offspring through the effects of the individual alleles. The fraction of the total change that is transmitted, V_g/V_z, is a common measure of heritability.

Changes in Transmission and Total Change

We now have the tools needed to find the total evolutionary change when considered in terms of the parts of phenotype that are transmitted to descendants. Here, the transmitted part arises from the predictors in an explicit causal hypothesis about phenotype.

From eq. 9.18, $\bar{z} = \bar{g}$, because the average residuals of a regression, $\bar{\delta}$, are zero. Thus, when studying the change in a character, we have $\Delta\bar{z} = \Delta\bar{g}$, which means that we can analyze the change in a character by studying the change in the average effects of the predictors of a character. From eq. 9.4, we can write the total change in terms of the coordinates of the average effects of the predictors, g, yielding:

$$(9.22) \qquad \Delta\bar{z} = \sum \Delta q_i g_i + \sum q'_i \Delta g_i = \Delta_g \bar{z} + \Delta_t \bar{z},$$

in which $\Delta_t \bar{z}$ is the change in the average effects of the predictors during transmission (Frank 1997, 1998). The total change divides into two components (proposition 4): the change caused by the part of selection that is transmitted to descendants plus the change in the transmitted part of the phenotype between ancestors and descendants. Alternatively, we may write $\Delta_g \bar{z} = \Delta_s \bar{g}$, the total selective component expressed in the coordinates of the average effects of the predictors, and $\Delta_t \bar{z} = \Delta_c \bar{g}$, the total change in coordinates with respect to the average effects of the predictors.

Choice of Predictors

If natural selection dominates other evolutionary forces, then we can use the theory of natural selection to analyze evolutionary change. When does selection dominate? From eq. 9.22, the change in phenotype caused by selection is Δ_g. If the second term Δ_t is relatively small, then we can understand evolutionary change primarily through models of selection.

A small value of the transmission term, Δ_t, arises if the effects of the predictors in our causal model of phenotype remain relatively stable between ancestors and descendants. Many factors may influence the phenotype, including alleles and their interactions, maternal effects, various epigenetic processes, changing environment (Scheiner, chap. 13), and so on. Finding a good causal model of the phenotype in terms of predictors is an empirical problem that can be studied by testing alternative causal schemes against observation.

Note that the equations of evolutionary change do not distinguish between different kinds of predictors. For example, one can use both alleles and weather as predictors. If weather varies among types and its average effect on phenotype transmits stably between ancestors and descendants, then weather provides a useful predictor. Variance in stably transmitted weather attributes can lead to changes in characters by selection. Calling the association between weather and fitness an aspect of selection may seem strange or misleading. One can certainly choose to use a different description. But the equations themselves do not distinguish between different causes.

Discussion

The Uses of Inductive versus Deductive Approaches

Sometimes it makes sense to think in terms of deductive predictions. What do particular assumptions about initial conditions, genetic interactions, and the fitnesses predict about evolutionary dynamics? For example, if we know the current frequency of genotypes, the fitnesses of those genotypes, and the pattern of mating between genotypes, then we can deductively predict the dynamics of change in genotype frequencies between the original population and their descendants.

Sometimes it makes sense to think in terms of inductive analysis. Given the observed changes between ancestor and descendant populations, how much do different causes explain of that total distance? For example, if we know the current phenotypes of individuals, and we observe the phenotypes of offspring, then we can inductively estimate the causes of the observed changes in terms of the partitioning of fitness into different estimated strengths of selection acting on the individual phenotypes.

Accomplishments of Inductive Theory

We illustrate the value of inductive theory with three examples. First, inductive approaches provide empirical methods for the study of natural selection in populations. Typically, one begins with data about the reproductive success of individuals and about measurements of various characters of those individuals. One then asks questions such as: How much does an increase in body weight enhance reproductive fitness? How much does stress measured by cortisol level reduce reproductive fitness?

Although these are simple questions, one has data only about the correlations between various characters and fitness. Teasing out estimated causal relations from such correlational data can be difficult. In other words, it is not so easy to inductively arrive at the relative causal strengths for the various characters in the explanation of variation in observed values of fitness. For example, suppose that larger body size is correlated with both reduced cortisol level and increased fitness. How do we explain the causes of increased fitness? It could be that large body size directly increases fitness and that reduced stress is correlated with large body size. Or it could be that reduced stress reflects good physiological health and immune system status, which directly enhance both fitness and body size.

Distinguishing between these alternative hypotheses requires a careful approach to the inductive analysis of natural selection. Lande and Arnold (1983) initiated modern approaches to inductive methods. Many subsequent approaches to inductive analysis have been developed, including techniques such as path analysis (Crespi 1990; Frank 1997, 1998;Scheiner et al. 2000) and an analytical approach known as Aster (R. G. Shaw and Geyer 2010).

Second, the theory of kin selection has developed complementary deductive and inductive approaches. The original deductive theory by Hamilton (1964a, 1970) made assumptions about the frequencies and fitnesses of alternative genes. Those genes were associated with altruistic behaviors that benefit relatives at a cost to the actor that performs the behavior. For example, a bee in a social colony might help her mother to reproduce rather than reproduce herself. That altruistic behavior benefits her mother's reproduction and simultaneously imposes a cost on her own reproduction. When would natural selection favor such an altruistic behavior that reduces the actor's own direct fitness?

From assumptions about the direct cost of altruistic behavior and the benefit to the recipient of the altruism, Hamilton used population genetics to deduce the conditions under which increased altruism would evolve by natural selection. Put another way, he analyzed the conditions that favor an increase in the frequency of genes associated with altruism. He found the condition $rB-C > 0$ for the increase in altruism, in which C is the direct cost in fitness of the altruistic behavior, B is the recipient's benefit from the altruistic behavior, and r is the relatedness between actor and recipient.

Hamilton's original theory assumed a given partition of the causes of fitness into a part attributed to the cost of the behavior and a part attributed to the benefit of the behavior. That deductive theory makes predictions about behavior. By contrast, actual studies of natural populations often obtain data about observed behaviors, relatedness, and reproductive fitness. From those data, one inductively estimates the partition of fitness into costs to the individual that expresses the character and benefits to the individuals that receive the consequences of the character.

Queller (1992) recognized the identical structure of Hamilton's original deductive theory and the inductive methods of Lande and Arnold (1983). Following Queller's insight, the modern theory of kin selection unified deductive and inductive theories into a single approach that focuses on the partitioning of fitness into causal components associated with various characters and their associated costs and benefits (Frank 1997, 1998). In this context, group selection is an alternative way to partition the causes of fitness into components (Goodnight, chap. 10; Hamilton 1975; Frank 1986, 1998).

Third, inductive approaches provide methods for the analysis of molecular genetic data. Before extensive molecular data were available, almost all population genetic theory was deductive. After molecular data became common, inductive theory dominated (Ewens 1990). Classically, one began with alleles and fitnesses, and then deduced gene frequency changes (Crow and Kimura 1970). Since the molecular revolution, one typically begins with current samples of alleles and then tries to induce the historical states and processes of the past (Graur 2016). For example, given an observed sample of DNA sequences in a population, one may compare a variety of alternative processes that might have generated the observed sample. One can ask which of the alternative processes is most likely to generate the observed pattern, an inductive perspective that begins with the observed data.

Suppose we have a sample of nucleotide sequences obtained from influenza viruses over a series of annual epidemics. We can reconstruct a phylogenetic history of the viruses from those nucleotide sequences. Within that history, we can estimate how particular nucleotides and associated amino acids changed over time. We can then ask: How has natural selection acted on particular amino acids that coat the surface of the virus? We may inductively conclude that certain amino acids changed in a manner correlated with the virus's escape from recognition by host immunity and subsequent spread in the next epidemic, suggesting that natural selection favors rapid evolution of those particular amino acids (R. M. Bush et al. 1999). Once again, we have inductively assigned a potential causal role of natural selection to explain the pattern of changes we observe in populations.

Status of the Theory

The theory of natural selection provides many of the key insights for understanding how organisms evolve. Several chapters in this volume illustrate the primacy of selection, including chapter 11 on the evolution of life histories (Fox and Scheiner), chapter 12 on ecological specialization (Poisot), chapter 13 on phenotypic plasticity (Scheiner), and chapter 14 on recombination (Orive).

As the theory of natural selection matured in the 1970s and 1980s, empirical studies showed it to be one of the best-supported theories in science (Endler 1986). The advent of modern computers and modern statistical methods led to extensive reviews of this empirical support, including quantification of such things as the strength of selection (Hoekstra et al. 2001; Kingsolver et al. 2001; Nielsen 2005). Nevertheless, the subtlety of the concepts suggests a need for evolutionary biologists to pursue deeper understanding of the theory and its implications.

As an example of deeper conceptual issues, we have emphasized that Price's formulation provides a useful way to understand the relations between deductive and inductive approaches to selection. Both deductive and inductive approaches play key roles in efforts to understand the diverse evolutionary patterns discussed in chapters 11–14 below. The approach taken in this chapter helps to clarify the points of connection (and of contrast) between the inductive and deductive approaches. Because of the recent increase in scientists' interest in and capability of collecting large datasets, we expect that inductive approaches to understanding natural selection will become increasingly important.

There has also been a trend in many information sciences to develop new methods of learning and inference that can be applied to large datasets beyond biology. The conceptual challenges in those various subjects often hint at the need to understand more deeply how information accumulates by various trial and error algorithms. Our understanding of natural selection will likely contribute to and gain from those broader developments in modern science.

Acknowledgments

A more comprehensive version of this chapter is in Frank (2014), parts of which were taken from a series of articles on natural selection published in the *Journal of Evolutionary Biology*. NSF grant DEB 1251035 supports SAF's research. GAF was supported by NSF grant DEB 1120330.

TEN

The Theory of Multilevel Selection

CHARLES GOODNIGHT

It is frequently observed that organisms behave in ways that are difficult to reconcile with selection acting only in the best interests of the organism, that is, acting at the individual level, and yet only recently have we begun to develop a general consensus on how these group-level traits evolve. In this chapter I focus on three main topics concerning multilevel selection: (1) the relationship between the Price equation, group selection, multilevel selection, and kin selection, (2) the underlying causes of the effectiveness of higher levels of selection, and (3) the implications of a multilevel perspective of evolution. By multilevel selection, I specifically am referring to selection acting at more than one level. If I am referring to selection at a single level, I will call it by an appropriate name, such as cell-level selection, individual selection, or group selection. Most of the older literature on "group selection" is a component of the multilevel-selection literature, since in this literature group selection is almost always compared in some manner with individual selection. Thus, multilevel-selection theory is not limited to selection acting simultaneously at the individual and group level, but applies equally well to selection at other levels, such as at the cell and individual level. That said, most theoretical and experimental studies have considered selection acting at the individual and group levels, so that will be the focus of this chapter.

Domain and Structure of the Theory

One of the fundamental features of the living world is that it is structured into groups from cells up to ecosystems, and potentially the entire biosphere (proposition 1, table 10.1). However, systems need not have this discrete, nested structure. For example, groups may consist of overlapping sets of kin

Table 10.1.   The Domain and Propositions of the Theory of Multilevel Selection

Domain: Selection acting at multiple levels of biological organization.

Propositions:

1. Living things are organized into natural groups that are often nested but may be overlapping or have indistinct boundaries.
2. It is possible to get a response to selection at any level of biological organization in which there is heritable variation in fitness.
3. Levels of organization may be defined by different criteria.
4. The direction of selection at one level is not dependent on the direction of selection at other levels.
5. The nature of and genetic basis for the response to selection are qualitatively different at different levels of organization.

Corollary 5A. Measures of heritable variation made on one level of organization may not provide information on the heritability of traits measured on other levels of organization.

Corollary 5B. Selection at one level cannot be described in terms of selection acting at a lower level.

groups when matings take place outside the family. Alternatively, boundaries may not be clear in continuous populations where mating varies by distance (Hamilton 1964b). Nevertheless, as pointed out by Lewontin (1970), evolution by natural selection will occur at any level of biological organization that satisfies three criteria: (1) there is phenotypic variation, (2) the phenotypic variation correlates with fitness, and (3) the phenotypic differences are heritable. Thus, in principle, evolution by natural selection can occur at any level that satisfies Lewontin's criteria (proposition 2, table 10.1).

Multilevel selection occurs when selection is simultaneously acting at more than one of these levels of biological organization. If selection is acting at only one level, then it is best to refer to it as selection at that level, such as cellular selection, individual selection, or group selection. A multilevel-selection perspective works well for understanding many forms of adaptive evolution, especially when selection processes at different levels are acting in opposition (proposition 4, table 10.1). For example, kin selection can be shown to be equivalent to an opposing combination of family selection and individual selection. Soft selection can be shown to be an opposing combination of group selection and individual selection. Many forms of frequency-dependent selection can be shown to be the result of selection acting simultaneously at more than one level. From a multilevel-selection perspective, cancer can be studied as the opposing forces of cellular selection favoring the rapid reproduction of the cancer cells, and organismal selection favoring those individuals that can suppress cancer growth and survive and reproduce.

Definitions of Group Selection and Multilevel Selection

Although the modern discussion of the efficacy of group selection was introduced by Wynne-Edwards (1962), the early debate was framed by Maynard Smith (1964, 1145). Maynard Smith felt that group selection could occur "[i]f all members of a group acquire some characteristic which, although individually disadvantageous, increases the fitness of the group." This is a very narrow vision of group selection based on three premises: that it occurs only if it leads to an adaptation that is clearly opposed by individual selection, that individual selection is far stronger than group selection (Wade 1978), and that it occurs only when there is no variation within groups (A. Gardner 2015).

A better definition was presented by Wade (1977), who defined group selection to be the differential extinction and proliferation of groups. This is a good working definition because it defines group selection in terms of the selection process rather than the adaptation, and it removes the narrow limits placed on the process in Maynard Smith's definition. Wade's definition is adequate for all of the experimental studies of group selection, and it continues to be widely used.

In the majority of laboratory studies, group selection has been imposed by differential extinction of groups. For example, in Wade's (1977) study in the group selection for high population size treatment, selection was imposed by choosing the population with the largest size and using it to set up as many new populations as possible. When the largest population was exhausted, he chose the second largest, and so forth. As a result only the largest populations contributed to the next generation, and the smallest were discarded without founding any new populations.

In other studies this definition has been stretched, suggesting that an expanded definition may be more appropriate (proposition 3, table 10.1). When populations are not clearly structured, it is not necessarily clear what a group is or what "differential extinction and proliferation of groups" means. For example, Wade and Goodnight (1991) observed a response to group selection caused by differential migration from populations, indicating that group selection can occur through differential proliferation of groups in the form of the spread of some groups at the expense of others.

There have been a number of experiments and models examining continuous populations that have demonstrated behavior that is best termed group selection, even though the "groups" are actually regions of a continuous population (e.g., L. Stevens et al. 1995; Goodnight et al. 2008) or self-organized interacting groups similar to D. S. Wilson's (1980) "trait groups" that persist only transiently (Eldakar et al. 2010). Finally, theoretical work has demonstrated that multilevel selection is mathematically equivalent to many forms of frequency-dependent selection (Goodnight et al. 1992). As a result, a reasonable definition of group selection is that it occurs when the fitness of an individual is a function of group membership. This still leaves the question of what a "group" is. Maynard Smith (1964) appeared to define groups as discrete entities with clear boundaries. However, recent work has emphasized, that groups can be regions in a continuous population, and that groups can persist even in the face of high migration rates (Wade and McCauley 1984). Thus, an individual's group is best considered to be the set of individuals with whom it interacts. This greatly broadens the definition of "group." For example, in a continuous population every individual may be at the center of its own group of interacting individuals, and as a result each individual's group might be unique and slightly different from that of its neighbors. Many consider this definition of group selection overly broad and as a result problematical (e.g., A. Gardner and Grafen 2009).

The Price Equation

The Price equation is a means of partitioning covariance first suggested by G. R. Price (1970). In the hierarchical setting of multilevel selection the Price equation (Frank and Fox, chap. 9, eq. 9.4) can be used to partition the covariance between relative fitness and a trait into within-group and among-group components:

$$(10.1) \qquad \Delta \bar{z} = cov(w_{ij} z_{ij}) + E[cov(w_{ij} z_{ij})] + E[w_{ij} Dz_{ij}],$$

where Δz is the change in the mean of a trait due to selection; w_{ij} and z_{ij} are the relative fitness and phenotype of the ith individual in the jth group, respectively; $cov(w_{ij} z_{ij})$ is the covariance between the group-mean relative fitness and the group-mean phenotype; $E[cov(w_{ij} z_{ij})]$ is the expected within-group covariance between individual relative fitness and individual phenotype; and $E[w_{ij} \Delta z_{ij}]$ is the fitness-weighted change in phenotype due to factors other than selection. See Frank and Fox (chap. 9) for a derivation. The relative fitness is equal to the absolute fitness divided by the mean absolute fitness, and thus mean relative fitness is 1. The term $E[w_{ij} \Delta z_{ij}]$ is often called "transmission bias," but any change in the trait that is not due to selection will be included in this term (Frank 2012a). I will follow the tradition of assuming that the environment is constant and this term is zero (e.g., Wright 1942; Frank 2012a).

The multilevel Price equation effectively divides a covariance into within- and between-group components; however, its relationship to group selection is problematical. Several authors have equated the covariance between group-mean relative fitness and group-mean phenotype with group selection (e.g., Wade 1985; A. Gardner 2008, 2015); however, the group-mean phenotype can change as a result of selection acting at other levels. Consider the situation of pure individual selection in a subdivided population. If the groups are finite in size there will inevitably be groups that have larger numbers of individuals with the favored phenotype, and thus a higher group-mean relative fitness. If we use the Price equation, this will show up as a covariance between group-mean relative fitness and group-mean phenotype, even though the relationship is entirely due to individual-level differences in fitness. For this reason Wade (1985), using the Price equation, concluded that hard selection had a group selection component even though hard selection is the situation where fitness is determined entirely by an individual's phenotype and is not influenced at all by the characteristics of the group.

Contextual Analysis

Modern studies of selection in natural populations have generally relied on the regression approaches first pioneered by Lande and Arnold (1983; Frank and Fox, chap. 9). In this approach, a number of phenotypic traits are measured along with at least one fitness trait. The phenotypic traits may be morphological, physiological, or behavioral, and the fitness trait is typically a component of fitness such as survival, mating success, or reproductive success. A multiple regression is then performed of the phenotypic traits on the fitness trait. A significant partial regression of a phenotypic trait on the fitness trait is considered to indicate significant selection acting on that trait. A trait can change either owing to direct selection acting on that trait or owing to selection acting on a trait with which it is correlated. For example, trait 1 can change owing to selection acting directly on trait 1, or owing to selection acting on trait 2. The (within-generation) change due to direct selection acting on trait 1 is:

$$(10.2) \qquad \Delta \bar{z}_{1\,(direct\_selection)} = V_p(z_1)\beta_{wz1},$$

and the change due to selection acting on trait 2 is:

$$(10.3) \qquad \Delta \bar{z}^*_{1(indirect\_selection)} = Cov_p(z_1 z_2)\beta_{wz2}.$$

Using linear algebra, this becomes:

$$(10.4) \quad \Delta \bar{z}^* = \begin{bmatrix} \Delta \bar{z}_1^* \\ \Delta \bar{z}_2^* \end{bmatrix} = \begin{bmatrix} V_p(z_1) & Cov_p(z_1 z_2) \\ Cov_p(z_1 z_2) & V_p(z_2) \end{bmatrix} \begin{bmatrix} \beta_{wz_1} \\ \beta_{wz_2} \end{bmatrix},$$

where $\Delta \bar{Z}^*$ is the within-generation change in the mean phenotype vector of the population, in this case the change in traits z_1 and z_2; $\Delta \bar{z}_1^*$ and $\Delta \bar{z}_2^*$ are the change in the mean of the traits making up the $\Delta \bar{Z}^*$ vector; β_{wz_1} and β_{wz_2} are the slopes of the partial regressions of the respective traits on relative fitness; $V_P(z_1)$ and $V_P(z_2)$ are the phenotypic variances for the two traits; and $Cov_P(z_1, z_2)$ is the phenotypic covariance between the traits.

Heisler and Damuth (1987) extended the phenotypic selection models of Lande and Arnold (1983) as "contextual analysis." The basic idea is that a "contextual" trait can be a summary statistic such as $\bar{z}_1$, the mean of the individual-level trait, or it may be a trait that is not expressed by the individual organism, such as population size. These are incorporated into the multiple regression in exactly the same manner as the individual traits:

$$(10.5) \quad \Delta \bar{Z}^* = \begin{matrix} \Delta \bar{z}_1 \\ \Delta \bar{z}_2 \\ \Delta \bar{\bar{z}}_1 \\ \Delta \bar{\bar{z}}_2 \\ \Delta c_1 \\ \Delta c_2 \end{matrix} = P \begin{matrix} \beta_{wz_1} \\ \beta_{wz_2} \\ \beta_{wz_1} \\ \beta_{wz_2} \\ \beta_{wc_1} \\ \beta_{wc_2} \end{matrix},$$

where $\Delta \bar{z}_1$ and $\Delta \bar{z}_2$ are the changes in the mean of the traits averaged across all individuals in all groups; $\Delta \bar{\bar{z}}_1$ and $\Delta \bar{\bar{z}}_2$ are the changes in the average group mean of each trait; c_1 and c_2 are contextual traits; Δc_1 and Δc_2 are changes in the mean taken across groups for the contextual traits; β_{wz_1}, β_{wz_2}, β_{wc_1}, and β_{wc_2} are the partial regressions of relative fitness on the group means of the group traits and the contextual traits; and P is the phenotypic covariance matrix (analogous to the matrix in equation 10.4).

Because contextual traits can be included in a selection analysis in the same manner as individual-level traits, contextual analysis can easily be extended to any linear model-based selection analysis, such as path analysis (Li 1975; L. Stevens et al. 1995). Similarly, theoretical models using contextual analysis (e.g., Goodnight et al. 1992) have generally focused on the group mean as the appropriate summary statistic. This choice is governed by the details of the individual models, not by any constraint of the approach. For example, in many settings it may be valuable to use a contextual trait that leaves out the focal individual, such as the mean of the focal individual's neighbors (Okasha 2006), or the within-deme variance that reflects the dispersion within subpopulations.

It is important to recognize that all regression-based selection analyses, including contextual analysis, are based on correlations. From a statistical perspective the Lande and Arnold (1983) approach, with its contextual analysis, provides the best prediction of the response to selection based on the measured traits and fitness. However, this does not necessarily imply that selection on those traits is causing the observed evolutionary change. Selection analysis based on regression cannot rule out the possibility that the observed evolutionary change is a correlated response to selection acting on different, unmeasured traits. For this reason, Wade and Kalisz (1990) argue that the results of a regression-based selection analysis should be considered a hypothesis that ideally is tested using manipulative experiments (e.g., Kelly 1996).

Contextual analysis, and all regression-based selection analysis methods, have been developed for measuring the phenotypic effects of selection. Thus, even if a selection analysis identifies selection acting on a particular trait, there may be no response to that selection if the trait is not heritable. This is generally understood for individual selection models but applies equally well to analyses involving contextual traits.

In contextual analysis, fitness is assigned at only one level. For example, fitness may be measured on the organism in the form of the number of offspring produced. The multiple regression is then performed, and if there is a significant regression of a contextual trait on fitness, then group selection is said to be occurring. Group selection shows up as a change in the fitness of the individuals that make up the group. This is different from the Price equation where group selection is inferred from changes in the group-mean fitness.

The Price Equation and Contextual Analysis

Using equation 10.1, and remembering that I am assuming $E[w_{ij}\Delta z_{ij}] = 0$, based on the standard methods of multiple regression (Li 1975), it can be shown that

$$(10.6) \quad \begin{aligned} \Delta\bar{z} &= \text{cov}(w_{\cdot j}z_{\cdot j}) + E[\text{cov}(w_{ij}z_{ij})] \\ &= \text{cov}(w_{ij}, z_{\cdot j} \bullet z_{ij}) + \text{cov}(w_{ij}, z_{ij}(z_{\cdot j})) + E[\text{cov}(w_j, z_{ij} \bullet z_{\cdot j})] \end{aligned}$$

where $\mathrm{cov}(w_{ij}, z_{ij} \bullet z_{\bullet j})$ and $\mathrm{cov}(w_{ij}, z_{\bullet j} \bullet z_{ij})$ are the partial covariances for the trait within groups and among groups respectively, $\mathrm{cov}(w_{ij}, z_{ij}(z_{\bullet j}))$ and $\mathrm{cov}(w_{ij}, z_{\bullet j}(z_{ij}))$ are the covariances at one level that are due to changes at the other level, and the $E[\ ]$ denotes the expected value. The partial covariance among groups is the covariance between the group mean of the trait and the group-mean relative fitness that is independent of within-group changes in relative fitness, and the term $\mathrm{cov}(w_{ij}, z_{\bullet j}(z_{ij}))$ is the covariance between the group mean of the trait and the group-mean relative fitness that is caused by changes in fitness at the individual level. Using the partial covariances it becomes apparent that the Price partitioning includes both the direct effects of selection at one level and the indirect effects of selection at the other level.

To express this equation in terms of partial regressions, note that in equation 10.6 the first two terms correspond to the mean within-group change in phenotype ($\Delta \bar{z}_{within}$), and the last two terms correspond to the among-group change in the phenotype ($\Delta \bar{z}_{among}$). The equivalence to the contextual analysis multiple regression equation is then obtained by expressing this in a linear algebraic format (Goodnight 2013b):

$$
\Delta \bar{Z} = \begin{array}{c} \Delta \bar{z}_{within} \\ \Delta \bar{z}_{among} \end{array} = \begin{array}{cc} \mathrm{Var}(z_{within}) & \mathrm{Cov}(z_{within}, z_{among}) \\ \mathrm{Cov}(z_{within}, z_{among}) & \mathrm{Var}(z_{among}) \end{array} \begin{array}{c} \beta_{within} \\ \beta_{among} \end{array}
$$

$$
(10.7) \qquad = \begin{array}{c} \mathrm{Var}(z_{within})\beta_{within} \\ \mathrm{Var}(z_{among})\beta_{among} \end{array} + \begin{array}{c} \mathrm{Cov}(z_{within}, z_{among})\beta_{among} \\ \mathrm{Cov}(z_{within}, z_{among})\beta_{within} \end{array} .
$$

This formulation is identical to the Price equation, demonstrating that the partial regression of contextual analysis successfully decomposes the change within and among groups into the effects of selection acting directly at that level and the correlated effects of selection acting at a different level.

The Special Case of Soft Selection

Soft selection has received special attention in the multilevel-selection literature because of a seeming contradiction. Soft selection is the situation in which every group produces the same number of offspring, but the individuals that are produced are the result of selection acting within groups. The fitness of an individual is a function of both its own phenotype and the mean phenotype of its group. For example, an intermediate phenotype individual would have a high relative fitness in a group of low fitness individuals, and a low relative fitness in a group of high fitness individuals. This suggests that there is group selection acting. However, because every group produces the same number of offspring, there is no variation among groups, suggesting that group selection is not acting.

The first to examine soft selection from a multilevel perspective was Wade (1985). He noted, based on the Price equation, that because the variance among groups is zero, the covariance between mean fitness and mean phenotype must also be zero, and thus there could be no group selection in soft selection. Goodnight et al. (1992), using contextual analysis, noted that while the simple regression of the group-mean phenotype on relative fitness was zero, the partial regression was nonzero, and came to the opposite conclusion that soft selection did have a component of group selection. This debate has continued, as a number of researchers have considered the contextual analysis result to be counterintuitive, and as evidence that contextual analysis does not always provide the correct answer (e.g., Okasha 2006).

One of the truisms of mathematical modeling is that often the results of a model are at odds with our intuition. Sometimes this is the result of a mistake in the model, but after that possibility has been eliminated, we are left with the possibility that our intuition is incorrect. This is exactly what has happened with contextual analysis and soft selection. To illustrate the logic behind the contextual analysis conclusion that group selection is acting, consider two traits that have a positive correlation, say, body length and body weight. If we were to select for increased body length, we would expect body weight to also increase because of the positive phenotypic correlation. Thus we expect longer animals to weigh more than shorter animals. If we wanted to select for longer animals and not have the weight change, we would need to specifically select for long skinny animals. That is, we would need to select for lower weight to counter the indirect selection caused by selection for increased body length.

It is the same for a trait and the group mean of a trait. The group mean is a contextual trait that is correlated with the trait in the individual. Thus if we select for an increased value of the individual trait, the group mean will naturally increase as well. If we want to select for an increased value of the trait without changing the group mean of the trait, we need to exert counterselection on the contextual trait, in this case whatever "process" keeps the group sizes exactly equal. This is the group selection that is countering the individual selection that prevents the group mean fitness from changing. Thus, for there is to be soft selection, there must be both selection on the individual level trait and selection that keeps the mean fitness among groups equal. Of course, it is also important to remember that "soft selection" is a theoretical model and our selection at the group level a model assumption. The real world will never conform to this exactly, although in the one example where it has been examined, the results were remarkably close to the theoretical expectation (L. Stevens et al. 1995).

The Effectiveness of Group Selection

The earliest modern models of multilevel selection using the evolutionary-change approach are those of Griffing (1977). In these models he explored the observation that when individual selection for increased yield is applied to crop plants, the response is often negative. If the heaviest-yielding individuals are selected and used to plant the next generation, in many cases the yield will decrease. Griffing explored this in great detail (e.g., Griffing 1981, 1982, 1989) and identified "associate effects" as being responsible for this effect. Associate effects, now called indirect genetic effects (Wolf et al. 1998), are heritable effects that a plant (or other organism) has on its neighbors. Griffing found that there tended to be a negative genetic correlation between the direct effects an individual had on itself and the indirect effects it had on other individuals. Plants that were genetically predisposed to produce high yields also tended to have a strong negative effect on the yield of neighbors. In common words, selection can result in increased yield either by increasing the efficiency of resource use or by increasing the ability to gather resources. Most plants, including crop plants, are already very efficient at using resources, and the majority of increased yield is attained by more aggressive root growth that allows the plant to more effectively take resources from its neighbors. Thus in these agricultural settings, individual selection for increased yield resulted in a field of aggressive plants with increased competition and overall lowered yield.

Griffing concluded based on these models that individual selection can act only on the effects of an individual on itself, that is, on direct fitness effects. In contrast, group selection would act on both direct and indirect effects, and would result in the greatest response to selection (Griffing 1977). Because group selection could act on both direct and indirect genetic effects, he recommended that it should be the primary method for selecting for increased crop yield.

Perhaps because of its origins in the agricultural literature, Griffing's work went largely unappreciated, and the general consensus, based on models such as Maynard Smith's (1964) "haystack" model, was that group selection should be less effective than individual selection. Thus, it came as a surprise when Wade (1977) did the first group selection experiments, and the response to group selection was much greater than expected. Wade's original experiment has been replicated and extended many times, with all studies seeing a significant response to group selection (Goodnight and Stevens 1997). These studies have confirmed many of the predictions of Griffing's (1977) models. I (Goodnight 1985) compared group and individual selection; in addition to a significant response to group selection, I observed the negative response to individual selection predicted by Griffing. I also demonstrated that the response to selection in two-species communities of *Tribolium* flour beetles depended on the interaction between the two species, and the response to selection disappeared when the community structure was disrupted (Goodnight 1990a, 1990b). Studies of egg laying in caged chickens have shown that group selection for increased egg production in chickens results in chickens that have slightly lower egg production per chicken when tested in single hen cages, but much higher egg production per cage when housed in groups. This increased egg production is a result of group selection associated with changes in other traits, indicating that the birds become less aggressive toward each other as total egg production increases (Muir 1996). Indeed, the breeding program pioneered by Muir has become a standard tool for animal breeders in recent years. The theory has been well worked out (e.g., Bijma et al. 2007a; Bijma et al. 2007b) and has become an increasingly important tool for animal breeding (Wade et al. 2010). In recent years multilevel-selection methods in agricultural settings have become quite sophisticated. Advances include the development of optimal multilevel-selection indices that maximize the expected response to selection (Griffing 1969; Muir 2005) and the use of best linear unbiased parameter estimates of breeding values in a group context (Bijma 2010; Muir et al. 2013).

These experimental studies of group selection have shown not only that group selection works, but also that it works far better than predicted by classical models that assume only additive effects. Group selection is more effective than predicted because it can act on indirect genetic effects. Individual selection cannot, although negative correlated responses to individual selection are often observed. As a result, until recently these indirect genetic effects have been ignored and were not included in the classic models of group selection. Nevertheless, it is probably these indirect genetic effects that are primarily responsible for the effectiveness of group selection (Wade 1978; Goodnight and Stevens 1997).

Measuring Multilevel Selection in Natural Populations

Given the laboratory and theoretical demonstrations of multilevel selection, an equally important question is whether selection commonly acts at multiple levels in nature. The earliest study addressing this question was Breden and Wade (1989). They did a series of laboratory and field experiments demonstrating that in the imported willow leaf beetle, *Plagiodera versacolora*, selection among individuals within groups favored cannibalistic individuals, whereas selection among groups favored groups with a lower proportion of cannibals. Although the experiments used in this study were appropriate, they did not provide a general protocol for studying multilevel selection in nature. Such a general protocol is provided by contextual analysis, a method that is particularly useful in experimental settings.

The first study to use contextual analysis to study multilevel selection was L. Stevens et al. (1995), which examined multilevel selection in the orange jewelweed, *Impatiens capensis*. Although the study population was continuous, it was clear that neighbors interacted to affect each other's fitness. Thus, this example of selection does not fit clearly into either a classic individual selection situation in which the fitness of an individual is strictly a function of its own phenotype, or a classic group selection situation in which there are clearly delineated groups that are isolated from other groups. Rather, each individual is surrounded by a neighborhood of others with which it interacts. Because of these complications, contextual analysis becomes the best framework for examining selection because it can be used for any population structure that can be clearly defined. In this study, the mean phenotypes of neighbors within a small (0.5 m) radius provided appropriate contextual traits. In effect each individual was treated as if it were the center of its own group that consisted of the neighbors with which it interacted. Additional studies have used contextual analysis to study multilevel selection in plants (Aspi et al. 2003; Donohue 2003, 2004; Weinig et al. 2007), arthropods (Tsuji 1995; Herbers and Banschbach 1999; Eldakar et al. 2010; Pruitt and Goodnight 2014), birds (Laiolo and Obeso 2012), and humans (Moorad 2013). All have found evidence of multilevel selection, although the current number of studies is not sufficient to draw any general conclusions about how common multilevel selection is in nature.

Multilevel Selection and Kin Selection,
Differences and Similarities

A long-standing controversy in evolutionary biology is whether altruistic traits—traits that increase the fitness of other individuals at a cost to the individual expressing the trait—evolve via kin selection or group selection. That this controversy has persisted so long is surprising given that since as early as 1970 (G. R. Price 1970) it has been clear that kin selection and group selection are describing similar if not identical processes. This controversy between the two approaches is likely due to the very different historical roots of the two approaches (Goodnight and Stevens 1997). Kin selection grew out of what I will call the "adaptation" approach, whereas group selection and later multilevel selection grew out of what I will call the "evolutionary change" approach.

The adaptation approach that grew into kin selection thinking is the better known of the two. In this approach, an adaptation is identified and plausible scenarios for its evolution are constructed. The investigator uses current patterns to attempt to infer past processes. In all but the simplest cases, there will be multiple processes that could produce the same pattern or adaptation, and distinguishing among those processes can be difficult. However, the investigator can design experiments and use rules for deciding which of the explanations is the most likely. Some explanations can be eliminated by the rules of science. For example, only explanations relying on known natural processes are considered acceptable. If multiple explanations for the evolution of a trait remain, it is often possible to remove some by theoretical modeling or experimentation. Ultimately, however, there may be several explanations for the evolution of a trait that are valid and cannot be eliminated by experimental or theoretical examination. There is no objective means of distinguishing among these remaining explanations, and as a result, agreed-upon rules must be established to choose among them. One such famous rule is G. C. Williams's (1966a, 5) principle of parsimony:

> In explaining adaptation, one should assume the adequacy of the simplest form of natural selection, that of alternative alleles in Mendelian populations, unless the evidence clearly shows that this theory does not suffice.

This is the most well known rule for deciding among explanations, but it is by no means the only possible rule. Two things to notice about this rule: First, the first phrase, "In explaining adaptations" clearly defines the adaptation approach and distinguishes it from the evolutionary-change approach. Second, it provides a clear rule as to ranking explanations and defines "simplest" as the selection of alternative alleles. The principle of parsimony is a clear rule, but it is also arbitrary; there are no objective criteria for deciding that this rule is correct.

In contrast, the evolutionary-change approach of multilevel selection stems from the quantitative genetics tradition originally developed for individual selection by Fisher (1930) and Wright (1931), and later extended to a multilevel setting by Griffing (1977) and others (e.g., Bijma et al. 2007a; Bijma et al. 2007b). As a result, in its early development the multilevel-selection approach was dominated by laboratory and agricultural experiments in which group selection was applied as an experimental treatment. Wade's (1977) group selection experiments measured population sizes of *Tribolium castaneum*, applying group selection favoring the populations with the highest population size in one treatment, and group selection favoring the populations with the lowest population size in a second treatment. At the end of the experiment the significant difference between these two treatments was attributed to a response to group selection. The trait of population size was chosen as much for its experimental tractability as for any inherent interest in population size as a group-level adaptation. For field research, methods for measuring group selection in natural populations, such as contextual analysis, were developed (Heisler and Damuth 1987; Goodnight et al. 1992). The observed change in the means of traits within a generation and the correlation between the traits and fitness are used to infer multilevel selection.

This difference in traditions leads to a number of important differences in how these two approaches are applied and in the language that they use. In the adaptation approach, group selection is invoked when other explanations (individual selection or genic selection) are inadequate. In contrast, in the evolutionary-change multilevel-selection approach, group selection is applied as a treatment in laboratory or agricultural settings or measured as an ongoing process in nature. Only recently have adaptations that clearly evolved as a result of multilevel selection been identified (Pruitt and Goodnight 2014). Because the multilevel-selection approach is generally used to measure ongoing processes, there is no need for rules for selecting among explanations. Rules such as Williams's principle of parsimony are not necessary and are not part of the multilevel-selection literature. Similarly, because the adaptation approach generally uses rules that favor lower levels of selection as explanations, kin selection has nearly exclusively focused on the evolution of altruism, or the situation where the interests of the group are in opposition to the interests of the individual. In contrast, with the multilevel-selection approach, group selection can be detected whether it acts in concert with or opposition to individual selection. As a result, practitioners of the multilevel-selection approach have tended to focus on traits such as population size, migration rate, and leaf area that are experimentally tractable but not particularly associated with altruism. Indeed, outside of very general discussions, it is rare to find mention of the evolution of altruism in the multilevel-selection literature.

While kin selection and multilevel selection come from very different traditions, they are actually addressing the same phenomena. It can be shown that standard methods for modeling kin selection use equations that are identical to those used for measuring multilevel selection in natural populations (Goodnight 2013a). Interestingly, the comparison between these two approaches shows that while kin selection and multilevel selection are indeed similar, they should not be considered identical. In particular, the kin-selection approach is an equilibrium approach that is used to identify the degree of altruistic versus selfish behavior that maximizes overall fitness, whereas multilevel selection is a nonequilibrium approach that is used to identify the rate of change in traits resulting from selection acting at multiple levels at the same time. The result is that kin selection and multilevel selection should be considered complementary approaches rather than being either identical or in conflict (Goodnight 2013a).

General Conclusions

Early models of multilevel selection led to the general dismissal of group selection as an important force in evolution (Wade 1978). Group selection was viewed as a conceivable but unlikely force both because group selection would generally be weaker than individual selection, and because genetic variation at the group level was expected to be smaller than genetic variation at the individual level (Maynard Smith 1964, 1976a). In view of those models, the large response to group selection in experimental studies was quite surprising (Goodnight and Stevens 1997). The conclusion that can be drawn from these laboratory experimental studies is that selection at higher levels of organization is potentially a major evolutionary force.

There were two related errors in early models that led to the conclusion that group selection would be ineffective (Wade 1978). The first was the assumption that heritable effects at one level could be translated to heritable effects at higher levels. For example, Maynard-Smith's (1964) haystack model was based solely on a single Mendelian locus with only recessive homozygotes behaving altruistically. He then assumed that the heritability of the group-level trait (growth rate) could be described in terms of single-locus genetic effects. In other words, he was assuming that the behavior of the group was a simple sum of the behaviors of the underlying genes. The second was failing to recognize the importance of indirect genetic effects and other interactions among individuals within a group. Those interactions are the primary explanation for why heritability at one level does not translate into heritability at higher levels. When those early models were developed, we did not recognize that higher levels of organization incorporated this additional source of heritable variation. Such interactions were always present. However, at the individual level they are expressed as environmental variation and do not become heritable until there is group structure and selection acting on that group structure. Thus, the effect of a neighbor on an individual's phenotype is not heritable at the individual level but becomes heritable at the group level if the offspring of that neighbor has a high probability of being the neighbor of the focal individual (Goodnight and Stevens 1997). Such indirect genetic effects represent a form of group-level heritable variation that is environmental variation at the individual level and would probably not be measured in any standard genetic breeding design.

Models that acknowledge the existence of indirect genetic effects and other forms of gene interaction provide a very different view of multilevel selection. From these models it becomes readily apparent not only that higher levels of selection have the potential to be highly effective, but also that they will often lead to adaptations that are genetically and phenotypically qualitatively different from adaptations due to selection at lower levels.

The hierarchical structure of life and the theory of multilevel selection permeate nearly all aspects of biological inquiry. Most obviously they have an impact on the theory of natural selection (Frank and Fox, chap. 9) in which classical individual selection is but one of the possible levels at which selection can occur. Because selection at different levels can lead to qualitatively different adaptations, the possibility of selection at different levels of organization can fundamentally alter our understanding of evolution. Other areas are less obvious, and in many cases not well studied; however, multilevel selection can interact with them as well. Good examples of such areas are the evolution of sex (Orive, chap. 14), where multilevel selection can affect sex ratios and presumably other aspects; life histories (Fox and Scheiner, chap. 11), where multilevel selection could act to favor those that promote stable demographies; speciation (Edwards et al., chap. 15), which could potentially occur through selection at a higher level; and macroevolutionary patterns (Jablonski, chap. 17), where the probability of extinction can be affected by both individual- and species-level traits. The implications of multilevel selection have not been adequately explored, so the contribution of higher-level selection to these other areas must remain speculative, but it is nevertheless intriguing. Unfortunately, our understanding of multilevel selection has been hindered because various approaches to studying it have developed independently. Because these approaches have arisen from different traditions, they have not been well integrated into a single body of theory. When this integration has been achieved, incorporation of multilevel selection theory into a general understanding of evolution will be possible.

Acknowledgments

This review was written while in residence at the Universidade Federal de São Carlos in Brazil, working in the laboratory of Reinaldo Brito. I thank Reinaldo Brito, Andrea Peripato, and the members of their lab for their support during this time. This research was supported by FAPESP grant number 2014/04455-5 to Reinaldo Brito.

ELEVEN

The Demography of Fitness

Life Histories and Their Evolution

GORDON A. FOX AND SAMUEL M. SCHEINER

The "king clone" of a creosote bush (*Larrea tridentata*) in the Mojave Desert is estimated to be over eleven thousand years old; after taking decades to reach maturity, creosote bushes flower (if water is available) at least yearly (Vasek 1980). Cicadas in the genus *Magicicada* spend some years (typically 13 or 17) as juveniles, feeding on root xylem; they emerge as adults synchronously over large regions, reproduce, and die (A. Martin and Simon 1990). House mice (*Mus domesticus*) reach maturity a month or two after birth, have multiple litters within a year, and senesce and die; in wild populations mean age at death is one hundred days or so, but some individuals survive more than a year (Berry and Bronson 1992). There is much variation among species in the pattern of how long organisms live and when and how often they reproduce. Life history theory is aimed at understanding this trait variation, in the context of trade-offs. Consider offspring size. If having larger offspring simply improves their survival, then we understand the ensuing dynamics without any need for a special theory of life history evolution: under these circumstances, larger offspring are always favored by selection. However, if there are trade-offs among traits (such as offspring size and number), we require a special theory of the demography of fitness to consider the outcome of the trade-off.

In this chapter we present a constitutive theory about the evolution of these patterns of survival and reproduction: the theory of life history evolution. By life histories we mean the schedule of births and deaths in organisms and major factors (like growth and metamorphosis) that affect that schedule (Cole 1954; Lewontin 1965; Roff 1992; Stearns 1992). Schedule refers not only to the timing of events but also to their magnitude; for example, given that an individual begins reproducing at some size, how

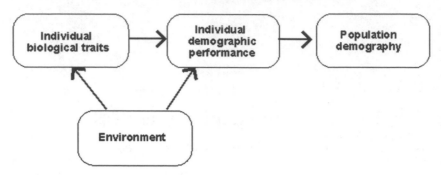

Figure 11.1. The relationship between individual biological traits, individual demographic performance, and population demography. An example of a trade-off between reproduction and survival involves the individual biological trait age at first reproduction, which affects the number of eggs at any age (individual demographic performance), and then has consequences for the probability of survival to subsequent ages (population demography).

much does it reproduce? But this definition fails to make clear the central role of life histories; the theory really concerns the demography of fitness and the causes underlying variation in fitness. This terminology makes clear the demographic content of life history theory and distinguishes the theory of life history evolution from descriptions of life cycles (e.g., alternation of generations in plants or single- versus multiple-host life cycles in parasites).

We make a crucial distinction between the biological traits of individuals, the demographic performance of individuals, and the demography of populations (fig. 11.1). Life history traits are measurable biological traits like size at a given age, or the fraction of new photosynthate allocated to flower production (G. C. Williams 1966a; Gadgil and Bossert 1970). These traits affect the demographic performance of individuals, which is also affected by many other things that for convenience we lump together as "environment." Individuals might be thought of as having characteristics like survival probabilities, but if so, they are not estimable; we can measure only the times of birth and death for individuals. Quantities like mean survival probabilities or mean fertility at a given size are estimable properties of populations, not of individuals.

There are limits to the applicability and utility of life history theory. First, although life history theory concerns trade-offs (see below), not all trade-offs that involve life history traits are trade-offs among life history traits. Consider a trade-off between two defense mechanisms affecting the manner of death prior to reaching the age of first reproduction. The overall probability of survival depends on how resources are allocated between the two defenses, but we have defined the theory of life history evolution as a theory for the evolution of trade-offs between life history traits; thus we would not consider a model of a defensive trade-off to be within the domain of life history theory. Second, research interest in some trade-offs may be focused on issues other than life history theory. For example, Irwin et al. (2003) studied the consequences of flower color variation in wild radish, *Raphanus sativus*, which depends largely on anthocyanins. These molecules affect not only flower color, but herbivory, so there are trade-offs involving reproduction, growth, and survival. But Irwin et al. asked how the flower color polymorphism is maintained (and thus focused on the mechanisms involved in the trade-off), not on how life history traits themselves evolved. This is a fairly common occurrence: many studies of trait evolution might be cast in terms of life history theory, but if the focus is not on the life history traits themselves, doing so may not be useful.

Thinking about life history evolution dates to at least Fisher's (1930) discussion of reproductive value and its possible relationship with senescence (B. Charlesworth 2000), but the bulk of the literature is more recent. The two principal roots of the theory are in evolution (e.g., Medawar's [1952] discussion of the causes of the evolution of senescence) and ecology, especially Cole's (1954) exploration of the consequences of variation in fertility and survival. These threads came together during the invention of evolutionary ecology (Odenbaugh 2011). One outcome of that effort was MacArthur and Wilson's (1967) concept of r- and K-selection, which was a major impetus to the growth of interest in life histories. Some of these models have been both influential and controversial. One of the most influential, still promulgated in textbooks, is Pianka's (1970) expansion of r-K theory. A controversial model was that of Grime (1977), whose C-S-R theory was developed in direct response to r-K theory. A more recent variant of r-K theory is the notion of a fast-slow continuum (Promislow and Harvey 1990; Salguero-Gómez et al. 2016). Some multispecies datasets have been analyzed to argue that slower-growing species tend to reproduce less and at later ages, but no argument has been advanced to show why this must be so, nor have models been published describing how populations would evolve to show this pattern. Finally, reanalysis of at least one influential dataset concluded that evidence for a fast-slow continuum may be weaker than has been claimed (Bielby et al. 2007).

Structure of the Theory

Domain

The domain of the theory of life history evolution is evolutionary change in the pattern of the timing of events concerning demographic performance (i.e., survival and reproduction), including the magnitudes of those events. The theory is directly related to the theory of evolution by natural selection (Frank and Fox, chap. 9), and could be considered a subdomain of that constitutive theory. It is related to multilevel selection theory (Goodnight, chap. 10) because some models of life history evolution concern evolution among related individuals.

The theory consists of seven propositions (table 11.1). Propositions 1, 2, and 4 hold for all models; the last three propositions apply to specific models. Proposition 3 is often ignored, leading to general heuristic models that may not always be adequate tools for understanding selection and its responses in particular populations. We distinguish this theory from others that are often considered to be theories of life history evolution, notably the r-K theory of Pianka (1970) and the C-S-R theory of Grime (1977). Those theories are not about the demography of fitness, the way in which individual demographic performance (births, deaths, growth) affects individual contributions to population growth. Rather they are about the characteristics of organisms likely to be found in particular types of habitats. They are constitutive theories concerned with how organismal biology creates or constrains specific trait combinations and, thus, exist within the domain of the theory of organisms (Zamer and Scheiner 2014). While the characteristics of organisms are necessary conditions for the theory of life history evolution (see propositions), the focus of life history theory is the demography of fitness. There are demographic models of life histories involving density-dependent regulation that use r-K terminology but, unlike Pianka's r-K theory, are based on demographic theory (e.g., Lande 1982; Boyce 1984; Engen et al. 2013). Because their domain and assumptions are quite different from those of Pianka's theory, calling them r-K models is confusing. Instead, they should simply be called models of life history evolution.

Trade-offs among Traits

The first proposition is that there exists a trade-off in one or more traits that affects the demography of fitness. This proposition defines the very nature of life histories and is used by all models of life history evolution. But

Table 11.1. The Domain and Propositions of the Theory of Life History Evolution

Domain: Evolutionary change in life history traits in response to natural selection.

Propositions:

1. There is a trade-off in one or more than one trait affecting the demography of fitness.
2. Trade-offs come from resource limitations, which are assumed to be temporarily fixed.
3. Demographic performance depends on the characteristics of organisms.
4. The traits affecting the demography of fitness meet the conditions required for evolution by natural selection.
5. Allocations may be hierarchical.
6. Allocations may be among individuals.
7. Timing may affect the consequences of variation in life histories.

there is more to it than simply a definition. To see this, consider the simplest possible model of variation in such traits, in which we fix all of the demographic characteristics but one. That model would look like a single-locus model of gene frequency change. In those models, all individuals in a population are demographically identical except in their chance of surviving to reproduction. If an individual survives, it reproduces once with the same timing and number of offspring as the rest of the population. Such a model involves variation in demographic performance, but only in the sense that the components of fitness are survival and reproduction.

Trade-offs such as those between survival and reproduction, or between reproduction at different times, are thus the focus of life history theory. The trade-off can also manifest as the expression of the same trait in different, related individuals (e.g., seeds from the same mother that vary in their germination behavior in dry versus wet years). This is a case where the unit of selection is the group of siblings, rather than a single individual (Goodnight, chap. 10). The existence of trade-offs leads to the notion of optimality: What is the best that can be done, given a trade-off? What values of each trait would yield the greatest fitness? While it is often reasonable to consider this kind of question, several issues are important. What is optimized? Some form of fitness? One of its components? Is anything necessarily optimized? How does the genetic architecture of the relevant traits affect predictions?

This proposition encompasses two of the requirements for evolution by natural selection—phenotypic variation in traits and a link between phenotype and fitness. What differentiates this constitutive theory from the more general theory of evolution by natural selection (Frank and Fox, chap. 9) is the specification of a class of traits (those affecting demography) and relationships among them (trade-offs).

Resource Limitations

The second proposition is that trade-offs come from resource limitations. Joint evolution of traits is an issue only if the traits somehow affect one another. Two nonmutually exclusive mechanisms have been proposed for such affects: resource limitation and genetic correlation. The latter mechanism lies outside the domain of life history theory. Nor are all trade-offs that are due to resource limitations within the domain of this theory; the trade-offs must affect life history traits. The idea itself, like so much in evolutionary biology, comes from R. A. Fisher (1930, 43–44):

> It would be instructive to know not only by what physiological mechanism a just apportionment is made between the nutriment devoted to the gonads and that devoted to the rest of the parental organism, but also what circumstances in the life-history and environment would render profitable the diversion of a greater or lesser share of the available resources towards reproduction.

This notion was developed considerably by G. C. Williams (1966a, 1966b), who tied reproductive allocation explicitly to demographic performance and popularized the notion of reproductive effort in the sense of proportional allocation.

A thought experiment makes clear the motivation behind this proposition. Consider a "Darwinian demon," an organism that begins reproducing as soon as it is born, survives indefinitely, and produces an unlimited number of offspring (Law 1979). Such an organism—which has maximized every component of fitness without constraint—would have infinitely large fitness and an infinitely large rate of population growth, and is clearly impossible. At some point, resources become limiting, so a finite total must be divided among different functions: a trade-off is born.

Resources may be treated implicitly or explicitly. Implicit assumptions are common in many models. For example, a model may simply assume a trade-off between present and future reproduction. That is, the model assumes that total reproduction is constrained, presumably because it requires some resource like energy. Another implicitly assumed trade-off involves time; for example, it may take some time to find suitable places to deposit eggs. An individual could lay many eggs in a single location, or spread those eggs over many locations resulting in different reproductive schedules. For models with explicit assumptions about resource trade-offs, the most common involve caloric energy (e.g., Hirshfield and Tinkle

1975; Schaffer and Rosenzweig 1977; Chiariello and Roughgarden 1984; Sibly et al. 1985; Fox 1992). There are also models of trade-offs in other resources, like nitrogen (e.g., McGinley and Charnov 1988).

An important advance in the study of life history evolution was the paper by de Jong and Noordwijk (1992), now referred to as the Y-model. They addressed a problem in how we measure trade-offs by looking at correlations among traits and pointed out that variation among individuals in resource acquisition can lead to positive correlations between traits even when those traits are governed by trade-offs. If resources limit allocation to both body mass and reproductive output, for example, individuals in favorable sites will have larger bodies and more reproductive output. A trade-off among traits within an individual, which is expected to manifest as a negative correlation, is overwhelmed by variation among individuals. The assessment of trade-offs thus depends on understanding both organismal characteristics and their ecological context.

Demography and Organismal Characteristics

The third proposition is that demographic performance depends on the characteristics of organisms. Besides external constraints imposed by finite resources, the evolution of life history characteristics is also determined by biological limitations on what an organism can do at a given time and level of resource availability (fig. 11.1). Consider events that must occur sequentially over time: an individual must first develop its reproductive structures before reproduction can begin. As another example, some breeds of domestic chickens can lay about one egg per day, and selective breeding has been unable to increase that rate. Such limitations may not be absolute in the sense that other species may be able to lay more than one egg per day, but such limitations may be constraining for a given species or clade. Factors that set these limitations are outside the domain of the theory of life history evolution; they belong within the domains of constitutive theories in the theory of organisms (Zamer and Scheiner 2014).

Different models of life history evolution can be built based on different organismal limitations. Given an assumed set of limitations, the resulting model is relevant for the set of species matching those limitations. Thus, there can be no universal life history model. One can construct a model that assumes general limitations, but the resulting model will not make predictions about any particular species. If the model is to be used for predictions and hypothesis testing, more specific limitations must be added.

Organismal limitations are frequently not addressed explicitly in models; they are either implicit or absent. Indeed, many influential models of life history evolution have been constructed entirely from demographic considerations (Cole 1954; Gadgil and Bossert 1970). Important as these models were in establishing the foundations of life history theory, it can be challenging to use them in interpreting empirical data. For example, Charnov and Schaffer (1973) established the demographic circumstances (for a population with two classes of individuals, juveniles and adults) under which selection favors reproducing only once. Does this explain all observed patterns of single versus multiple reproductive bouts? Perhaps, but it is also possible that in some cases biological limitations on the amount of reproduction possible within a year are the limiting factor. Trait evolution that can eliminate or alter such trade-offs (e.g., placentas in mammals) are studied within the theory of evolutionary developmental biology (Love, chap. 8). An important advance to life history theory would be the building of models at its intersection with developmental biology.

Models that do not address organismal limitations must be interpreted conditionally. Informally one might say, "This set of traits is predicted to evolve in a certain manner, if such evolution is biologically possible." Consider the use of matrix and integral projection models (Easterling et al. 2000; Caswell 2001) for interpreting the sensitivities of growth rate parameters (derivatives of the long-term growth rate with respect to terms in the model) as estimates of the potential strength of selection on each model term. Caution is needed for this interpretation because these sensitivities actually estimate the potential importance of selection on each term only if it is biologically possible to change that term while holding all others constant (Easterling et al. 2000; Caswell 2001; Metcalf and Pavard 2007).

Understanding the biological limitations of life history evolution is a major challenge. Some researchers have labeled organismal limitations as "constraints" without considering how such constraints might operate. Without further specification of its meaning, the term "constraint" is not readily interpretable (Antonovics and van Tienderen 1991; Van Tienderen and Antonovics 1994; Roff and Fairbairn 2007). For example, constraint has been taken to mean: (1) that a trait has never evolved in a specific manner within a particular lineage, (2) that it cannot currently evolve owing to a lack of genetic variation, (3) that trade-offs are involved, (4) that the trait is physically impossible, and (5) that its evolution would be slowed owing to genetic correlations. Nor is this list exhaustive (Antonovics and van Tienderen 1991). All of these factors may play a role in trait evolution, but lumping them all under the vague term "constraint" results in a usage that fails to convey meaning.

Genetic correlations are often measured in an attempt to estimate both trade-offs and limitations. However, this use of genetic correlations is incorrect (B. Charlesworth 1990; Houle 1991; Roff and Fairbairn 2007). Trade-offs and limitations are a function of the biological properties of individuals, while genetic correlations are measures of variation among individuals. Furthermore, a lack of variation (e.g., a genetic correlation of 1) is not necessarily an indication that such variation cannot exist but that it does not exist in that population at that moment. Conversely, the lack of a correlation (e.g., a genetic correlation of 0) is not necessarily an indication of no limitation. Genetic correlations are bivariate measures, while limitations may be the outcome of multivariate interactions within an individual.

Requirements for Natural Selection

The fourth proposition is that the traits affecting the demography of fitness meet the conditions required for evolution by natural selection. This proposition is a catch-all that covers any other conditions necessary for natural selection to occur. For example, it subsumes the requirement that life history traits be heritable (Mousseau and Roff 1987; Vaupel 1988; Price and Schluter 1991). Because life history traits are closely related to fitness, we might expect their heritability to be low and, therefore, the response to selection on life history traits to be slow. While it is true that on average life history traits typically have lower heritabilities than morphological or physiological traits (Mousseau and Roff 1987), that average is substantially above zero and there is extensive variation around that average. One reason that we might expect genetic variation for life history traits to be maintained is precisely the trade-offs that are the focus of life history theory. On the other hand, these measures of heritability may not be for the correct traits. Variation in demographic outcomes means that even if biological traits have high heritabilities, demographic performance may not (fig. 11.1). Vaupel (1988) showed that even if the tendency to survive is highly heritable, longevity itself is expected to have a low heritability.

Selection on life history traits can take multiple forms, and these differ in their long-term consequences. Population growth can be modeled as density-independent or density-dependent, models may assume either an unchanging environment or one that varies, and that environmental variation can be stochastic or deterministic. A key factor is the rate of change of the environment relative to the life span of the organisms. Under density-independent growth in a constant, deterministic environment, fitness (equivalent here to r, the intrinsic rate of increase) is maximized (B. Charlesworth 1970; Taylor et al. 1974). Density-dependent growth leads to a maximization of $\hat{n}$, the equilibrium population size (B. Charlesworth 1994). In a randomly varying environment, there may be additional selection for a minimization of extinction risk. Extinction risk depends on two quantities: the population size N and the stochastic growth rate $(\lim_{t \to \infty} (1/t) \log[N(t)/N(0)]$, Tuljapurkar 1990). Even populations with stochastic growth rates above zero have some probability of extinction, which is greater for small populations or those with low growth rates (Haccou et al. 2005). Thus, density dependence can act to increase the risk of extinction by lowering population sizes or growth rates. Conversely, under density-independent growth in a randomly varying environment, a population can avoid extinction by achieving a large size owing to the random occurrence of favorable environments (Ludwig 1996). Finally, selection can also be frequency dependent, most notably for traits like offspring sex ratio. Under frequency-dependent selection, equilibria are generally not maxima of any function, but game theory and related areas provide a rich set of tools for analysis of these equilibria.

The assumption that traits meet the conditions for evolution by natural selection refers to the evolutionary process, and need not hold at equilibrium. For example, models focusing on the outcome of selection need not assume current additive genetic variance for the life history traits—only that such variance was present when the traits evolved.

Might a theory of life history traits be built without invoking the process of natural selection? A scientist might develop a constitutive theory within the domain of the theory of organisms (individuals and the causes of their structure, function, and variation; Zamer and Scheiner 2014) that predicted possible variation in life history traits based on rules of developmental biology. Such a theory could act as a complement to theories of life history evolution by providing limitations and constraints on the range of those traits (see discussion of proposition 3). However, such a theory would be incomplete with regard to predicting patterns in nature. While it could set boundary conditions on trait values, actual patterns would result from the outcomes of natural selection.

Hierarchical Allocations

The fifth proposition is that allocations may be hierarchical (De Jong 1993; Worley et al. 2003; Roff and Fairbairn 2007). For example, consider an individual that is simultaneously male and female. It may allocate some energy to growth, and some to producing offspring. The energy allocated to offspring might be further divided between male and female function. Such a hierarchy does not imply that the decisions are necessarily sequential, only that decisions concerning one partition create constraints on other partitions.

Such hierarchical sets create a key evolutionary outcome. Both genetic and phenotypic correlations between allocations to traits at the end of the hierarchy depend strongly on correlations between allocations to traits early in the hierarchy (De Laguérie et al. 1991; Björklund 2004; Roff and Fairbairn 2007). De Jong (1993) pointed out that hierarchical allocation to pairs of traits leads to repeated cases of the Y-model (Van Noordwijk and De Jong 1986) and thus a tree-like structure. Just as variation in resource availability can create positive between-individual correlations between traits trading off within individuals, the trade-off between traits low on the tree (e.g., reproductive versus body mass) can lead to positive correlations in allocations between traits high on the tree (e.g., male versus female function; De Jong 1993). Complicated patterns also can occur in genetic correlations where there is a strikingly analogous result. Using quantitative genetic models, Worley et al. (2003) found that if the amount of genetic variance in traits low in a hierarchy is large, positive genetic correlations can evolve between traits high in the hierarchy, even in the presence of trade-offs.

Allocations among Individuals

Allocations can also occur among individuals, the sixth proposition. Such allocations always involve partitions among kin, either between parents and their offspring or among siblings. For example, bird eggs are energetically expensive. Allocating resources to more eggs, or larger eggs, can reduce resources needed for parental survival (C. C. Smith and Fretwell 1974). A classic example of such an allocation is selection on the sex ratio of offspring (Darwin 1871; Fisher 1930; A. W. F. Edwards 1998). In an outbreeding population, an evolutionary equilibrium exists when individuals allocate half their reproductive resources to producing sons, and half to daughters.

Life history models of bet-hedging are based on such among-individual allocations (Slatkin 1974; Philippi and Seger 1989; Simons and Johnston 1997; Starrfelt and Kokko 2012). Bet-hedging can be favorable only in unpredictable environments (say, wet versus dry years, or good versus bad patches) because then fitness tends to vary over time or space. For a given individual there can be a trade-off between the fitness measured for a set of offspring in a given environment and fitness measured over all environments (Lewontin and Cohen 1969). Consider the evolution of seed dormancy in an annual plant. In a constant environment, fitness is maximized when all seeds germinate the following year. But in a varying environment, long-term fitness may be greater if a plant produces some seeds that delay germination until later years, avoiding the possibility of many (or all) descendants having low (or zero) fitness in some years. Although a delay in germination (and therefore reproduction) reduces the mean fitness in any given year, having some seeds remain dormant reduces the variance in fitness among years. Similarly, bet-hedging can occur through dispersal: if some patches are favorable and others unfavorable in a given year, offspring dispersing to a mixture of patch types can increase long-term fitness by reducing the risk that all offspring will be in unfavorable patches (Venable and Lawlor 1980). Different bet-hedging models have treated conservative (risk-avoiding) and diversified (risk-spreading) strategies as separate phenomena (e.g., Childs et al. 2010). Starrfelt and Kokko (2012) showed that by decomposing the variance in fitness into variation at the individual level and a correlation among individuals, these two seemingly disparate strategies are actually ends of a continuum.

Timing

The seventh proposition is that timing may affect the consequences of variation in life histories. The amount of resources allocated to a trait is not all that matters; the timing of allocation is also important. The effects of timing can be seen by considering what happens at the level of the population: all else being equal, populations grow faster if reproduction occurs at an earlier age. Consider two genotypes that reproduce once then die: both produce an average of three offspring but A reproduces after two years while B reproduces after only one. Because population growth is multiplicative, after four years an A individual will have nine descendants while a B individual will have eighty-one. More generally, if the expected (over the entire lifetime) number of offspring is R_0 and a genotype begins reproducing after τ years, the annual growth increment for a lineage is $R_0{}^{1/\tau}$; a shorter generation time means faster growth. Of course, all else is not always equal. Demographically, the question is whether by delaying reproduction, organisms can produce enough additional offspring to offset the longer generation time. Most research on the evolution of delayed reproduction has focused on increases in reproduction due to increased parental body size. However, other conditions can also select for delayed reproduction, for example among-year variation in conditions resulting in bet-hedging.

Timing is often considered with respect to events among years, either for individuals with an annual life cycle (among generations), or for individuals with yearly bouts of reproduction. Timing also matters within years (or other time units, as appropriate) or reproductive bouts in response to variation in the availability of resources, competition or facilitation by conspecifics, or the activities of predators, pathogens, mutualists, dispersers, and the like (Mahall and Bormann 1978; Fox 1989; Lyons and Mully 1992; Stanton and Galen 1997).

Many studies eliminate issues of timing by considering selection only on composite quantities that can be considered either over an entire lifetime (e.g., lifetime reproductive success: the total number of offspring during a life), or for a single bout of reproduction (e.g., reproductive effort: the amount of resources allocated to reproductive versus nonreproductive traits). This reduction in the complexity of life histories to single quantities is understandable, but it is also, in most cases, a mistake. The problem is not so much that information is lost by compressing the time course of events into a single quantity; rather, such quantities can be quite misleading. In the case of reproductive effort, information about the standing crop of reproductive and nonreproductive biomass provides no information about which phenotypes have greater fitness, or about how different phenotypes achieved their biomass; single frames do not provide much information about entire movies. Similarly, one cannot conclude anything about fitness itself from comparisons of lifetime reproductive success, because, as they say in comedy, timing is everything. Two individuals can produce the same total number of offspring but have differences in fitness depending on when those offspring are produced. In addition, such compression may obscure unrealistic model assumptions.

Models and Their Development

A Non-Life-History Model

To see where life history models come from—and the boundary of the domain of life history theory—we begin with a simple model for population growth. If there is no age or stage structure in a population, we can write

(11.1) $$n(t+1) = \lambda n(t),$$

where n is population size and λ is the number of individuals that a single individual gives rise to; this model can include haploid or asexual populations, or females if they are not mate-limited. This is not yet a life history model because there are no trade-offs possible. By assuming that there is no age structure, we have assumed that a newborn has the same properties as an adult; a birth then has exactly the same effect on population size as a death, and the deaths of all individuals have the same effects. An increase in λ is always favored by selection.

More formally, we can make this model into the standard one-locus haploid selection model by assuming that genotypes vary in survival to reproduction s, and produce the same number of offspring if they do survive. This gives

(11.2) $$p_i(t+1) = \frac{n_i(t)s_i}{\sum_j n_j(t)s_j},$$

where n_i and p_i are the number and frequency of the ith genotype. We can think about selection (proposition 4) because we are considering multiple types. On the other hand, the most obvious change—dividing by the denominator in equation 11.2—has the effect only of changing the model from one of absolute population size to one of relative frequencies. But equation 11.2 underlines a key point: it is a selection model, but not a life history model, because there are still no trade-offs possible.

Introducing Trade-offs

Trade-offs (proposition 1) might be introduced into the model in equation 11.1 in several ways. One could separate survival and reproduction in the λ term and include an assumption about how they trade off. One might allow the organisms to have overlapping generations and thereby allow trade-offs between present reproduction and future survival or reproduction.

As an example of the importance of including trade-offs, consider

Cole's (1954, 118) comparison of the fitness of an annual plant (without a seedbank) with that of a perennial:

> The most extreme case of iteroparity, and the one exhibiting the absolute maximum gain which could be achieved by this means, would be the biologically unattainable case of a species with each individual producing b offspring each year for all eternity and with no mortality. . . . Thus we have $r = \ln(b + 1)$, which is to be contrasted with $r = \ln(b)$ for the case of an annual. For an annual species, the absolute gain in intrinsic population growth which could be achieved by changing to the perennial reproductive habit would be exactly equivalent to adding one individual to the average litter size.

Thus, on Cole's account, the selective advantage of iteroparity is at best small, inviting the question of why there are so many iteroparous organisms. This issue was clarified when Charnov and Schaffer (1973) pointed out that Cole's conclusions flowed from his assumption that surviving adults and newly germinated plants have the same chance of surviving another season, i.e., that there are no trade-offs. They allowed for trade-offs using a general model that made survival and reproduction age-specific. Annual growth rate was changed from b to $B_a C$, where C is the fraction of annuals surviving to reproduce and B_a is their output given that they survive, and perennial growth rate was changed from $b + 1$ to $B_p C + P$, where B_p is the perennial reproductive output given survival and P is the fraction of perennials surviving between years. This reduces to Cole's model if $C = P = 1$. Under the more reasonable assumption of trade-offs, Charnov and Schaffer showed that there are many circumstances under which iteroparity can be strongly selected for.

Resource trade-offs (proposition 2) can be modeled by explicitly indicating how the organism gains and uses resources. While easy to say, developing such a model can be difficult because it requires substantial knowledge of the ecology and underlying biology of the organism. The earliest models of resource allocation in life histories were stimulated by G. C. Williams's (1966a, 1966b) argument that an optimal life history entailed allocating resources such that the organism could maximize both current and future reproductive value. This argument was further developed by several influential models (Schaffer 1974; Taylor et al. 1974; Schaffer and Rosenzweig 1977; Schaffer 1981). These models all assumed that the problem was one of resource allocation and examined the evolution of the fraction of available resources allocated to reproduction at each age, i.e., age-specific reproductive effort.

Making the Models More Specific

These general models were extended and made somewhat more specific by Gadgil and Bossert (1970), who added fitness costs and gains, and Schaffer (1974; Schaffer and Rosenzweig 1977), who considered fertility and postbreeding survival and growth as functions of age-specific reproductive effort. These models were still general, as rules for acquiring resources (including the rates at which they were acquired) were not included. Although the models were age structured, often they were not explicit about what time intervals were involved. Hirshfield and Tinkle (1975, 2227) defined reproductive effort as "the proportion of total energy, procured over a specified and biologically meaningful time interval, that an organism devotes to reproduction." Unfortunately, timing can matter substantially (proposition 7), so that the total fraction of energy devoted to reproduction over a given interval does not uniquely define the amount of reproductive tissue (or fitness consequences) that results.

Despite the general nature of these models, a considerable effort was made to test them with data. For plants, these efforts were based on the claim by Hickman and Pitelka (1975) that the energy content of each part was linearly related to its dry weight, unless the tissues differed substantially in lipid concentrations. However, simple measurement of dry weights quickly proved to be a quagmire (Evenson 1983); one had to decide how much of each tissue was "for reproduction," and such measurement could be made on individuals only once. Indeed, since organismal rules were unstated, researchers found themselves debating what "reproductive effort" actually meant, whether energy was always the appropriate currency, and how to measure it (e.g., K. Thompson and Stewart 1981; Bazzaz and Reekie 1985). That debate was focused largely on plant studies, perhaps because there was no result analogous to that of Hickman and Pitelka (1975) for animals; testing reproductive effort models in animals may have seemed considerably more daunting as a result. This speculation aside, it seems clear that reproductive effort models suffered from at least one key difficulty: in attempting generality they failed to incorporate enough information about organismal limitations or about the timing of allocation (propositions 3 and 7) as to be readily interpretable (Fox 1992). While the term "reproductive effort" may be permanently established in the lexicons of evolutionary biology and ecology, it is so vague as to be meaningless. Stearns (1977) suggested that a more biologically meaningful framework was needed.

Initial steps toward such a framework were made in the 1970s and 1980s by incorporating rules governing resource acquisition and allocation, typically using dynamic programming—optimal control theory—that allowed optimization over an entire lifetime (Vincent and Pulliam 1980; D. King and Roughgarden 1982a, 1982b; Schaffer et al. 1982; Chiariello and Roughgarden 1984). A key assumption of these models is a lack of constraint in the allocation between compartments (typically, vegetative, reproductive, and storage biomass). These models were used to predict optimal allocation schedules such as the optimal timing of flowering. However, as these compartments were treated as independent of one another, optimization frequently led to solutions that were not biologically reasonable, such as an instantaneous switch from 100 percent allocation to growth to 100 percent allocation to reproduction (Fox 1992; Horn 1992). The recent development of dynamic energy budget models (Noonburg et al. 1998; Nisbet et al. 2000; Lika and Kooijman 2003; Kooijman 2010) suggests that it is possible to build mechanistically defensible models that make predictions about the consequences of varying the schedules of growth and reproduction.

Types of Allocations

One way to make models more explicit is to consider the various ways that resources can be allocated (propositions 5 and 6). To our knowledge there are no models that deal with hierarchical allocations among different classes of life history traits. This is unsurprising, given the general result that variation early in a hierarchy can determine the signs and magnitudes of correlations later in the hierarchy (De Jong 1993; Worley et al. 2003) and the challenges of simultaneously modeling sequential trade-offs. In contrast, it is much easier to model hierarchical trade-offs within a single trait class. For example, there is a rich literature—both theoretical and empirical—on trade-offs within reproductive tissues (C. C. Smith and Fretwell 1974; Charnov et al. 1976; Charnov 1982; D. G. Lloyd 1987; McGinley et al. 1987; Venable 1992). In particular, there has been much study of the trade-offs between size and number of offspring, and between male and female function.

Models for selection on the trade-off between seed size and number all flow from the work of C. C. Smith and Fretwell (1974). Their key assumptions were that offspring fitness increases with size, but at a decreasing rate, resulting in a parental fitness peak at some intermediate offspring size. This sort of model assumes a single offspring size/number relationship, which was borne out in a study of *Uta stansburiana* lizards (Sinervo et al. 1992). However, in most species there is considerable variation in offspring size, and many studies found a positive correlation between the number and the size of offspring (Roff 1992).

To account for biological phenomena not captured by the model of C. C. Smith and Fretwell (1974), other models have been developed that include variation in maternal resource availability (Venable 1992), sib competition after germination or birth (Godfray and Parker 1992), sib facilitation (McGinley 1989), nonsib competition (Einum and Fleming 2004a), multiple resource limitation on offspring size and number (McGinley and Charnov 1988), differences in the ability of offspring to extract resources from the mother (Sakai and Harada 2001), and bet hedging (McGinley et al. 1987; Einum and Fleming 2004b). Theory appears to be considerably ahead of empirical evidence here, and it may be difficult to evaluate the relative importance of some of these mechanisms since they are not mutually exclusive. Understanding of organismal limitations may again prove useful here.

The basic insight for models of sex allocation comes from Darwin's (1871) observation that parents very frequently produce approximately the same numbers of male and female offspring. He believed that this was likely due to selection but did not suggest a mechanism by which this could occur. In 1884 Carl Düsing (Edwards 1998, 2000) proposed a mathematically explicit mechanism for a tendency to equal production of the two sexes. Half a century later, Fisher (1930), using essentially the same argument as Düsing, proposed that if production of the two sexes varies, those producing an excess of the rarer sex have, on average, a tendency to leave more descendants, and so the system tends toward a 1:1 sex ratio. Thus sex allocation is inherently subject to frequency-dependent selection. Darwin's basic model has been modified in many ways, to account for phenomena such as a differential cost of male and female offspring (Trivers and Willard 1973), allocation under varying degrees of selfing versus outcrossing (Queller 1984; Charnov 1987; D. G. Lloyd 1987), and many of the complexities of plant mating systems (West 2009).

Phenotypic models of selection are often analyzed using optimization approaches and are often called informally "optimization models." Optimization, however, is not a modeling technique but a possible outcome. Populations may not reach optima for many reasons, even under conditions of density independence and constant selection (Akin 1979; Hofbauer and Sigmund 1988). But phenotypic models can lead to valuable insights on selection, even if predictions of optima are best regarded with some caution.

Genetic models of the evolution of sex allocation have been developed (D. Charlesworth 2006), because many questions about sex allocation necessarily entail genetic issues such as the effects of inbreeding when allocation affects the degree of selfing or when we consider the evolution of self-incompatibility. Under some circumstances, the details of the relevant mutations can have strong impacts on the dynamics. For example, in plants cytoplasmic mutations that cause male sterility (Schnable and Wise 1998) spread more readily than nuclear mutations with the same effect. Modeling the population genetics of sex allocation can provide a readily interpretable framework for asking questions about the circumstances under which a mutant type can increase in frequency (D. Charlesworth 2006).

Other life history characters can also be subject to frequency-dependent selection (e.g., insect larval resource acquisition [Kerswell and Burd 2012]). However, there have been few models of frequency-dependent selection on life history traits other than those related to sex allocation. Some papers on adaptive dynamics theory (Hofbauer and Sigmund 1988; Geritz et al. 1998; Waxman and Gavrilets 2005) mention life history traits as motivation (De Mazancourt and Dieckmann 2004; Ferriere and Legendre 2013). Geritz et al. (1999) modeled the joint evolution of seed size and competitive ability, using an adaptive dynamics approach, and found conditions under which polymorphisms in seed size are maintained.

Outlook for Life History Theory

The theory of life history evolution is mature—and quite successful—at some levels, but still developing at others. The theory has been most successful in explaining variation in life history traits among species and higher taxa. For example, it has provided predictions about such problems as the circumstances under which iteroparity or semelparity are favored, why between-year propagule banks evolve in some organisms, and why selfing organisms allocate fewer resources to male function than their outcrossing relatives. There is no doubt that trade-offs based on resources occur, but as one changes focus from comparing species or populations to understanding the response of a population to selection, additional factors can become quite important. For example, if an artificial selection experiment is conducted on seed size, the population response may not involve the predicted changes in seed number as well, because genetic correlations need not have the same sign as correlations based on resource trade-offs. Tradeoffs may exist, but in any particular population the genes under selection may have effects beyond those trade-offs. Combining ecological trade-offs (which set the context for selection) with genetic information and developmental patterns (which set the context for selection response) is a central challenge for further progress in understanding responses to selection on life history traits. Are there general patterns that we can predict, or must all predictions be specific to the circumstances of a given population?

While the theory is quite successful in guiding research, it has been much less so in educating biologists. Indeed, Pianka's r-K theory—which we noted is not really a theory of evolution at all—continues to be presented as such, and often as the only model of life history evolution. More than a decade ago, Reznick et al. (2002) pointed out that while r-K theory was no longer influential in research, it continued to be presented in textbooks as one of several useful frameworks for understanding life history evolution. This is still the case. Consequently, one of us (GAF) has encountered new graduate students who were astonished when told that r-K theory is not influential among researchers. Moreover, we see casual references to r-K theory in a number of recent research papers. Why is there such a disconnect between contemporary research and education? Probably there are several reasons. The Pianka theory provides a quick and simple explanation of life history patterns even though life histories are obviously anything but. This complexity may confound teachers (including those who write textbooks) who may not be aware of, or understand, this complexity. That explaining this complexity requires the use of mathematics creates further teaching challenges, especially if students demand simple explanations. We have the opportunity to correct these problems, however, by presenting life history evolution through the use of the theory developed in this chapter. Our list of propositions, and an exploration of their implications, can provide a simple entry into the complexities of life history evolution.

Intersections with Other Components of Evolutionary Theory

The present volume suggests several potentially fruitful directions for future development of life history theory, by considering the intersections between this chapter and others. We have underlined the notion that trade-offs among biological traits are what necessitates a distinct theory of life history evolution. Without those trade-offs, the theory reduces to the theory of natural selection: the expected fitness differences between values of a trait that affects survival or reproduction drive changes in the frequencies of the different phenotypes in a population, if there is a heritable basis for the variation (Frank and Fox, chap. 9). A key theme for research in life history evolution has centered on adaptations to unpredictable environments, often entailing phenotypic plasticity (Scheiner, chap. 13) or bet-hedging via variation in allocations among individuals. These studies are challenging mainly because of technical difficulties, like the problem of capturing enough environmental variation in the time available for study, including variation in population density as one of the variables (Gremer and Venable 2014). The theory here is relatively mature.

On the other hand, approaches to inductive inference about the role of natural selection in shaping observed life histories are not well studied. They present challenges to evolutionary theory because most populations are structured, and inferences about them typically require attention to the role of population structure in their dynamics (Metz and Diekmann 1986; Tuljapurkar 1990; B. Charlesworth 1994).

To some degree, life history traits all depend on developmental biology. Many models assume that trait values are continuous and can be changed in infinitesimal increments, but the developmental-genetic processes underlying trait variation are unlikely to make this true in general. Similarly, we know little about the biology underlying most trade-offs and most processes of hierarchical allocation. Perhaps the largest challenge to life history theory, then, is coupling it with our increasing knowledge of evolutionary developmental biology (Love, chap. 8). Providing a stronger biological basis to our understanding of how life history traits vary, and how they are constrained, will provide considerable depth to our understanding of life history evolution.

Formalizing life history theory leads to some valuable insights. If we may be allowed a personal comment, we had always been skeptical of models like r- and K- selection or Grime's C-S-R selection but before writing this chapter had not recognized that fundamentally these are not theories about evolutionary process and are thus distinct from the theory that considers the way selection is shaped by demographic trade-offs. It was our efforts at formalization that led to the insight that that theory describes the demography of fitness, which we think aptly summarizes this valuable body of work.

Acknowledgments

GAF was partly supported by grant DEB-1120330 from the US National Science Foundation. Two anonymous reviewers made valuable suggestions. This chapter is based on work done by SMS while serving at the US National Science Foundation. The views expressed in this chapter do not necessarily reflect those of the National Science Foundation or the United States Government.

TWELVE

The Theory of Ecological Specialization

TIMOTHÉE POISOT

"Ecological specialization" is a catch-all term to describe a series of evolutionary mechanisms by which a given species, taxon, or group of organisms undergoes an improvement of its performances in a subset of the environments it can potentially occupy. Conversely, specialization can happen if the availability of these environments changes, with or without concomitant evolutionary change. Observed at any given time, the distribution of performances of an organism across different environments is its *specificity*. *Specialists* are organisms whose distribution of performances is biased toward higher performance in a comparatively small subset of possible environments; conversely, *generalists* have equivalent performance across most of the environments they exploit.

The goal of this chapter is to provide a constitutive theory of ecological specialization, including an overview of the mechanisms involved in the evolution of specialization and how specialization proceeds in nonindependent ways for biotic (species-species interactions) and abiotic (species-habitat interactions) components. In particular, I examine the role of trade-offs in restricting or allowing the evolution of different degrees of specificity, the importance of considering the spatial and temporal heterogeneity of the environment, and the role of lifestyles and species interactions in determining the degree of both abiotic and biotic specialization. I also describe how the current specificity of an organism constrains its ecology and future evolutionary dynamics. The literature on specialization has been extensively reviewed over the last years, and for this reason, this chapter is primarily concerned with large-scale issues and outstanding research questions.

Because specialization is such a pervasive feature across evolutionary biology, it comes with a few lexical ambiguities. I use the following conventions: "Activity" or "environment" is any state or feature exploited by an organism, or any action that this organism performs. In the classical perspective of the niche as an n-dimensional hypervolume (Hutchinson 1957), each position on a niche-forming axis is an environment or activity. "Performance" is any measure of the benefit derived by the organism from being exposed to a given environment, most often any estimation or empirical measure of fitness (Forister et al. 2012).

Domain

The domain of the theory of ecological specialization lies at the intersection between the domains of the theory of natural selection (Frank and Fox, chap. 9) and the theory of ecological niches (Chase 2011). Whereas the theory of natural selection deals with general changes in fitness in response to the environment, the theory of ecological specialization deals explicitly with how these changes operate when the environment is heterogeneous. That heterogeneity can be either spatial or temporal; in this chapter I focus on models dealing with spatial heterogeneity. This theory is complementary to the theory of the evolution of phenotypic plasticity (Scheiner, chap. 13) that examines the consequences of varying phenotypic expression of a single genotype. In the theory considered here, genotypic expression is assumed to be fixed.

I consider two complementary subdomains: (1) specialization occurs as a consequence of natural selection and shapes species niches, and (2) specialization triggers and constrains future natural selection by expanding or restricting the range of situations to which organisms can be exposed. Although the second subdomain deals primarily with ecological mechanisms, it must be included in a comprehensive theory of ecological specialization. The consequences of evolutionary mechanisms are mainly observable in their ecological consequences (P. W. Price 2003), and these ecological consequences themselves constrain future evolutionary trajectories (Futuyma and Moreno 1988).

The term "specialization" is, to some extent, misleading. Studies on the specialization of organisms describe the direction of change in the distribution of performances across different environments (fig. 12.1), whether the endpoint of this change is an increase or a decrease in specificity. Performance can be measured in evolutionary terms (i.e., fitness) or in other (e.g., demographic) terms, with equal validity. Throughout this chapter, "specialization" should be understood as an abbreviation of "any change in the distribution of performances, induced by either evolutionary or ecological change." "Performance" will refer to a variety of measures selected

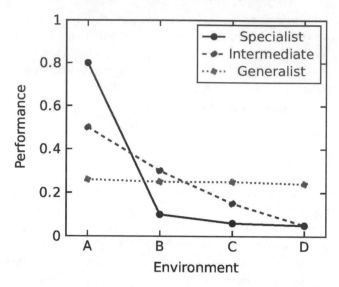

Figure 12.1. Hypothetical distributions of performance of three organisms across four environments. The specialist has a high performance in environment A and very low performances in the others. The intermediate species has a more gradual decline of performance across the gradient. Finally, the generalist has roughly equal performances throughout the gradient. Note that the performances have been standardized so that they sum to unity. This allows comparisons of the specificity of organisms with different cumulative performances. The distribution of performance values in these three situations would range from extremely left-skewed (specialist) to not skewed (generalist).

by the experimenter or modeler, and I will use "fitness" for the cases in which the performance measure has a clear evolutionary significance. As an example, if performance is defined as the growth rate of a population, this has evolutionary consequences and can be termed "fitness"; conversely, the number of prey eaten is not a direct measure of the evolutionary success of the population and will be termed "performance." Fitness is often the integration of performance across many axes. Consequently, because the theory of specialization deals with the distribution of performances along single axes, it gives insights about the components of fitness but very rarely about fitness itself.

Here I summarize the main elements of the constitutive theory, in particular how specificity and niche are related. One of the most important distinctions is between fundamental and realized specificities. Fundamental specificity is the expected distribution of performances across environments in the absence of biotic or abiotic constraints for the focal organism.

It is defined in the absence of extrinsic boosts to growth rate that can be conferred by mutualisms. This definition does not imply that fundamental niches are infinite; rather, they reflect the physiological, biochemical, and other limitations of where an organism can live. The fundamental specificities (with regard to temperature) of a cold-adapted bacteria from the Arctic and a heat-adapted bacteria from a hydrothermal vent differ in that they are adapted to opposite thermal stress; one would likely instantly deteriorate when exposed to the other's environment. The realized specificity is the effective distribution of performances, one that accounts for both biotic and abiotic constraints and advantages. Most often, realized specificity is measured from empirical data. This distinction is important because realized specificity stems from the integration of constraints along and between several niche axes. As such, any theory of specialization should aim to predict realized and potential specificities and to uncover the mechanisms that allow linking one to the other.

Basic Propositions

Increased specificity (i.e., higher performances in a narrower subset of environments) is selected for when the preferences of an organism (which environment or activity it actively seeks out) and its resulting performance positively covary. This correlation creates a positive feedback loop in which suboptimal environments are decreasingly exploited, triggering specialization. This situation may result in speciation (Edwards et al., chap. 15), in which the incipient species exhibit increased specificity. This dynamic has been established for habitat choice by organisms with active dispersal (Ravigné et al. 2009), and for feeding-source selection by mothers for their juveniles, the so-called "mother knows best" hypothesis (Jaenike 1978b). There have been a number of reviews of how specificity evolves in recent years. Ravigné et al. (2009) provide an excellent argument about how both specificity and diversity increase jointly when habitat selection and local adaptation happen jointly. They also synthesize most of the modeling approaches used to address the evolution of habitat specificity.

Devictor et al. (2010) proposed a more integrative view of specificity, about which I will comment extensively in subdomain 2. They suggest that specificity can be coarsely partitioned into its Grinnellian and Eltonian components. The Grinnellian specificity, by similarity to the Grinnellian definition of the niche, represents the requirements of the species in terms of its habitat. This includes the abiotic conditions and the presence of populations of other species that are, for example, part of its diet. The Eltonian specificity on the other hand, describes the effects, or functional roles, that a species has in a given community. For example, species that have different proportions of prey items in their diet have different functional roles. More plainly the two components can be expressed as: What does the species need? What does the species do?

This representation is a first attempt at tackling two key issues: What are the feedbacks between biotic and abiotic levels of specificity? How can we build on theory surrounding specificity to move closer to an integration with niche theory? Poisot, Verneau, et al. (2011) suggest that even though specificity is expressed in a large variety of contexts, it happens through a very few key evolutionary mechanisms. These mechanisms fall into three broad categories: constraints, life histories, and environmental features, the latter two being mechanisms involving gene-environment covariances. These three mechanisms are the focus of subdomain 1.

Forister et al. (2012) provide an insect-focused overview of specificity and its ecological and evolutionary dynamics. They establish that performances are distributed in an environment-dependent way, even when we focus purely on biotic components of the environment. Vamosi et al. (2014) discuss the phylogenetic distribution and dynamics of specialization, with specific attention given to the fact that species may exhibit different levels of specificity on different niche axes.

Overview of Modeling Efforts

It should be no surprise that a topic as vast and fuzzily defined as specificity has been modeled from a variety of angles. D. S. Wilson and Yoshimura (1994) famously listed forty-three models addressing just the study of co-existence between specialists and generalists. More than twenty years later, it is dubious that a comprehensive review would be feasible. Those forty-three models can be grouped into four classes, depending on the number of factors they accounted for: the shape of the fitness set (Levins 1962, 1965), whether habitat or resource selection was active or passive, negative density dependence, and environmental stochasticity. This last factor was considered only for intergenerational heterogeneity, excluding spatial stochasticity.

The framework of adaptive dynamics (e.g., Doebeli and Dieckmann 2005) is a natural candidate to study specialization because it makes explicit predictions about performance alongside a niche axis defined by continuous trait values, the measure of performance can be directly related to fitness, and the framework can be amended to allow for explicit spatial structure and heterogeneous environments. These features result in the generation of a large number of predictions about local adaptation and dispersal that sometimes have been interpreted as predictions about specialization (Kisdi and Stefan 1999; Troost et al. 2005).

Alongside their review of modeling efforts, D. S. Wilson and Yoshimura (1994) introduced a simple model for the coexistence of habitat specialists and habitat generalists in a spatially heterogeneous landscape. This model combines explicit rules for fitness with chance events in survival and falls within the broader family of lottery models. Lottery models have been used increasingly in recent years, notably to study the dynamic of specialists and generalists in metacommunities (Büchi and Vuilleumier 2014, 2016). Because of their temporal scale, both adaptive dynamics and lottery models of specialization make predictions on an ecological and microevolutionary timescale, rather than a macroevolutionary timescale (Jablonski, chap. 17). Very long timescales have mostly been addressed through simulations of traits on phylogenetic trees, rarely with dynamic models of populations or species.

Overview of Outstanding Challenges

There are two outstanding challenges when dealing with a broad theory of specialization, identified in the seminal paper by Futuyma and Moreno (1988). Although largely refined since then, they have not been clearly resolved. First, although species are often classified as specialists or generalists, this is a false dichotomy. Even when not accounting for the quantitative performance in each environment, most measures of specificity are continuous values (Poisot et al. 2012), because specificity is a continuous phenomenon. Examples of perfect specialist or perfect generalist species are rare, and most ecological communities show the coexistence of species with intermediate levels of specificity (Poisot et al. 2015). These observations imply that specialization rarely has a clear-cut outcome and should instead be considered as a process that can lead toward smaller or larger skewed distributions of performances.

Second, specialization is defined for both biotic and abiotic environments. Host specificity is just as important as habitat specificity, and both should be accounted for by a general theory. Fortunately, previous literature surveys suggest that both are driven by the same set of core mechanisms (Devictor et al. 2010; Poisot, Bever, et al. 2011; Forister et al. 2012). Defining these mechanisms is the main purpose of subdomain 1; understanding how the biotic and abiotic levels of specificity are intertwined

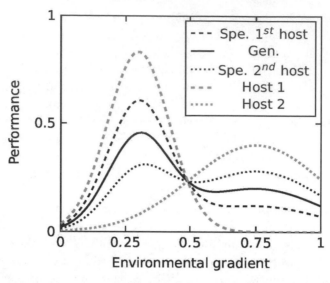

Figure 12.2. Perceived specificity of three symbionts (black lines) on two hosts (gray lines) on an environmental gradient of abiotic resources. Each host has a performance $\omega_i(g)$ at position on the gradient, and the performance of symbiont k is $\omega_k(g) = \Sigma(a_{ki} \times \omega_i(g))$, where a_{ki} is the preference of symbiont k for host i. While it is possible to measure the specificity of each symbiont with regard to the gradient, doing so would be meaningless, as it would only reflect the interaction between their host preferences and the performances of hosts. This illustration emphasizes how crucial the selection of the correct axes for measuring specificity is.

is addressed in subdomain 2. Notably, these two levels are entangled, as specificity in one family of axes affects specificity in the other (fig. 12.2).

The Constitutive Theory

The nine propositions of the theory of ecological specialization are presented in table 12.1. The first five propositions deal explicitly with how distributions of performance evolve toward greater or lower specialization. Specialization tends to occur because of the existence of constraints at different levels. At the organismal level, evolution is constrained by physiological or biochemical limitations. The environment adds another layer of constraints through limited availability of suitable habitats, finite resources, and so forth. The lifestyles and behaviors of a species can also drive it toward specialization (e.g., active versus passive foraging, migration). Finally, direct and indirect biotic interactions will push species toward more or less specificity.

Table 12.1. The Propositions of the Theory of Ecological Specialization

Domain: Patterns of adaptation and its consequences in response to natural selection in a heterogeneous environment.

Propositions of subdomain 1: Specialization in response to natural selection.

1.1. Specialization occurs when there is environmental heterogeneity that results in fitness trade-offs in trait values across environments.
1.2. These traits meet the conditions required for evolution by natural selection.
1.3. Specialization occurs as a response to patterns of dispersal and environmental heterogeneity.
1.4. Specialization is determined by interactions between the biotic and the abiotic components of the environment.

Propositions of subdomain 2: Natural selection in response to specialization.

2.1. Specificity is defined along many axes.
2.2. Biotic and abiotic specificities interact.
2.3. Niche dimension is the result of specificities.
2.4. Specificity affects emergent community properties.

The last four propositions highlight how the evolution of ecological specialization underpins many ecological mechanisms and should be investigated in a broad context. These propositions concern how the current specificity of a species constrains or allows its future evolution because the overall niche of the species is the resultant of its specificity along many entangled axes. I first consider subdomain 1: specialization in response to natural selection.

Proposition 1.1: Environmental Heterogeneity and Fitness Trade-offs

At its most fundamental, specialization happens because differences in fitness effects of trait values are such that improving net performance in one environment decreases performance in others (Kassen 2002). Therefore, specialization is expected to happen whenever there is selection for locally higher performances. Débarre and Gandon (2010) showed that in a spatially continuous landscape, whether one generalist or two specialists evolved from an initially monomorphic population depends mostly on the curvature of the fitness trade-off. If ρ_1 and ρ_2 are the performance in two environments, and $\omega(\rho_1,\rho_2)$ is the organism's fitness as a function of both performances, a concave trade-off permits maximizing ω by maximizing both ω_1 and ω_2 (one generalist), while a convex trade-off maximizes ω only when either ω_1 or ω_2 is maximized (two specialists).

Such trade-offs can occur in a variety of ways. The fitness trade-off can occur owing to selection on a single trait. In this case, the trade-off is intrinsic to the fitness function such that the optimal value for that trait differs among environments. Alternatively, the fitness trade-off can occur because fitness is determined by a combination of two or more traits that are constrained in their joint phenotype such that phenotypic combinations that would result in high fitness in both environments are not possible. The most commonly invoked genetic mechanism (antagonistic pleiotropy) is known to generate unbreakable constraints (Laubichler and Maienschein 2009), although these constraints can involve so many traits that they end up being difficult to measure (Hereford 2009).

These constraints are also expressed at the population level, notably when performance is expressed as the consequence of a complex trait architecture. There have been recent calls to rebuild our understanding of community properties using traits (McGill et al. 2006), but these calls tend not to account for the fact that traits can rarely, if ever, be mapped directly unto performance in a given environment as such mapping can be done many ways. Most evolutionary (and ecological) constraints emerge because traits contribute differently to components of performances, and the success of a given activity results from the interaction of trait values. Malcom (2011) showed that characters that are regulated by a smaller network of associated genes show greater adaptability, suggesting that evolutionary dynamics can be increasingly constrained by past innovations.

Perhaps unsurprisingly, then, constraints and specificity are nonrandomly distributed in the phylogenetic structure of clades. In systems where this has been studied, up to 45 percent of the variance in specificity can be explained by phylogenetic constraints (Desdevises et al. 2002). Constraints can be hard to get rid of once they appear (Diniz-Filho and Bini 2008), so that constraints that appear at deep nodes in the phylogeny can still be strongly expressed in extant taxa. Although specificity can be measured with reference to currently experienced environments, understanding the drivers of specialization requires accounting for events at large temporal scales.

Along with fitness trade-offs, all of the other conditions necessary for evolution by natural selection (Frank and Fox, chap. 9) must be met by these traits (proposition 1.2, table 12.1).

Proposition 1.3: Dispersal and Environmental Heterogeneity

The model of Débarre and Gandon (2010) showed that steep environmental gradients led to specialization, although higher dispersal (being exposed to more habitats over a lifetime) counterbalanced this effect. It makes intuitive sense that species dispersing across different environments must be able to at least persist in most of them (ignoring habitat sinks), and higher rates of dispersal would lead to decreased specificity. This prediction has been demonstrated experimentally by Venail et al. (2008); high rates of dispersal over heterogeneous environments reduced the specificity of bacteria for different carbon sources. Kawecki (1994) argues that this dispersal effect is limited by the accumulation rate of deleterious mutations whose expression varies across environments. However, the ability to distantly disperse implies that an individual or its offspring may experience a highly heterogeneous habitat, which should be associated with generalist species.

Papaïx et al. (2014) showed that in host-pathogen interactions, short dispersal range in the host favors pathogen diversity, and pathogen specificity is determined by both host and pathogen dispersal. When the scale of dispersal increases, specialized pathogens are lost first, followed by moderately specialized ones. The most likely hypothesis is that high dispersal both synchronizes the dynamics of different populations in space, thus making the environment of the pathogen more or less temporally heterogeneous but spatially homogeneous, and triggers competitive exclusion of specialized strains whose optimal host/environment is in low abundance.

The mechanisms described above are inherently spatial. When there is temporal heterogeneity, the longer the duration of the environment that is less optimal for specialists, the higher the potential for their exclusion. Temporal heterogeneity does not allow for refuges through dispersal, unless there is some form of dormancy or quiescent life history stage (e.g., a seed bank), a form of dispersal in time (Chesson and Huntly 1997).

Intriguingly, Büchi and Vuilleumier (2014) show that specialists are more likely to coexist when their dispersal ability is high, independently of whether they preferentially disperse toward favorable habitats, for all levels of spatial autocorrelation of habitat suitability; they dub this effect the "specialist's paradox." All else being equal, species with higher dispersal rates should encounter a broader range of environments, and therefore tend toward less specificity. Although it makes sense that high dispersal ability coupled with active habitat search would reinforce specificity, it is puzzling that high dispersal ability alone is beneficial to specialists in heterogeneous environments.

Empirical data (Morris 1996) may hold part of the solution to this paradox. Specialist and generalist rodents exhibit different modes of habitat selection as a function of the density of conspecifics. While both tend to avoid overcrowded habitats, generalists experience their surroundings at a larger scale than specialists do. While generalists have the ability to persist in suboptimal environments, since by definition they pay less of a price in their resulting performance, they can disperse only locally. Specialists, on the other hand, are under a stronger pressure to find suitable habitat, making a high dispersal ability beneficial.

These considerations about the role that dispersal plays in shaping specificity over ecological and evolutionary times hearkens back to the more fundamental issue that, depending on whether ecological or evolutionary timescales are considered, either specificity is driven by different mechanisms, or the same mechanisms can have different outcomes.

Proposition 1.4: Interactions between Biotic and Abiotic Components of the Environment

Thrall et al. (2007) identified a key gap in our knowledge of the evolution of specialization: the "biotic complexity" of the environment—whether the focal species is embedded in a complex web of interactions—makes specialization unpredictable. Most of this unpredictability stems from the fact that antagonistic interactions tend to decrease performance and can act to shrink the niche space, mutualistic interactions tend to improve performance and can act to expand the niche space, and most, if not all, species are involved in both types of interactions (Fontaine et al. 2011; Kéfi et al. 2012). The realized niche of a species can be distorted by these interactions to the point where it no longer resembles the fundamental niche. The realized niche of a species, and hence its realized specificity, will be contingent upon the combination of species and the way they interact in a local environment. Recent models of mutualistic interactions (Bastolla et al. 2009) highlight how these interactions are themselves intertwined; the mutualists that pollinate a plant are themselves in competition for access to pollen and the plants are in competition for access to pollinators; the high dimensionality of this problem makes predictions extremely difficult.

Specificity may also reflect the broader environmental context of the community. Specialists are more likely to undergo extinction at the lower bound of productivity, unless they disperse to less crowded patches, thereby decreasing community specialization. For this reason, environments with high productivity can sustain both high diversity and a broader range of levels of specificity, as exemplified in tropical systems. This productivity effect implies that the relationship between diversity and specialization can reflect not only evolutionary constraints but also the ability of specialists to persist alongside generalists. In a recent analysis of 175 bipartite networks representing different types of ecological interactions (Poisot et al. 2015), we report that the coexistence between organisms with different specificities is maximized when the average specificity is medium to low. This pattern suggests that a broader range of specificities can coexist when increasingly specialized species are able to persist in the same system as a core of generalists. Whether this relationship emerges through mechanisms that are ecological (e.g., extinctions are more common in specialist-rich communities, competition is stronger in generalist-rich communities), evolutionary (e.g., a wider range of specificities offers more variability), or both remains an open question.

Subdomain 2: Natural Selection in Response to Specialization

Natural selection can lead to specialization, but specialization can also lead to changes in patterns of natural selection. Subdomain 2 deals with those reciprocal processes.

Proposition 2.1: Many (Nonindependent) Axes

Devictor et al. (2010) propose that specificity can be viewed as the intersection of biotic and abiotic specificity, which they interpret as a one-to-one mapping between, respectively, the niche in the sense of Elton, and the niche in the sense of Grinnell. These dimensions have different functional consequences: what a species does within a community (biotic specificity, the Eltonian niche), and where it may be expected to do it given environmental requirements (abiotic specificity, the Grinnellian niche). This distinction is useful because it refocuses the discussion about how specificity evolves on classical niche theory. However, this shift fails to take into account that both of these dimensions of the niche are composed of a large number of relevant axes that are unlikely to be orthogonal (Holt 2009) and unlikely to have similar distributions of performances.

From a biotic standpoint, a species is expected to be involved in several types of interactions. For example, the same symbiont can be, depending on its host, mutualistic or antagonistic (Thrall et al. 2007) and exhibit different specificities in these different types of interactions. Kéfi et al. (2012) made the point that although each species is involved in few trophic interactions, the number of competitive interactions can be much larger. As a result, although predators are relatively specialized, species of all trophic ranks are essentially competitive generalists. The same situation holds for abiotic environments. Keeler et al. (2014) report that outside their hosts, avian influenza viruses can tolerate a broad range of pH values but require low salinities and concentrations of ammonia; these viruses differ in their degree of specificity when different axes of their Grinnellian niche are considered. These cases point to the fact that Eltonian specificity and Grinnellian specificity are somewhat artificial notions because within these two dimensions organisms may be specialists on one axis and generalists on another, and there are interactions between the biotic and abiotic niche axes, the consequences of which are manifest in proposition 2.2 (see below).

One complication of measuring the degree of specificity on a single axis is that niche-forming axes are not expected to be orthogonal, and performances along different axes are expected to covary. In a classical example, Stanley and Morita (1968) showed that the temperature at which bacteria reach their peak fitness changes as a function of salinity. Thus temperature specificity would likely be evaluated differently if another component of the environment changes. Such nonindependence of performances is expected to be the rule rather than the exception, since there is not a strict correspondence between genes, traits, and the environment in which they are expressed (Forister et al. 2012). Even without the involvement of complex genetic architectures, biotic niche axes are not independent. Hochachka and Dhondt (2000) reported that the infection of house finches by *Mycoplasma gallisepticum* resulted in a decline of abundance in natural populations, resulting in changes in interaction with other species, thus making several biotic niche axes covary.

Proposition 2.2: Biotic and Abiotic Specificities Interact

Even though it makes sense to consider the overall specificity of an organism as an interaction between the dimensions of its Eltonian and Grinnellian niches, this distinction overlooks the fact that axes forming these two niches can have some collinearity, especially biotic interactions. Although it is feasible to measure the abiotic specificity of an organism, this specificity would reflect the interactive effects between its biotic specificities and the abiotic specificities of all possible interaction partners (fig. 12.3).

To some extent, such interactions have already been considered by community ecologists. In their trophic theory of island biogeography, Gravel et al. (2011) propose that the spatial distribution of predators is at best nested within the cumulative spatial distribution of their prey. The realized Grinnellian niche of any species that require interactions to persist is contained within the sum of the realized Grinnellian niches of the species it interacts with. This means that even though these species have an abiotic

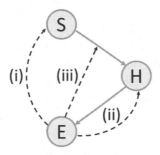

Figure 12.3. Biotic and abiotic specificities are entangled. In this simple situation of a system with a single host (H) and symbiont (S), the environment (E) can have effects either directly on the symbiont (i), directly on the host (ii), or on the host-symbiont interaction (iii). This results in a situation where the Eltonian and Grinnellian specificities are not discrete entities but rather interact in (potentially) complex feedback loops.

specificity, they do not interact directly with the environment; they do so indirectly, through biotic interactions.

This is not to say that species relying on biotic interactions have no relevant dimensions to their Grinnellian niche. For example, for avian influenza viruses (Keeler et al. 2014), some abiotic factors will affect virus performance directly (e.g., salinity, temperature), whereas others will affect it indirectly through their impact on the performance of suitable hosts. More broadly, levels of specificity are expected to be labile through time. Khidr et al. (2012) showed that monogeneans (*Microcotyloides* sp.) infesting *Terapon puta* vary in their infection success as a function of water pH. Changes of pH over seasons drives the seasonal dynamics of parasite performance. Similarly, Poisot, Verneau, et al. (2011) showed that the ability of phage to infect the same bacterial strains of *Pseudomonas* varies as a function of the availability of abiotic resources, not because the hosts or phage are directly affected by the environmental change, but because decreased resource input limits host growth, which in turn decreases the strength of infection by the viral population. Arneberg et al. (1998) conducted a survey of how host abundance affected the density of gastrointestinal strongylid nematodes in mammals over time. They show that variations in the abundance of hosts had strong effects on the performance of parasites, even after controlling for other covariates. Temporal variation of host populations was in response to various environmental factors, and the biotic performance of the parasites varied without evolutionary change, owing to the cascading effect of environmental change through the host populations.

Thus, the nonindependence of Eltonian and Grinnellian niches in species interactions can happen when the performance of either interacting species is tied directly to the state of the environment, either when one species is affected by the environment through cascading effects, or when the environment affects the existence or outcome of the interaction directly. Understanding these feedback loops is critical, since they determine which niche axes should be considered when measuring specificity.

Proposition 2.3: Niche Dimension Is the Result of Specificities

The interaction between levels of specificity has consequences for overall niche dimensionality. For example, S. M. Wagner et al. (2015) manipulated polyphagous herbivores by adding endosymbionts to test the effect of a mutualistic interaction on plant use specificity. In the presence of a positive insect-symbiont interaction, the insect was able to use a broader variety of plants, even becoming a pest by gaining the ability to feed on commercially relevant crops. The situation is, nonetheless, not as clear-cut as when positive interactions increase niche breadth and negative interactions decrease it. Elias et al. (2008) showed that mutualistic interactions are a highly sought-after resource. In a situation where mutualists share environmental requirements, this can create niche convergence in species seeking these mutualisms. The ultimate consequence is that the necessity to maintain mutualisms results in higher competition. Because of the ubiquity and nonindependence of biotic interactions, how they drive specificity and niche breadth will not be easily predicted.

At a broader scale, Slatyer et al. (2013) reported a seemingly general relationship: species with a wide diet (biotic generalists) are present over a wider spatial range. Thus, they are apparent habitat generalists. This relationship makes sense, in that the primary determinant of a consumer's niche is its resource. Species with a wider diet have more items to choose from and by chance are going to find suitable environments over a larger spatial scale. The interaction between biotic and abiotic components, although hard to predict, drives the overall niche dimension.

Proposition 2.4: Specificity Affects Emergent Community Properties

One of the key big-picture issues is how specificity and its evolution will cascade up to the level of community structure and whether specialization that involves niche differentiation has consequences on the emergent properties of the community. If species specialize and do so on different parts of the axis, this will give rise to niche complementarity, which is known to promote, for example, ecosystem functioning (Loreau 1998) and stability (Loreau 2004). On the other hand, the clustering of species on similar parts of the axis—a failure to specialize—will result in a high redundancy. Such redundancy is not per se bad as it provides an insurance effect (Yachi and Loreau 1999). If a species occupying one part of a niche axis goes extinct, functionally redundant species are still present.

This redundancy is also true of biotic interactions. Carvalho et al. (2014) showed that in an agricultural landscape, bees varied in both levels of floral specificity and niche complementarity; bees visited different sets of flowers, decreasing intraspecific competition. Interestingly, temporal variation in resource use in this system does not appear to be primarily driven by environmental variation. Avoiding niche overlap is an active process and can help maintain specialized bees, not because the resource is optimal, but because it would not be worth bearing the cost of competition. In a field-isolated bacteria-phage system, Poisot et al. (2013a) reported a slightly different situation: generalists performed better in that they are more efficient at infecting bacteria. All else being equal, this difference should have the result that specialists are competitively excluded. However, specialists predominantly used a distinct subset of hosts, exposing them to less competitive pressure, and therefore allowed persistence. That the hosts used by specialists were phylogenetically distinct from the hosts of generalists implies a long-term co-evolutionary relationship. Depending on whether specificity happens alongside niche differentiation, communities may be driven toward more or less compartmentalized states (Espinosa-Soto and Wagner 2010). Specificity also appears to be a strategy driven by the avoidance of competition, and not necessarily one that maximizes performance on optimal or preferred resources.

Chesson and Kuang (2008) provide an excellent demonstration of why biotic and abiotic specialization must be accounted for simultaneously. Coexistence of primary consumers (e.g., herbivores, grazers) is facilitated when they consume different sets of primary resources or producers while being consumed by different sets of predators; intermediate situations resulted in intermediate levels of coexistence. Not only does specificity and its distribution within a community affect emergent properties, but specificity can also act at a distance and have effects that cascade across trophic and organismal levels.

Implications of the Constitutive Theory

Different Lifestyles Offer Different Opportunities for Specialization

Specialization is also driven by the fact that an organism's lifestyle can decrease or increase constraints on its need to match a current environment. In an analysis of spatially replicated data on rodents and their ectoparasites, Poisot et al. (2013b) showed that obligate parasites (species that require a host interaction to complete their life cycle) are consistently more specialized than facultative parasites (species that live on hosts when available but can have a free-living period). Whereas obligate parasites are highly adapted to their hosts, and therefore more specialized, facultative parasites have moderate adaptations to a parasitic lifestyle that are not accompanied by host specificity. Moreover, facultative parasites showed less spatial consistency in their host use, suggesting less stringent adaptations to their hosts. Systems in which related taxa can have different lifestyles are good models for understanding the phylogenetic dynamics of specialization and how it ties to the evolution of lifestyles (Vamosi et al. 2014). For example, J. A. Smith et al. (2013) report that the joint evolution of specialization and social parasitism in allodapine bees is a derived trait that emerged several times in the phylogeny and was followed by subsequent co-speciations. This evolutionary pattern suggests that because specializations often appear together with a change in lifestyle, specialization can be triggered by other evolutionary changes.

More broadly, the interplay between lifestyle differences and specializing adaptations can create a positive feedback. Smorkatcheva and Lukhtanov (2014) show that the preference for subterranean environments in rodents evolved more readily when females were social and gregarious; it rarely evolved when females were solitary. They further suggest that this joint evolution drove the overall phylogenetic structure of surface-living and subterranean clades. Adaptation to the subterranean environment, and the morpho-functional adaptations that followed, were prompted by the shift in group dynamics from solitary to gregarious. Habitat shifts can create a similar feedback. Amson et al. (2014) and Reese et al. (2013) examined cases of the evolution of skeletal traits in response to physical environmental factors and associated resource-gathering strategies. In both cases, traits that conferred specialization allowed the species to shift to a new environment that created further selection on those traits that enhanced specialization.

Natural Selection Can Result in Generalization

The considerations outlined in the previous sections lead some to suggest that specialization is an evolutionary dead end. There are two ways this point is usually argued. First, since adaptations to a new environment are usually self-reinforcing, it would take tremendous amounts of gene flow and dispersal to maintain a generalist population. Instead, it is expected that generalism is a transient state in which not yet diverged species are following an increase in their fundamental niche (Loxdale et al. 2011). Second, regaining a status as a generalist would require going against accumulated mutations that are neutral in the current environment but deleterious elsewhere (Kawecki 1997, 1994). This question needs to be addressed in a phylogenetic context. If specialization is a dead end, then specialists should be the most derived taxa. The problem with testing this prediction is that generalists that are derived taxa can be dismissed as "undergoing speciation."

These questions have been well studied in fish and their monogenean parasites. These are a good system to study these questions owing to their long-standing association, wide geographical distribution, and the availability of molecular tools to provide robust phylogenies (Pariselle et al. 2011). Simková et al. (2006) show that in the genus *Dactylogyrus* (ectoparasites of Cyprinid fishes), high specificity (i.e., having a single host) is an ancestral state, and most derived species are generalists. Similar observations have been made in other genera of monogeneans (Desdevises et al. 2002; Morand et al. 2002). Yet the existence of extant generalists does not rule out the transience of generalism. In the genus *Lamellodiscus*, in particular, having several hosts is associated with higher genetic and morphometric variability (Kaci-Chaouch et al. 2008), although this variability is structured by the host species (Poisot, Verneau, et al. 2011), indicating that generalism can be lost (through speciation) if gene flow across different subpopulations is weak.

Specificity Is a Constituent of an Integrated Niche Theory

At its most fundamental, the niche of a species is defined as the parameter space in which it can maintain a nonnegative growth rate (thus enabling its persistence without immigration). It is worth noting that this definition says nothing about the performance on each niche axis, since the cumulative contribution of each to overall growth rate is the quantity of interest. In short, the niche of a species is the addition of its specificity on each axis, including the interactions between them. This is true for both fundamental and realized niches. Usual definitions of the niche (Chase and Leibold 2003; Chase 2011) frame it as both the requirements of a species (Grinnellian) and its effect on the community (Eltonian). Both of these aspects are captured by an integration of biotic and abiotic specificities. As mentioned previously, the dimensions of the niche respond to the interactions of fundamental specificities, the availability of environments, and biotic interactions; on the other hand, the effect of a species is purely a function of its realized specificity.

An important question is how the breadth of requirements affects a species' effects within the exploited breadth. Using data from lizards, S. Edwards et al. (2013) suggested that this relationship can be driven by evolutionary constraints. The acquisition of morphological or physiological adaptations that allow a species to have a larger effect can result in further evolution of a wider niche. Thus, specialization should be viewed in the light of the phylogenetic structure of a focal species and the distribution of its relevant traits within this phylogeny (Vamosi et al. 2014).

Missing Links in the Constitutive Theory

Biotic-Abiotic Integration with Nonindependent Traits

As Hardy and Otto (2014) pointed out, it is difficult to define specialization precisely, not only because of the blurred lines between what is abiotic and what is biotic, but also because performance along each axis is the product of complex, interacting traits. It is unlikely that a single trait can be mapped directly onto performance in a given environment, or even that a single trait would be involved in performance in only a single environment. Paterson et al. (2010), using experimental evolution between *Pseudomonas fluorescens* and a lytic phage, showed that while coadaptation left clear genomic imprints in the host genome, it did not modify the pathogen genome in a measurable way, although the phage genome was small and composed mostly of coding sequences whose products have well-known roles on the infection pathway. This is a puzzling result, since it suggests that even with the high resolution offered by genomic approaches, identifying the traits and underlying genetics involved in determining performance in a given environment may be prone to failure. This search is further complicated by the fact that the expression of biotic specificity varies across abiotic environments and vice versa. Common-garden experiments (Nuismer and Gandon 2008) or time-shift experiments (Gaba and Ebert 2009; Blanquart et al. 2012) can help in this regard. Scenarios for environmental change that are coupled with full genome sequencing might provide a better understanding of the underlying genetic mechanisms of specialization. Nevertheless, for the time being such scenarios are limited to microbial systems, with comparative methods being the fallback for other types of organisms.

The Paradox of Specialists

As highlighted by proposition 1.3, there is an apparent paradox in the maintenance of specialists with high dispersal abilities. Dispersal, all else being equal, is supposed to expose individuals to more heterogeneous environments, thus promoting generality. This statement must be tempered by two things. First, the distance of dispersal matters only as a relative value to the environmental grain. Dispersing even over a short distance in a highly fragmented environment means more perceived heterogeneity, as opposed to the same distance in a homogeneous environment. Second, dispersal might not be random; active habitat selection might mean that species disperse over long ranges because they settle only once they have found an optimal habitat or resource. In species with active dispersal, this hypothesis is easily tested. Specialist species should have a greater variance in how far offspring disperse from parents, and increasingly so in landscapes with intermediate levels of fragmentation. Species with a narrow tolerance should have a positive skew in how far they disperse: the most "unlucky" individuals will have to cover a large distance to find a suitable patch. Interestingly, this conjecture provides another way to reconcile specificity and high dispersal abilities. The latter may have been selected because of the imperative of finding a suitable habitat, whereas generalists, having by definition a broader tolerance, should have to disperse less to find suitable habitats. A relatively simple comparison can be used to test this idea. For equal dispersal mode and ability, species that produce more offspring should be able to be more specialized. The cost of specificity is offset by the fact that more offspring explore a larger part of the landscape, increasing the likelihood of finding a suitable patch for the next generation.

Ecological and Evolutionary Induced Changes in Specificity

Because specificity depends on the availability of each environment or resource in which performance occurs, there can be rapid, strong, and consistent changes in performance even in the absence of evolutionary change. An example is the apparent increase in generality of European birds, coupled with a decrease in the abundance of specialists, leading to a widespread biotic homogenization. Birds that are habitat specialists are more affected by disturbance and fragmentation (Devictor et al. 2008) because they exploit a subset of habitats that is both more sensitive to disturbance and distinct from the habitats of generalists (Julliard et al. 2006). Over the last two decades, the habitats of specialists have been affected in a way that reduced the prevalence of specialists (Le Viol et al. 2012), to the benefit of generalist species. The same trend is reported at the worldwide scale (Barnagaud et al. 2011; Clavel et al. 2011); structured habitat disturbance coupled with structured habitat choice is leading to an increase in the distribution of generalists. Yet the notion that these communities are moving towards greater generalism should be careful considered. From the perspective of the community weighted average (specificity weighted by the abundance of each species), the trend is clear. Yet this aggregate measure is masking the reasons for the change. An increase in the abundance of generalists and a decrease in the abundance of specialists can drive its value towards more generalism without any change in species' specificity. Fundamentally what ultimately matters is not the specificity of the species, it is the distribution and availability of its most suitable environments. For European birds, if the habitats more suitable for generalists (pastures and agricultural systems) have been more subject to fragmentation or disturbance, the aggregated trends would have been greater specificity. Since environmental disturbance is expected to keep accelerating, measures that correct for, or at least account for, the availability of resources, should be a primary research focus in coming years.

Conclusion

The specificity of organisms, whether with regard to abiotic or biotic components of their environment, is a highly dynamic state that can change rapidly over ecological timescales in response to changes in habitat or resource availability, population densities, or dispersal opportunity. These changes feed back into longer-term changes on evolutionary timescales. Understanding the level of specificity displayed by an organism is necessary to account for ecological and evolutionary constraints and opportunities. This specificity should be analyzed in an integrative manner, as specificity along one niche axis can be a function of specificity along other axes. An understanding of how species traits, levels of specificity, and niche-forming axes are entangled is one of the most important conceptual gaps in this constitutive theory and will require a careful integration of models and data. A remaining challenge is understanding how specialization can affect niche evolution.

THIRTEEN

The Theory of the Evolution of Phenotypic Plasticity

SAMUEL M. SCHEINER

Individual adaptability is, in fact, distinctly a factor of evolutionary poise. It is not only of the greatest significance as a factor of evolution in damping the effects of selection and keeping these down to an order not too great in comparison with 1/4N and u, but is itself perhaps the chief object of selection.

—(Wright 1931, *147*)

A central component of the theory of evolution is the theory of natural selection (Frank and Fox, chap. 9). That theory explicitly recognizes the critical role of the environment in determining the relationship between phenotype and fitness. Missing from that theory, however, is the other critical role of the environment, determining the relationship between genotype and phenotype. The theory of the evolution of plasticity deals explicitly with that second role. Within the framework of the theory of evolution and its constitutive theories, the theory of plasticity evolution could be considered a subset of the theory of natural selection or its own constitutive theory that represents an intersection of the theory of natural selection and the theory of the evolution of development (Love, chap. 8), depending on the processes that determine the phenotype. In practice, this distinction does not change the structure and content of plasticity theory.

The role of phenotypic plasticity in evolutionary processes was recognized as early as the nineteenth century, although it was relegated to the sidelines for most of the twentieth century. It is only since the 1980s that it has risen to prominence. In considering the myriad ways that plasticity has been viewed over the past century and a quarter, a key distinction must be recognized. The role of plasticity in evolution has been considered from two perspectives. One perspective addresses how plasticity affects particular kinds of traits (e.g., dispersal, Scheiner et al. 2012) or processes (e.g., speciation, Schlichting 2004). The other perspective addresses the conditions that favor plasticity as a mode of adaptation to heterogeneous environments versus other modes (e.g., genetic differentiation, developmental stochasticity, Tufto 2015). My theory of the evolution of phenotypic plasticity deals with this second perspective. The other perspective is within the domains of other constitutive theories, although particular models may reside at the intersection of one of those theories and the theory of plasticity evolution.

In this chapter, I first provide a brief history of the concept of phenotypic plasticity in evolutionary thinking. I confine my history to plasticity as other chapters in this book provide detailed examinations of other aspects of the history of evolutionary thinking. For a detailed history of plasticity theory, see Sarkar (2004). Second, I present the domain and propositions of the theory of the evolution of phenotypic plasticity and the status of those propositions with regard to empirical evidence. Third, I examine issues affecting model construction and the intersection of plasticity theory with other theory domains. Finally, I consider the current status of the theory and future research directions.

A Brief History

Early attempts to incorporate plasticity into evolutionary theory were made by Baldwin (1896a, 1896b), Osborn (1896, 1897), and Morgan (1896). The interests and ideas of the three differed to some degree, although all were motivated by a shared concern that neo-Lamarckian ideas about the inheritance of acquired characteristics were challenging Darwin's theory of evolution. Today we primarily remember Baldwin's contributions, although those were certainly influenced by the other two. Baldwin was interested in the effects of learning on the evolutionary process. His goal was to show how traits that were modified during development could be favored by evolution and how apparent neo-Lamarckian evolution could be accommodated within a Darwinian framework. To use modern terminology, he postulated that trait plasticity itself would be favored by evolution when those plastic responses increased individual fitness. His primary interest was in learning as a form of plasticity, although he pitched his idea so as to encompass all traits. His theory encompassed both the idea that plasticity could enhance the process of trait evolution and the idea that plasticity itself could be an object of selection. However, because this theory was being articulated prior to the rediscovery of Mendel's theory of genes, from a modern perspective it is incomplete and largely devoid of mechanisms.

For the most part geneticists and evolutionary biologists ignored this work, possibly because it was framed in the context of learning. The ur-text for empirical studies of phenotypic plasticity was Woltereck's (1909) study of *Daphnia*, which is notable for coining the term "reaction norm," the pattern of change in trait value across environments. It contains no citations to Baldwin, Osborn, or Morgan, although this may be because Woltereck was German and the earlier papers were all published in English. Wright's (1931) foundational paper on his shifting balance theory does not reference either Baldwin or Woltereck, although it discusses plasticity (referred to as individual adaptability), listing it among both the factors that promote genetic homogeneity and those that promote genetic heterogeneity. Despite his statement that individual adaptability may be the chief object of selection (see epigraph), he never attempted to incorporate plasticity into any of his evolutionary models. The only citation of Baldwin among the primary architects of the modern synthesis was by Huxley (1942). Schmalhausen's (1949) treatise is a notable exception to this neglect of plasticity, but it also was mostly ignored when finally published in English, possibly because its author was in the Soviet Union and so apart from the European and American modern synthesis.

Some of the neglect of these early theories of plasticity evolution can be blamed on G. G. Simpson's (1953a) paper in which he mischaracterized Baldwin's ideas under the term "the Baldwin Effect." While not denying the possibility that plasticity might play a role in evolution, Simpson dismissed it as at most a minor phenomenon. That mischaracterization was repeated by Waddington (1953a), who was pushing his own theory of genetic assimilation (Waddington 1942, 1953c) that he wanted to distinguish from the ideas of Baldwin. The result is that Baldwin's contribution to these ideas went unrecognized until recently (Scheiner 2014). Until the 1980s, Waddington's theory stood as the most completely articulated one that incorporated plasticity. It was a theory about the role of plasticity in adaptation, rather than about the evolution of adaptive plasticity, as plasticity is only a transitory state in his model.

What became the seminal paper about the heritability and evolution of plasticity was that of A. D. Bradshaw (1965). Bradshaw was reacting, in part, to a different concept of how organisms adapt to environmental variation embodied by the term "phenotypic flexibility" (Thoday 1953), which referred to the overall ability of an organism to maintain high fitness in different environmental conditions. Bradshaw contrasted that with his trait-based concept of phenotypic plasticity that emphasized that plasticity was trait specific rather than a property of the organism as a whole and was subject to natural selection, allowing for adaptive evolution of plasticity (Nicoglou 2015; Peirson 2015). Although Bradshaw's paper recognized the importance of plasticity in evolution, it did not present a formal theory of plasticity evolution. Nor did he cite Baldwin, although this oversight may be due to Simpson's dismissal of Baldwin's theory a decade earlier. That Bradshaw's review paper was explicitly about plasticity in plants is notable. For most of the twentieth century, evolutionary biologists who focused on animals tended to view plasticity as nuisance variation. Much of the focus of genetic studies in the first half of the twentieth century took place in tightly controlled laboratory settings with such model species as *Drosophila melanogaster* and focused on alleles that were assumed to have a single phenotypic expression. In contrast, those who studied plants saw plasticity as central to adaptation. The most extensive early work on genotype-environment interactions was done with the decidedly non-model-plant species *Achillea millifolium* and *Potentilla glandulosa* (Clausen et al. 1940). In the 1960s, Bradshaw and his students were some of the pioneers of the newly emerging field of plant evolutionary ecology. It was not until the 1980s that animal evolutionary ecologists began to pay substantial attention to the plant evolutionary ecology literature.

The first mathematical model of the evolution of plasticity is that of Levins (1963), one of the founders of evolutionary ecology (Odenbaugh 2011). In his paper, phenotypic flexibility is recognized as an evolutionary endpoint, but only under specific conditions. "An irreversible modification of development at any early stage will in general be advantageous only if the environmental factor evoking the modification is correlated with the environment at later stages when selection is operating" (Levins 1963, 75). This requirement for cue reliability is central to the theory of plasticity evolution (see below).

It was another twenty years before the next model of plasticity evolution was published (Via and Lande 1985). That model was the first to formally examine the evolution of plasticity per se and is notable in citing Schmalhausen and Bradshaw. In contrast, a contemporaneous model (Lively 1986) cited Levins but neither Schmalhausen nor Bradshaw, while Gavrilets (1986) was developing his model independently of all of those efforts. All of these threads became woven together amongst the explosion of models over the next decade (see review in Berrigan and Scheiner 2004), the huge increase in empirical studies beginning in 1991 (Scheiner and DeWitt 2004, fig. 13.1), and several major review articles (Schlichting

1986; West-Eberhard 1989; Scheiner 1993a). In the meantime, an entirely independent literature was developing derived from Baldwin's paper that was completely divorced from biological models of the evolution of plasticity. This work appeared in psychology journals in reference to the evolution of learning (Burman 2013) and in computer science journals in reference to evolutionary computation and computer learning (Turney et al. 1996).

Over the past decade, the pace of model development has slowed but not stopped. Despite all of these models and an informal consensus about the form and assumptions of such models, a formal constitutive theory of plasticity that made explicit the domain and propositions did not appear until 2013 as a brief appendix to one of my papers (Scheiner 2013b). That presentation just listed the propositions of the theory. Here I provide a full exposition.

The Domain and Propositions

Domain

The domain of the theory of the evolution of phenotypic plasticity is evolutionary change in trait plasticity in response to natural selection (table 13.1). Plastic traits are those whose phenotypic expression is environmentally dependent and can include any type of trait, including learning. In the literature on biological evolution, the modeling focus has primarily been on morphological traits, and many models deal with fixed traits. Even models of labile traits often overtly or by implication refer to morphological traits (e.g., Nonaka et al. 2015) rather than physiological or behavioral plasticity.

Models of plasticity evolution sometimes conflate trait evolution and the pattern of adaptation of individuals. Trait plasticity can result in an individual or a lineage having high fitness across a range of environments. Such an individual or lineage would be considered a generalist. But there are other ways to be a generalist. Individuals can be genetically and phenotypically uniform so that they have high fitness averaged across environments—a jack-of-all-trades generalist. Or an individual can produce phenotypically variable offspring through some sort of random process of developmental instability—a bet-hedging generalist. The conflation occurs when what is modeled is fitness itself, not the traits that underlie that fitness (e.g., M. Lynch and Gabriel 1987).

Plasticity theory is drawn from the sixth principle of the theory of evolution (see table 1.3) and is directly related to the theory of evolution by

Table 13.1. The Theory of the Evolution of Phenotype Plasticity

Domain: Evolutionary change in trait plasticity in response to natural selection.

Propositions:

1. Environmental heterogeneity exists that affects the phenotypic expression of traits.
2. The optimal phenotypic value of these plastic traits varies in space and/or time.
3. Individuals or lineages must experience the environmental heterogeneity either within or across generations.
4. These plastic traits meet the conditions required for evolution by natural selection.
5. Nonoptimal plasticity may result from maintenance, production, or information-acquisition costs of plasticity.
6. Nonoptimal plasticity may result because the environment at the time that the phenotype is determined does not provide a reliable cue about the environment at the time of selection.
7. Nonoptimal plasticity may result from developmental limitations on plasticity.

natural selection (Frank and Fox, chap. 9). It could be considered a sub-domain of that constitutive theory. It is related to the theory of the evolution of specialization (Poisot, chap. 12) in that plasticity is an alternative outcome to specialization, and to the theory of evolutionary development (Love, chap. 8) as constraints on development can be a critical limitation on plasticity evolution, although not all plasticity is developmental. Plasticity theory also impinges on speciation theory (Edwards et al., chap. 15) as the inclusion of plasticity into models of speciation can change the dynamic and outcome of that process, an example of the role of plasticity in other evolutionary processes. The theory of plasticity evolution presented here focuses on plasticity as a mode of adaptation.

Genotypes can vary in their phenotypic expression for reasons other than the external environment. For example, dominance and epistasis change allelic phenotypes. Purely random variation during development can affect final phenotypes, although I leave for another time the debate over whether such variation is still the deterministic result of microenvironmental variation or truly due to stochastic events. While these other sources of variation may affect plasticity evolution (e.g., Scheiner 2014), they are not the sources of phenotypic variation that define the domain of the theory.

Environmental Heterogeneity and Fitness

The first four propositions of the theory establish the conditions for adaptive evolution of plasticity. The first two propositions are plasticity-specific versions of the requirements for evolution by natural selection (see table 9.1). The first proposition is that environmental heterogeneity affects the phenotypic expression of traits, and it is related to the requirement of natural selection that phenotypes vary. By "phenotype" I mean all of the physical attributes of an organism, everything beyond the information content of the genome. This proposition embodies two requirements for selection on plasticity. First, the environment has to be variable in space and/or time. Second, that environmental variation must affect phenotypes such that individuals with the same genotype express different phenotypes in different environments. If an individual can express multiple phenotypes within its lifetime, then that individual can be a unit of selection. When an individual expresses just a single phenotype, then the unit of selection is the lineage that encompasses the set of variable individuals (see the first principle of the theory of evolution, table 1.3).

That the environment is heterogeneous at nearly any grain of space or time that we can measure is undoubtedly true. Within the context of biological theory, this principle is related to the fifth and sixth principles of the theory of ecology (Scheiner and Willig 2011). (Environmental conditions as perceived by organisms are heterogeneous in space and time. Resources as perceived by organisms are finite and heterogeneous in space and time.) However, such environmental variation does not enter the domain of the theory of plasticity evolution until it affects genotypic expression, a condition that is based on the eighth principle of the theory of genetics (Scheiner 2010). (Information usage is context dependent.)

The second proposition is that the optimal phenotypic value of these plastic traits varies in space and/or time. This proposition is related to the second proposition of the theory of natural selection—phenotypic variation leads to fitness variation. The third proposition is that individuals or lineages actually experience that environmental heterogeneity. The third proposition is implied by the first two, but here is made explicit because it can be violated. For example, even with spatial heterogeneity, if dispersal rates are very low then individuals or lineages would never or rarely experience that heterogeneity. The fourth proposition is a catch-all that covers any other conditions necessary for natural selection to occur. For example, it subsumes the requirement that trait plasticity be heritable, which is drawn from the theory of genetics (Scheiner 2010).

Restrictions on Adaptation

The last three propositions in table 13.1 establish the conditions that prevent adaptive evolution. A key question in evolutionary theory is why species are not always perfectly adapted. For plasticity, this question becomes

"If plasticity can result in organisms that always match the optimal pheno-type, why is adaptive plasticity not more common?" A recent meta-analysis of plasticity in plants found that only 33 percent of traits showed adaptive plasticity, 43 percent were genetically differentiated among populations, and 14 percent showed nonadaptive plasticity (Palacio-López et al. 2015). The answers to this question were enumerated by DeWitt et al. (1998) in their list of costs and limitations of plasticity. A cost is a factor that results in a decrease in the fitness of an individual even when a trait's phenotype matches the optimum value; a limitation is a factor that prevents an indi-vidual from matching that optimum.

In my theory, costs are embodied in the fifth proposition that nonop-timal plasticity may result from maintenance, production, or information-acquisition costs of plasticity. Maintenance and production costs are those that occur because a trait develops through a plastic pathway or because ge-notypic expression is plastic. Such costs must be distinguished from those associated with simply building a trait (Murren et al. 2015). Distinguish-ing those two types of costs can be done by comparing the fitness of indi-viduals that have the same trait values but differ in the plasticity of their genotypes (Van Tienderen 1991; Scheiner and Berrigan 1998). Currently there is very little evidence for such costs (Murren et al. 2015) except in single-celled organisms (Callahan et al. 2008) or under very stressful con-ditions (Van Buskirk and Steiner 2009). However, the likelihood of such costs might depend on the type of trait, and they may be more prevalent for traits that involve continual change or learning-like development as op-posed to those that involve one-time developmental switches (Snell-Rood 2012). Information-acquisition costs are much more likely to be impor-tant. They are of two types. First, organisms need mechanisms to sense the environment. Eyes, ears, and so forth all require energy and materi-als to build and maintain. Second, for mobile organisms the gathering of information may involve exposure to predators or adverse environmental conditions.

Limitations are embodied in the sixth and seventh propositions. The sixth proposition relates to cue reliability, which is likely to be the major reason that plasticity is less than ubiquitous. One cause of cue unreliability is that the rate of change of a trait's phenotype is slower than the rate of change of the environment. This can occur in several ways. First, a trait may take only a single fixed value during an individual's lifetime. If the envi-ronment changes after development ceases, or after the course of develop-ment is set, selection happens in an environment different from that of development. The longer the time between trait determination and selection, the lower the correlation between the two environments. An extreme example is transgenerational plasticity in which the phenotype of an individual's state determines the phenotype of its offspring or grandoffspring (e.g., Magiafoglou and Hoffmann 2003; Galloway 2005; Lacey and Herr 2005; Kuijper and Hoyle 2015). Reduced cue reliability can also occur if the phenotype is determined by multiple environmental factors, which can result in apparent maladaptive plasticity when only one factor is actually measured (Chevin and Lande 2015).

Second, a trait's phenotype may be labile, but if the change in the trait's phenotype takes time, again a lag is created between the environment when the developmental trajectory is determined and when selection occurs. Learning, Baldwin's paradigmatic plastic trait, is an example of such delayed phenotypic change. For the purposes of modeling plasticity evolution, such delays can act similarly to having a single, fixed trait value if selection occurs in a single bout, but might involve more complex models if selection occurs multiple times over a life cycle (Lande 2014).

Third, cue unreliability can occur even if the environment is constant in time, if it varies in space and the organism moves between the time that trait values are determined and the time that selection occurs. A key characteristic of the sixth proposition is that limitations are set by the nature of the organism's external environment, the environment's spatial and temporal pattern of variation, and how that pattern of variation interacts with organismal development and movement.

The seventh proposition encompasses a different set of limitations that are determined by internal factors of developmental or physiological responses (Snell-Rood et al. 2010). Even if an organism can change its phenotype instantly, it may still have plasticity limitations, boundaries on the range of phenotypes that an individual can take. For fixed traits, it may be that the range of phenotypes expressed by a plastic genotype is less than the total range of phenotypes that various nonplastic genotypes can express. For example, we performed a selection experiment with *Drosophila melanogaster* on the plasticity of thorax length in response to temperature (Scheiner and Lyman 1991). We were unable to select for a genotype with a reaction norm that had a shorter thorax at lower temperatures; the minimal reaction norm that could evolve had a slope of zero. However, we could select for genotypes that were either small at low temperatures or large at high temperatures. The limit was not on the possible phenotypes, just on the phenotypes that could be produced in a plastic manner. Even nonfixed traits will have limits. For example, the heartbeat of a chicken can never be as fast as that of a hummingbird, even if a greatly speeded up heartbeat would be advantageous under some conditions. Such developmental limitations are likely an important constraint on adaptive plasticity. Within the context of theory, all such developmental limitations are contained within the principles of the theory of organisms (Zamer and Scheiner 2014).

Tonsor et al. (2013) found evidence for a correlation between trait plasticity and stochastic trait variation for various morphological, physiological, and life history traits of the plant *Arabidopsis thaliana*. This correlation might be due to a pleiotropic effect of trait plasticity weakening trait canalization. If there is very strong stabilizing selection within environments, such pleiotropy might limit plasticity evolution even if plasticity would be adaptive among environments. The Tonsor et al. study is one of the few to demonstrate such a correlation. Because technical aspects of demonstrating such a correlation are daunting, this pleiotropic effect may be a more important limitation than we know.

Models

Dozens of models of plasticity evolution have been and continue to be published. In this section I consider various issues concerning model construction and outcomes. For the most part I do not consider specific models or examine model details. Rather, I provide an overview of plasticity models and how they may inform other aspects of evolution.

Classes of Models

Plasticity models can be categorized in various ways. A previous review (Berrigan and Scheiner 2004) divided models based on the types of genetics assumed—optimality (i.e., no genetics), quantitative, genic (i.e., individual loci)—and patterns of environmental heterogeneity (discrete versus continuous, spatial versus temporal). In this chapter, I consider a different aspect of the models, their focus. On that basis, we can define two classes. The first class focuses on the evolution of phenotypic plasticity and asks the question "When is trait plasticity favored over other forms of adaptation such as genetic differentiation, a jack-of-all-trades phenotype, or developmental instability?" The other class of models focuses on some other process or endpoint that may be affected by trait plasticity such as adaptation to a new niche (e.g., Waddington 1942), speciation (e.g., Schlichting 2004), or the evolution of novel phenotypes (e.g., West-Eberhard 2003).

On the one hand, all such models could be considered to exist within the domain of a single constitutive theory. However, I propose that they do not. Understanding why illustrates the approach to theory in this book. Theories are forms of explanation. Constitutive theories are often most fruitful when they focus on one or a few phenomena in need of explanation. Thus, the domain of the theory of plasticity evolution presented here is how natural selection shapes trait plasticity. Plasticity per se is the focus. The second class of models resides at the intersection of the theory of plasticity evolution and other constitutive theories.

One method for deciding in which theory domain a particular model belongs is by determining the rules by which the model was built. A constitutive theory can be thought of as a rule set for model building (Mindell and Scheiner, chap. 1). If a given model includes aspects that cannot be related in some fashion to its purported constitutive theory, either it belongs in the domain of a different theory or it is a portmanteau model that draws on multiple constitutive theories. An example of such a portmanteau model is that of Scheiner et al. (2015). That model is an individual-based, computer simulation built to answer the complementary questions: How does adaptation by plasticity versus genetic differentiation change when the environment includes another, coevolving species? How is the process of coevolution altered by the presence of phenotypic plasticity? The first question resides firmly within the domain of the theory of plasticity evolution, while the second question is in the domain of the theory of coevolution. Recognizing the hybrid nature of a given model creates a larger domain within which to interpret results and points to additional parameters or parts of parameter space that remain unexplored.

The Cost of Plasticity

A maintenance or production cost of plasticity is often included in models (e.g., Sultan and Spencer 2002; Fischer et al. 2014; Lande 2014; Botero et al. 2015). The role of such costs is typically to constrain the evolution of plasticity so that other modes of adaptation are attained within some parts of the parameter space. I presume that costs were included because the modeler recognized that adaptive plasticity is not ubiquitous and needed some way to get model behavior that somewhat mirrored that reality. However, given the lack of evidence for such costs (see above), such inclusion is highly speculative. Parameter values for such costs are generally unconstrained by data. I challenge others to consider how their models can lead to less than ubiquitous plasticity without invoking such costs.

More empirical information on costs is needed. We might expect that maintenance and production costs would generally be small. If selection otherwise favors trait plasticity, there should be selection to minimize costs. Perhaps the problem is that we are looking for costs in the wrong places. We need to look in places where we do not expect to have adaptive plasticity. Costs might be found if we compared traits among populations of the same species that differ in their environments such that plasticity is favored in some populations and not others. We could also compare traits within individuals that differ in selection on trait plasticity. This latter comparison requires that we know enough about the genetic and developmental bases for the traits to know that they are completely independent of each other.

The Genetic Architecture of Plasticity

Many models contain hidden assumptions. One role of theory is to make such assumptions explicit. For example, optimality models have no explicit gene flow. That does not make such models wrong, but it means that the model applies to only a specific demic structure. Any tests of such models or reference to such models as explanations of a given empirical pattern need to consider whether the model is relevant to that situation.

Quantitative genetic models do not allow for examining effects of genetic architecture beyond those that are contained in quantitative genetic parameters (i.e., genetic variances and covariances). Analytic genic models can include other types of genetic effects (e.g., dominance, epistasis, linkage) but have their own limitations. Because of mathematical tractability, such models typically assume that traits or trait plasticity is controlled by just one or two loci. However, the number of loci matters for the evolution of plasticity. Using an individual-based simulation, Scheiner and Holt (2012) serendipitously discovered that the number of loci whose expression is independent of the environment (i.e., are nonplastic) affected selection on alleles whose expression is environmentally dependent (i.e., are plastic) and, thus, the tendency for plasticity to be favored by selection. Increasing the number of nonplastic loci acted to decrease the among-generation cue reliability of the (genetic) environment. In hindsight, it was obvious that the genetic environment can be just as important as the external environment for selection. Only an individual-based, genic simulation model could easily discover the effect of this type of environmental heterogeneity (e.g., Leimar et al. 2006).

The genetic basis of plasticity can be modeled in two ways: genes that are expressed in some environments and not others (conditional expression), or genes that are always expressed but differently in different environments (expression modulation). These different genetic bases can result in different predictions about the evolution for plasticity. For example, Van Dyken and Wade (2010) showed that loci that are conditionally expressed are subject to the accumulation of deleterious mutations, which they termed "plasticity load."

Model Synthesis

The message of the preceding section is that a variety of modeling approaches are needed when exploring theories. Models contain different assumptions or allow exploration of different parts of parameter space. If different models have the same qualitative behaviors, we have more confidence that the predictions are general. A given empirical system aligns in different ways with each model, so having more than one model to compare against is a strength. Such comparisons of model behavior are useful when the models are of very different types (e.g., analytic optimality versus genic simulation) and can be thought of as a form of model meta-analysis.

Comparisons among models are further facilitated by formal attempts at synthesis. Consider two historical examples. The first comes from the origins of the modern synthesis when Fisher (1918) showed that quantitative genetic models were compatible with Mendelian genetic models, thereby resolving a decades-long debate (Provine 1971). The second is more recent and involves debates over plasticity evolution. De Jong (1995) showed that the two types of quantitative genetic models, one focusing on reaction norm parameters and one focusing on cross-environment covariances, were simply two forms of the same phenomenon, thereby helping to clarify a somewhat contentious debate (Via et al. 1995). What sometimes appear to be incompatible approaches or results can be harmonized by constructing a more inclusive model or by recognizing that different models are contained within the same theoretical domain. My hope is that the constitutive theory presented here will facilitate such syntheses of models of plasticity evolution.

The Baldwin Effect and Genetic Assimilation

An important role of formal models is demonstrating whether verbal models are logically consistent or lead to the ascribed behaviors. An illustration of this role can be seen in the discussion of Simpson's Baldwin effect and Waddington's model of genetic assimilation. The Simpson/Waddington model is a three-step process: (1) a trait is under stabilizing selection in a homogeneous environment; (2) a stepwise change in the environment occurs that both shifts the optimal trait value and alters trait expression owing to plasticity; (3) nonheritable, plastic trait expression is replaced by fixed genetic factors. Waddington's theory differs from Simpson's in one crucial way: he postulated that selection will replace plasticity with a fixed response because nonplastic development will result in a more accurate phenotype (Waddington 1942).

Simpson and Waddington both make a Genetics 101 error, confusing heritability with whether a trait is inheritable. All phenotypes have a genetic basis of some sort, i.e., they are all inheritable. However, not all variation in phenotypes among individuals is due to genetic variation, i.e., some trait variation has a heritability of zero. So the plasticity required for step 2 must have a genetic basis (theory proposition 3), but there may simply be no genetic variation for that genetic basis. The existence of that plasticity in a population can be due to one of two explanations. First, trait plasticity could be the result of previous selection. However, the Simpson/ Waddington model assumes that prior to the shift in the environment, it was homogeneous and trait selection was stabilizing, which should select against trait plasticity. Second, the plasticity may exist in some sort of mutation-selection balance. For this explanation to be sufficient, mutation has to maintain sufficient standing variation for plasticity to create a sufficient plastic response to the change in the environment and yet allow that plasticity to erode away after the shift. A formal model of the Simpson/ Waddington theory (Scheiner et al. 2017) found that such erosion either did not occur, took much longer than the environment would likely remain stable, or required costs of or limitations to plasticity that would have prevented its presence in the original population.

The Evolution of Phenotypic Novelty

West-Eberhard (2003, 2005) presented her theory of phenotypic accommodation as an alternative to both Baldwin and Waddington. As discussed above, a correct reading of Baldwin shows that Waddington's ideas are a subset of Baldwin's; similarly West-Eberhard's theory can also be interpreted as another restating of Baldwin's (Crispo 2007). West-Eberhard expands on those ideas in one crucial respect. Both Baldwin and Waddington considered just a shift in trait values. West-Eberhard considered the additional possibility that trait plasticity can act as precursor to phenotypic novelty, completely new forms of trait expression. Currently lacking is a formal model of that process, so we do not know under what conditions it would occur. A starting point for such a model is the theory of phenotypic novelty of Scheiner (2009) as distilled from Gilbert and Epel (2009). An important limitation of West-Eberhard's ideas is explaining why the plasticity exists in the first place. She assumes that potentially adaptive plasticity is just inherent in developmental processes. Perhaps the theory of the evolution of development (Love, chap. 8) can provide this missing piece.

Speciation

Phenotypic plasticity comes with that two-edged sword noted by Wright (1931), that plasticity can both promote and inhibit differentiation. It promotes differentiation by permitting individuals to survive or reproduce in novel conditions, and inhibits differentiation by permitting individuals with the same genotype to survive or reproduce over a range of conditions. There is a subtle difference between those two conditions, whether the environmental variation is novel—or at least very rare—or is a (semi)regular feature. In general, models that focus on plasticity as a mode of adaptation assume that the environment remains heterogeneous in space or time; otherwise there is no selection to maintain plasticity (theory proposition 1). In contrast, many models that focus on the role of plasticity in other evolutionary processes assume that plasticity will be transitory or nonevolving (e.g., Lande 2009; Chevin and Lande 2010; Reed et al. 2010).

Consider the potential role of plasticity in speciation. Speciation requires that populations evolve to become reproductively isolated (Edwards et al., chap. 15). If speciation is being driven by divergent selection, plasticity will enhance speciation only if the Simpson/Waddington process occurs such that plasticity is replaced by genetic differentiation (Schlichting 2004). Otherwise, the original species will simply expand its ecological range. I know of no formal model of speciation that includes plasticity. The paper by Thibert-Plante and Hendry (2011), despite its title, does not include the evolution of reproductive isolation and confirms the general conclusion that plasticity will tend to inhibit genetic differentiation. On the other hand, the optimality model of Nonaka et al. (2015) shows how plasticity can enhance genetic differentiation in sympatry, if habitat choice is linked to assortative mating.

Hierarchical Selection on Plasticity

Knowledge gained from other constitutive theories can also enhance our understanding of plasticity evolution. For example, hierarchical selection theory (Goodnight, chap. 10) can help resolve the twenty-year-old debate about whether plasticity is ever the direct object of selection, or whether its evolution is always a by-product of selection on traits in alternative environments (Scheiner 1993b; Schlichting and Pigliucci 1993; Via 1993). At the time we simply agreed to disagree (Via et al. 1995). Of course, if traits are labile, then it is the summed effect on fitness due to the variety of trait values over an individual's lifetime that is the object of selection and clearly plasticity can be the direct object of selection. Baldwin's theory about selection on learning is the essence of this idea. The controversy was about fixed traits. In that case, plasticity is a trait not of an individual but of a lineage, either siblings or other relatives spread over different environments in space, or parents, their offspring, and their grandoffspring spread over different environments in time. Selection is then occurring at the level of the lineage, a different hierarchical level from selection that is occurring upon individuals. In that instance, there is both direct selection on plasticity at the level of the lineage and indirect selection at the level of individuals. Both sides in the debate were correct.

Status of the Theory

The theory of plasticity evolution can be categorized as mature. An extensive body of models has developed over the past thirty years. Those models can all be derived from its constitutive theory. Its propositions are well established. Despite the maturation of the theory and the large number of models, there is a surprising disconnect between theory and data (Hendry 2016). For the constitutive theory, empirical data are still needed to establish the relative importance of the various types of costs and limitations. More telling is that very few tests have ever been done of the models, and nearly all of those tests were done in a confirmatory mode in which data are shown to be consistent with a singular expected pattern. This lack of model testing is not unique to this constitutive theory (see many other chapters in this book) but may be particularly acute for plasticity models.

This lack is partially due to the difficulty of testing plasticity models (Wund 2012). Plasticity is a property of a genotype or lineage. To express the full range of phenotypes of a genotype requires exposing an individual to multiple environments (for labile traits), or raising clones or siblings in different environments (for fixed traits). Doing so for sample sizes sufficient to measure genetic variances and covariances can quickly bloom out of control. Add to that the need to sample from multiple metapopulations or species that experience different patterns of environmental heterogeneity and the logistics become enormous. I know of only one study to take up that challenge—the evolution of plasticity of life history traits in twenty-three species of frogs (Relyea et al. 2018).

Although direct tests of the theory are rare, there has been extensive empirical study of the evolution of phenotypic plasticity. In a recent review, Hendry (2016) posed eight questions concerning the evolution of plasticity (table 13.2). The answers, especially that of question 3, are broadly consistent with theoretical expectations. Several of the answers, however, were vague, and Hendry indicated that more precise questions need to be posed. Linking those questions to the theory of plasticity evolution would help that precision.

In a recent paper on the role of plasticity in coevolution (Scheiner et al. 2015), we suggested one solution to this logistical problem: compare trait

Table 13.2. Some Key Questions and Answers about Empirical Evidence concerning the Evolution of Phenotypic Plasticity (Hendry 2016)

Question	Answer
To what extent is plasticity adaptive?	Plasticity is sometimes adaptive, sometimes maladaptive, and sometimes neutral.
To what extent is plasticity costly or limited?	Plasticity must have costs and limits, but those contracts are highly variable, often weak, and hard to detect.
What environmental and organismal characteristics favor the evolution of plasticity?	Multiple lines of evidence support the expectation of greater trait plasticity in more variable environments, when environments cues are more reliable, and when costs are lower.
To what extent does plasticity aid colonization and responses to environmental change?	Plasticity sometimes aids colonization of new environments and responses to in situ environmental change. However, plasticity responses aren't always necessary or sufficient in these contexts.
Does plasticity constrain genetic evolution?	Plasticity will sometimes promote and sometimes constrain genetic evolution.
Does plasticity help or hinder ecological speciation?	Plasticity will sometimes help and sometimes hinder ecological speciation.
How fast does plasticity evolve?	Plasticity can show considerable evolution change on contemporary time scale, although the rates of this evolution are highly variable.
How might plasticity have community/ecosystem effects?	Plasticity likely has considerable influence on ecological dynamics at the community and ecosystem levels, with foraging traits, biological stoichiometry, and chemical production by plants being particularly promising candidates.

plasticities within the same individual. Our model predicted different amounts of plasticity for traits responsible for predatory interactions and those responsible for mutualistic interactions. A comparison of trait plasticities within a single individual has the virtue of using a single environmental effector and traits that have been subject to a single evolutionary history.

Tests of plasticity theory can also be done through the use of meta-analysis. A meta-analysis by Murren et al. (2014) showed that the evolution of reaction norm slope and curvature was common. That paper was not framed as a test of models; however, we were able to use those results to find support for some of the predictions of our coevolution model (Scheiner et al. 2015). A meta-analysis by Palacio-López et al. (2015) found that adaptive plasticity was less common than often assumed, and that maladaptive plasticity was not rare. Again, that paper was not framed as a test of any particular model, although the results show that the types of seemingly maladaptive plasticity generated by some models (Scheiner 2013b; Chevin and Lande 2015; Scheiner et al. 2015) may be worth exploring in more detail. At the very least, such meta-analyses help set boundaries on model parameters.

Such meta-analyses also can point to overarching questions that can be answered by a combination of modeling and data. Given that adaptive plasticity is less common than local adaptation (Palacio-López et al. 2015), can we come up with general rules as to why? Is it due to external factors (e.g., cue reliability) or internal factors (e.g., developmental constraints, costs)? Models address this question by determining what portion of their parameter space leads to either adaptive plasticity or local adaptation. Data then tell us how often those alternative conditions are met.

For such tests to occur, empiricists need to pay attention to models when designing their studies. Models provide guides for empirical work by indicating what types of data need to be collected. For example, what is the spatial grain of environmental heterogeneity relative to the grain of gene flow? What is the temporal rate and pattern of environmental change relative to the timing of developmental events? When during development are phenotypes determined?

The theory of the evolution of phenotypic plasticity is an interesting case of a mature theory where a related empirical literature is very large, yet where those theory and empirical realms interact hardly at all. My hope with this chapter is to substantially increase that interaction. A formal constitutive theory provides a framework for unifying a disparate array of models so that key theoretical questions can emerge that need empirical answers. Conversely, empirical data can be synthesized to constrain theoretical parameters and support or refute general qualitative patterns that emerge from the models. Advancing evolutionary theory requires such two-way engagement.

Acknowledgments

I thank Carl Schlichting, Luis-Miguel Chevin, and several anonymous reviewers for their cogent comments. This chapter is based on work done while serving at the US National Science Foundation. The views expressed in this chapter do not necessarily reflect those of the National Science Foundation or the United States Government.

FOURTEEN

The Evolution of Sex

MARIA E. ORIVE

The evolution of sex has arguably produced one of the most diverse and expansive bodies of theory within evolutionary biology, leading both to extensive verbal arguments and to a large quantity of mathematical models. It nevertheless remains an active area of theoretical and experimental research, implying that not all of the questions raised by the presence of sex across the tree of life have been answered. It is important, however, when discussing the evolution of sex to carefully delineate what is meant by "sex," since biologists often use the term to describe at least four different biological phenomena: (1) the existence of separate sexes (or dioecy); (2) anisogamy, or the fusion of two dissimilar gametes; (3) meiosis, a specialized form of cell division leading to the production of gametes, which may or may not include genetic recombination; and, finally, (4) genetic recombination itself. The vast majority of the theoretical considerations for the evolution of sex have focused on this final definition, and thus it is the evolution of recombination that I have chosen to focus on here.

We should also take care to differentiate between sex, which changes the genetic state of cells or individuals, and reproduction, which produces ecologically distinct individuals that are largely independent of one another (G. Bell 1982). There can be sex in the absence of reproduction—consider bacterial conjugation, where there is transfer of genetic material between two cells, but no increase in cell number. There likewise can be reproduction (an increase in the number of individual living organisms) in the absence of sex, if we consider all of the various forms of clonal reproduction that do not change the genetic makeup of clonal offspring (e.g., G. Bell 1982; Hughes 1989; Klimes et al. 1997). Finally, it is important to clarify that I am using the term "recombination" in the general sense of the bringing together of genes inherited from different parents; see Maynard Smith (1988a) for a discussion of the usage of "recombination" versus "crossing over." I am also focusing on lineages that recombine their entire genomes regularly, which necessarily excludes some forms of recombination, such as horizontal or lateral gene transfer and gene conversion. As such, the focus is on recombination of genomes within species (for a discussion of the concept of species as it relates to genetic interconnectivity, see Nathan and Cracraft, chap. 6; for a discussion of the role of lateral gene transfer in genealogical patterns of lineage diversification, see Kearney, chap. 7). Frank and Fox (chap. 9) argue that the overall process of evolutionary change can be partitioned into the processes of natural selection, where information about the environment is accumulated, and transmission of that information. The theory of the origin and maintenance of sexual recombination seeks to explain how natural selection shapes this transmission itself, and thus lies at the intersection of these two aspects of evolutionary change for sexual organisms.

Mechanistic Theories

The early evolutionary origins of recombination likely involve the repair of DNA damage, an idea tracing back at least as far as E. C. Dougherty (1955), wherein he considered two different aspects of sex—the transfer of DNA molecules between two or more cellular compartments, and the recombination of DNA molecules within a cellular compartment. This second step would be an advantage in the case of DNA damage and the evolutionary cause for the origin of DNA recombination; genetic transfer is more likely to have evolved secondarily. This mechanistic theory of the evolution of recombination was called the "repair hypothesis" by Bernstein et al. (1988), who distinguished it from theories that depend on the distribution of genetic variation (Bernstein et al. 1985). It is not clear, however, why genetic exchange (crossing over) of the DNA molecule beyond the repair site is necessitated by DNA repair. As an example, Maynard Smith (1988a) notes that a process equivalent to double-strand repair occurs in mating-type exchange in yeast, without crossing over beyond the repair site. Further, studies of natural transformation in prokaryotes show a lack of evidence for regulation of the process by DNA damage (Redfield 1993). Nevertheless, it seems likely that DNA repair played a role in the early evolution of recombination in the ancestor of modern eukaryotes, as evidenced by the relationship between the molecular machinery of DNA repair and that of recombination and crossing over (Redfield 2001; Lieber 2010).

Other mechanistic hypotheses for the evolution of recombination include those assuming that proper segregation of chromosomes during

Table 14.1. The Theory of the Evolution and Maintenance of Sexual Recombination

Domain: The evolutionary origin and maintenance of sexual recombination.

Propositions:

1. Sex and recombination incur costs when segregation within a locus or recombination between loci breaks up genotypes of high fitness.
2. Evolutionary explanations for the evolution of sex rest on the role of recombination in breaking up and forming nonrandom genetic associations (linkage disequilibria).
3. Negative linkage disequilibrium always favors recombination via its long-term effect on population fitness.
4. The linkage disequilibria underlying the long-term advantage of recombination can be formed via random, stochastic processes (drift processes) or via deterministic, nonrandom processes (epistasis, population structure).
5. Two differing evolutionary forces act on modifiers of recombination: short-term selection on individuals to have the highest mean offspring fitness and long-term selection to increase the genetic variation in population fitness.
6. Environmental or genomic heterogeneity in the form of selection fluctuating over time or space, the interaction between genetic drift and linkage, or structure imposed by epistasis and genomic architecture, expands the conditions under which increased recombination is favored.
7. Genomic processes such as gene conversion, horizontal gene transfer, and phenotypic assortment provide alternatives to sexual recombination for breaking up negative disequilibrium and allowing an escape from long-term population fitness decline in asexual lineages.

meiosis is dependent on the formation of chiasmata, resulting in recombination as a by-product of a system for the proper sorting of homologous chromosomes. However, efficient chromosomal disjunction occurs without the formation of chiasmata in the heterogametic sex of organisms with both male heterogamy (such as *Drosophila*) and female heterogamy (*Lepidoptera*) (Burt et al. 1991). Thus, this mechanistic explanation does not adequately explain the persistence of recombination for many taxa. The vast majority of the theories developed to explain the persistence of recombination have focused instead on the costs of sex and on the effects of recombination on genetic variation.

Costs of Sex

No discussion of the evolution of sex can be complete without clarification of what is meant by the "cost of sex" and how evolutionary theory has addressed this idea. That sexual reproduction involves a real cost is implied by the existence of life cycles where organisms (such as aphids and rotifers) reproduce asexually to take advantage of abundant resources and switch to sexual reproduction only when resources begin to disappear (for a review of such cyclical parthenogenesis, see De Meester et al. 2004).

Theory has focused on two different costs associated with sexual reproduction: the cost of producing males, and the cost of breaking up genotypes. The cost of producing males (or the cost of male function in hermaphrodites) depends on the existence of anisogamy. The key assumption is that two forms of gametes exist, and the number of more costly gametes (the eggs) limits the number of new individuals that can be produced each generation. Consider a population of sexually reproducing individuals of size N, half of which are females ($N/2$) and half of which are males (Maynard Smith 1978a). If the number of eggs that can be produced by a single female is k, and the probability that an egg will contribute to the next generation is s, this population will produce $Nsk/2$ new individuals in the next generation. Contrast this with an asexual (parthenogenetic) population that is similar in every way (including the number of eggs that can be produced and their probability of survival). Here all N individuals are parthenogenetic females, and thus the number of new individuals is Nsk. This contrast between $N/2$ versus N individuals producing the limiting gamete type is then the so-called "twofold cost of sex" (or more accurately, the twofold cost of anisogamy); the asexual population will, all things being equal, increase at a rate double that of the sexual population. The same argument holds for the comparison between a hermaphroditic or monecious sexual population and an asexual population, if we assume that half of the total reproductive contribution is allocated to eggs and the other half to sperm for the sexual population, while the full reproductive contribution is allocated to parthenogenetic eggs for the asexual population. So what keeps sexual species, once they have evolved anisogamy, from evolving parthenogenesis? Most answers have focused on contrasting the advantage of no longer producing males with advantages reaped from continuing recombination. What is clear, however, is that discussions of the cost of males are best placed in the context of the maintenance of sexual reproduction once it has arisen, since anisogamy is believed to have arisen from isogamous sexual reproduction, after the early evolution of sexual reproduction (Parker et al. 1972; Randerson and Hurst 2001).

A more subtle cost lies in the very aspect of sexual reproduction that is often seen as its chief advantage—the breaking up by recombination of parental genotypes. Any adult organism that survives to a reproductive stage has passed through the selective sieve of survivorship and is more likely to have an advantageous genotype than the average newly formed zygote. Why should it then subject that genotype to dissolution? Imagine a heterozygous genotype at a single locus with high fitness. Sexual reproduction will lead, on average, to a decrease in heterozygosity in its offspring, and fitness will decrease, a process termed "segregation load" (G. Bell 1982). Likewise, if we consider genotypes across loci, recombination will break up high-fitness multilocus genotypes. This cost of sexual reproduction underlies much of the theory that will be addressed here (proposition 1, table 14.1). The opposite side of this same coin, however, lies at the heart of many of the most well-developed theories of the evolution of sex—what recombination breaks apart, it also brings together. And it is the bringing together of either beneficial or deleterious mutations (the dissolution of negative linkage disequilibrium, in the language of population genetics) that will be the focus of this chapter. Evolutionary explanations for the evolution of sex and recombination rest on the role of recombination in breaking up and forming linkage disequilibrium (proposition 2, table 14.1). That this is the key to the "queen of problems in evolutionary biology" (G. Bell 1982) is widely agreed upon. But where researchers have disagreed is in the specifics of how best to model the interactions of mutation, selection, and finite population size in shaping the evolution of sex.

Optimality Theory versus Modifier Theory

Two contrasting theoretical approaches have been applied to the evolution of sex; here I will discuss the scope of each approach, along with important examples of each. The first is optimality theory, which is based on optimization of a specific criterion such as individual or population mean fitness, times to fixation of beneficial mutations, rates of evolution, or genetic loads. It is assumed that evolution proceeds in the direction that optimizes the criterion with respect to the evolutionary parameters of interest. Such models do not assume that organisms can or will be optimal. They are also agnostic regarding the ability of natural selection to optimize in any particular case, which will depend on the details of the population under natural selection (Parker and Maynard Smith 1990).

For the evolution of recombination, optimality arguments (also called equilibrium models, Otto and Lenormand 2002) define some central criterion to be optimized and consider the effects of changing recombination rate on this criterion, often employing (most often implicitly) group selection arguments (Goodnight, chap. 10). The success of increased or decreased rates of recombination is inferred from comparisons of the value of the optimization criterion, which is often a characteristic of a group. Groups with a more optimal value of the criterion are assumed to outcompete and replace other groups, pointing to the direction that recombination will evolve. These group-selection-type of arguments go back to Weismann

(1891), who argued that sex was selected for its effect on the genetic structure of populations or species, rather than individuals, acting to create the
"material upon which natural selection may work" (G. Bell 1982). These
models of the evolution of sex focus on how recombination redistributes
both beneficial mutations (the Fisher-Muller model) and deleterious mutations (Muller's ratchet) and how that redistribution by recombination in
turn sets the value of the optimality criterion.

In contrast to models based in optimality theory, a second large family
of models asks when and under what conditions sex and recombination
can evolve and spread. Many of these models are based on the idea of a
modifier locus that alters the level of recombination or segregation and
that interacts with loci determining fitness. These models thus constitute
a modifier theory of the evolution of sex and recombination (Feldman
et al. 1996). In contrast to optimality models, they are explicitly models
of individual selection that focus on defining the specific conditions under
which alleles acting to regulate the rate of recombination can increase in
frequency, and they are dynamic and nonequilibrial. In a single large population with a constant environment, modifiers of recombination (haploid
model) or modifiers of segregation (diploid model) can spread only under very restrictive conditions (Otto and Lenormand 2002). Both require
weak, negative genetic interactions (across loci for modifiers of recombination, across alleles at the same locus for modifiers of segregation) in a relatively limited parameter space. Only the addition of some sort of structure
(spatial, temporal, genetic architecture) permits less restrictive conditions
(proposition 6, table 14.1). Examples include structure generated by selection fluctuating over time for a single species, or for two or more interacting
species (Red Queen); structure imposed by selection that varies over space
in a system including migration; structure arising via the interacting effects
of genetic drift and linkage (Hill-Robertson effects); and structure generated by epistasis and genomic architecture.

Key Optimality Models of the Evolution of Sex

The Fisher-Muller Model

While some early arguments regarding the evolution of sex focused on the
generation of new genotypes (Weismann 1891; East 1918), it is clear that
in the absence of nonrandom associations of alleles across loci (linkage
disequilibrium), the creation of new beneficial combinations of alleles by
recombination is exactly balanced by their loss (Felsenstein 1988). It is

only in the presence of negative linkage disequilibrium (the presence of an excess of haplotypes containing both beneficial and deleterious alleles) that recombination gives a benefit. Both Fisher (1930) and Muller (1932) described a model where recombination was advantageous within a population because it brought together favorable mutations that initially arose on different genetic backgrounds (and so arose in negative linkage disequilibrium, to use modern terminology; proposition 3, table 14.1).

The Fisher-Muller model posits a population that is finite but large enough so that multiple advantageous mutations can arise during the same time period. In the absence of recombination, new mutations arising on different genetic backgrounds (in negative disequilibrium) cannot go to fixation at the same time. They necessarily compete against one another unless they arise in the same clonal lineage. The advantage to the population is in the rate of fixation of beneficial mutations, and so a "rate of evolution" optimality argument lies at the heart of this model. Finite population size is an important aspect pointed out by Felsenstein (1974, 1988). In an infinite population, two new mutations would each arise at a rate governed by the mutation rate, μ, and a proportion μ_2 of new individuals (in a haploid population) or new gametes (in diploids) would be double mutants. The population would then be in linkage equilibrium, and recombination would give no advantage to such a population over a population without recombination.

Related to the Fisher-Muller model is the Hill-Robertson effect or Hill-Robertson interference (Hill and Robertson 1966; Felsenstein 1974), which describes an interaction between genetic drift and selection caused by genetic linkage. In a finite population, as described above for complete linkage (for asexual lineages), linked genes interfere with each other's ability to fix. With free recombination, a favorable mutation that arises is "seen" by selection against many different genetic backgrounds. Good and poor backgrounds on average tend to cancel each other out, and thus the average fitness of individuals carrying the mutation will depend only on the selective advantage of the particular mutation. In the case of linkage, the chance association between the favorable mutation and the genetic background will tend to persist. The average fitness of individuals carrying the mutation will now also depend on the genetic background within which it arose. Beneficial mutations arising on good backgrounds will increase in frequency more than equally beneficial mutations arising on poor backgrounds. This leads to greater variation in fitness for individuals carrying beneficial mutations, and thus increases the random variation in the frequency of such mutations from generation to generation. In effect, linkage increases the amount of genetic drift accompanying selection and causes a reduction in the effective population size for the locus where the beneficial mutation arose. The chance that the favorable mutation will fix will thus be less, on average, under linkage than under free recombination. This is true even in the absence of any epistatic effects between loci; Hill and Robertson (1966) assumed additive fitness effects. Interference between beneficial mutations can reduce the rate of accumulation of beneficial mutations even in fully sexual populations, by decreasing the average probability that an individual new mutation will fix by a proportion that depends on the density of adaptive sweeps (Weissman and Barton 2012).

Muller's Ratchet

The modern interpretation of the model proposed by Muller (1964) focuses on deleterious mutations, but in truth Muller's ratchet is a variant of the earlier Fisher-Muller model (Felsenstein 1988). We imagine an asexual lineage with some number of loci where deleterious mutations may arise. In a population of finite size, these may occasionally fix owing to drift. Once a deleterious mutation is fixed, the number of deleterious mutations in the most fit haplotype (the haplotype with the lowest number of deleterious mutations) increases from n to $n + 1$. Correspondingly, the population mean fitness decreases, and the "ratchet" clicks forward. In the absence of back-mutation and recombination, there is no way to regain the more fit, n mutation haplotype, and so as is true of a physical ratchet, there is no way to move backward to higher fitness. How does recombination allow a population to escape the ratchet? Recombination between haplotypes containing deleterious mutations at two different loci leads to a haplotype containing both mutations and another free of both mutations. The haplotype with one fewer deleterious mutation can thus be reconstituted by recombination.

Felsenstein (1988) pointed out that Muller's ratchet is a variant of the Fisher-Muller model, if we focus on the unmutated, favorable variant at each locus. The difference between this formulation and the original Fisher-Muller model is that in considering Muller's ratchet, we are assuming that the beneficial alleles are at high frequency because the beneficial alleles are the preexisting wildtype alleles, whereas in the Fisher-Muller model the beneficial alleles are new mutations and thus start at low frequency.

The optimality criterion here is the expected number of individuals in the optimal, zero mutation class. A higher expected number for this class means a higher mean fitness for the population because it is assumed that fitness drops with the number of mutations; thus, this model focuses on genetic load. The relative fitness of an individual with k mutations is $(1 - s)^k$, where s is the selective disadvantage per mutation. For a finite population of size N undergoing deleterious mutations at a rate of U per genome per generation, the expected number of individuals in the zero class is given by Haigh (1978):

$$(14.1) \qquad\qquad E(n_0) = Ne^{-U/s}.$$

If $E(n_0)$ is large, Muller's ratchet moves slowly if at all, and deleterious mutations accumulate independently of one another. If $E(n_0)$ is small, the ratchet rapidly moves the population distribution along k, increasing the number of deleterious mutations in the average individual. The process is most important for slightly deleterious mutations and small populations, where drift can drive the ratchet. This process, all other things being equal, sets an upper limit on the genome size of strictly asexual lineages; if U increases as genome size increases, then $E(n_0)$ necessarily decreases as genome size increases, making the ratchet more important for larger genomes (Maynard Smith 1988a). This has been an argument for why lineages of asexual eukaryotes (with larger genome sizes) should not persist over long evolutionary timescales.

The Red Queen and Her Court

In addition to linkage disequilibrium generated stochastically, there are a large number of models of the evolution of sex where the linkage disequilibrium broken up by recombination is generated deterministically. In these models, epistatic interactions can be generated by selection fluctuating over time or space. Perhaps the most well-known members of this family of models are based on the Red Queen model (Van Valen 1973), so-named for a famous passage in Lewis Carroll's *Through the Looking-Glass* (1871), "it takes all the running you can do, to keep in the same place," an allusion to the need to constantly evolve in order to maintain current fitness in a changing environment. In the presence of environmental fluctuations (either abiotic or biotic), the optimal genotype changes from generation to generation (Jaenike 1978a; Hamilton 1980; G. Bell 1982). As pointed out by Maynard Smith (1988b, 1988a), this type of environmental fluctuation over time leads to selection favoring changes in the sign of linkage disequilibrium. For example, if the combination A_1B_1 is favored in one generation, linkage disequilibrium will build up owing to an over-representation of that allelic combination. When this combination is no longer favored, recombination breaking up linkage disequilibrium will be favored. Selection therefore leads both to cyclical changes in multilocus genotype frequencies and to changing linkage disequilibrium.

Another family of models that also depends on selection to generate linkage disequilibrium includes those where the important environmental variation is spatial rather than temporal. These often focus on the benefit of recombination in the context of local competition among relatives, contrasting the production of genetically more homogeneous offspring with genetically heterogeneous offspring (Lenormand and Otto 2000). In one such model, termed the "tangled bank" (G. Bell 1982), genetic diversity translates into the ability to occupy ecologically diverse environments. In a spatially complex environment with heterogeneous resources, sexually produced offspring can exploit the complex environment more completely than can asexual offspring, who must compete with one another for a limited portion of the resources; the benefit of recombination here rests in the decrease in within-brood competition (Ghiselin 1974a; G. Bell 1982; Koella 1988). Another type of model that focuses on environmental heterogeneity is the lottery model in which genetically diverse offspring are more likely to include individuals of high fitness (G. C. Williams 1975; Maynard Smith 1976b). It is clear that the relative grain size of environmental heterogeneity versus the dispersal distance for offspring has an important effect. Production of genetically diverse offspring is also beneficial in a lottery model when local conditions change over time so that, in effect, the environment becomes heterogeneous for a spatially static distribution of genotypes (resulting in a Red Queen scenario), rather than the genotypes moving across a heterogeneous environment owing to dispersal. For both, recombination in a lottery model leads to an increased spread of genotypes over a fitness landscape (Vos 2009).

For the tangled bank, lottery, and Red Queen models, linkage disequilibria are formed via deterministic, nonrandom processes involving epistasis and population structure. They are in effect infinite-population size models, in contrast to the Fisher-Muller model where the linkage disequilibria underlying the benefit of recombination are formed via random, stochastic processes such as mutation in a finite population or drift caused by finite population size (proposition 4, table 14.1). The interacting effects of finite population size and selection are a key component to current models of the evolution of recombination, considered below.

Modifier Theories of the Evolution of Sex and Recombination

Modifier Theory in Large Populations

Two differing evolutionary forces act on modifiers of recombination: short-term selection on individuals favoring those having the highest mean off-spring fitness, and long-term selection to increase the genetic variation in population fitness (Barton 1995; Otto and Lenormand 2002) (proposition 5, table 14.1). A theory considering the fate of a modifier allele (M) that alters the recombination rate between a set of loci must include analysis of both the association between those loci (linkage disequilibrium, D) and the amount and type of fitness interaction between the loci (epistasis, ε) (Nei 1967; Feldman et al. 1980; Feldman et al. 1996; Otto and Feldman 1997). If we define epistasis as the difference between allelic effects on fitness in unison and what we expect from the individual locus effects, we see that epistasis generates linkage disequilibrium of the same sign. Positive epistasis results in a greater improvement in fitness in unison for beneficial alleles and a less severe decrease in fitness for deleterious alleles (fig. 14.1), and thus leads to an overrepresentation of the allelic combination relative to that expected from the marginal frequencies (positive linkage disequilibrium, D > 0). Negative epistasis generates negative linkage disequilibrium (D < 0) and always favors recombination via its long-term effect on population fitness (proposition 3, table 14.1) as described above for optimality models. Recombination within allelic combinations of intermediate fitness (combining both beneficial and deleterious alleles and exhibiting negative D) results in haplotypes with high fitness (where beneficial alleles have been brought together) and haplotypes of low fitness (where deleterious alleles have been brought together), increasing the genetic variation in population fitness and allowing long-term selection to act. The short-term effects of recombination are more complex, however, and depend on the signs of both D and ε and on the form of selection on the loci.

In models for the evolution of recombination via modifiers, the focus is on the change in frequency of the modifier allele, ΔpM. Under weak selection, where alleles at two loci, A and B, change fitness by small amounts sA and sB individually and by sAB when present together, the change in frequency for a modifier allele changing recombination by a small amount ∂r is given by

(14.2)
$$p_M = -\frac{\partial r p_M q_M D}{r_{MAB}} \left[\varepsilon + s_A s_B \left(\frac{1}{r_{MA}} + \frac{1}{r_{MB}} - 1 \right) \right],$$

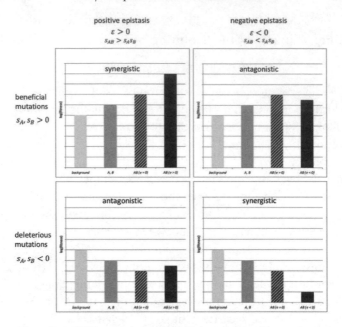

Figure 14.1. Relationship between positive vs. negative fitness epistasis, and antagonistic vs. synergistic fitness effects. Epistasis shown as deviation from multiplicative fitness ($\varepsilon = s_{AB} - s_A s_B$) with equal effects of the alleles A and B individually ($s_A = s_B$) to more easily contrast the individual effects with those of the alleles in unison (s_{AB}). Log(fitness) utilized so that multiplicative fitness effects are additive. Each panel shows: background fitness level (light gray), individual effect of alleles A and B (dark gray), joint effect of A and B in absence of epistasis ($\varepsilon = 0$, dark gray hatched), and joint effect of A and B in presence of epistasis ($\varepsilon \neq 0$, black). Note that positive epistasis implies a greater improvement in fitness in unison for beneficial alleles (synergistic fitness effects) while implying a less severe decrease in fitness for deleterious alleles (antagonistic fitness effects); negative epistasis results in the reverse relationship (smaller improvement in fitness in unison for beneficial alleles [antagonistic] and greater decrease in fitness for deleterious alleles [synergistic]).

where r_{MAB} is the rate of recombination for M, A, and B, r_{MA} is the rate for M and A, r_{MB} is the rate for M and B, and ε is the amount of epistasis, measured here as the deviation from multiplicative fitness, $\varepsilon = 2s_{AB} - s_A s_B$ (adapted from eq. A1.5e, Barton 1995). This change in allele frequency reflects both the short- and long-term effects of selection on the modifier allele. Negative D acts to make the change in the modifier frequency positive (as long as the term in the square brackets in 14.2 is positive) and is sensitive to the rate of recombination between the modifier locus and the fitness loci (r_{MAB}). The modifier allele must stay in association with the beneficial allelic combination it creates long enough for their increase in frequency to lead to an increase in frequency of M, a result of the long-term effects of selection. In contrast, the effect of short-term selection is more complex. The effect of changing the average fitness of offspring is greatest when D and ε have opposite signs, or more exactly, when $(\varepsilon + s_A s_B)D < 0$ (Otto and Lenormand 2002). Although epistasis generates disequilibrium of the same sign, other factors can generate either positive or negative disequilibrium; for example, spatial correlations in selection coefficients can generate positive disequilibrium, while random genetic drift can generative negative disequilibrium. These and other forces can influence the relative effect of epistasis on the short-term results of recombination.

In this model, recombination modifiers of small effect are favored in response to two forces: fluctuating epistasis and directional selection. Fluctuating epistasis may arise in the presence of biotic coevolution, as in Red Queen-type models, if the epistasis imposed by one species is in direct response to linkage disequilibrium arising in the other species (Nee 1989). Less restrictive are the conditions under which recombination modifiers are favored in response to directional selection: epistasis between loci must be both weak and negative (Barton 1995). Weak epistasis seems plausible under many biologically reasonable scenarios, but whether epistasis is generally negative is less clear. Theory on fitness interactions between loci has posited that negative epistasis arises from stabilizing selection on additive quantitative traits (Maynard Smith 1988a; B. Charlesworth 1993) or that negative epistasis is necessary to avoid excessive mutational load under deleterious mutations (Kondrashov 1988, but see MacCarthy and Bergman 2007). While work on artificial gene networks has suggested that negative epistasis can evolve as a consequence of sexual reproduction (Azevedo et al. 2006), it is not clear if natural gene networks have the same type of connectivity (Leclerc 2008). Empirical studies measuring epistasis have found mixed results, with some reporting $\varepsilon < 0$, some finding $\varepsilon > 0$, and some finding variable or no epistasis (summarized in de Visser and Elena 2007 and Kouyos et al. 2007, see references therein).

Modifier Theory in Small Populations: The Importance of Drift

The relative roles played by linkage disequilibrium and epistasis in promoting the spread of recombination modifiers becomes much easier to disentangle when drift acts as a stochastic force generating nonrandom associations. Directional selection acting on beneficial alleles in finite populations generates negative linkage disequilibrium (on average) because associations in positive linkage disequilibrium (either beneficial with beneficial or deleterious with deleterious alleles) are rapidly either fixed or lost when they arise by chance (Barton and Otto 2005). Combinations uniting beneficial and deleterious alleles persist the longest, causing recombination modifiers to increase in frequency. The strongest examples of this effect are seen in small populations (Otto and Barton 2001), in large populations with genetic drift imposed by spatial structure (G. Martin et al. 2006), or in populations subject to directional selection at multiple loci (Iles et al. 2003). This effect of genetic drift in conjunction with directional selection requires a high rate of beneficial sweeps acting to remove haplotypes in positive disequilibrium. In populations of small ($2N = 100$) to intermediate ($2N = 10^4$) sizes, Otto and Barton (2001) showed that this effect of linkage disequilibrium generated by random genetic drift was often stronger than the effect caused by selection for recombination generated by epistasis.

Drift is acting in the background of all populations, large and small. A truly synthetic theory for the fate of recombination modifiers needs to allow for the stochastic effects of finite population size. In discussing Hill-Robertson interference, I noted that linkage increases the amount of genetic drift accompanying selection near a selected locus, reducing the effective population size for the locus where the beneficial mutation arose (or, conversely, in the presence of purifying selection against deleterious mutations). Keightley and Otto (2006) showed that purifying selection against repeated deleterious mutations provided an advantage to modifier alleles and, what is more striking, that this advantage increased with increasing population size. The advantage arises because recombination frees the focal locus from Hill-Robertson interference, thus allowing deleterious mutations to be purged by selection. This advantage was greater than the force of epistasis in generating disequilibrium, and thus the form of epistasis (its magnitude and sign) is not critical in determining the advantage to the modifier allele. The surprising result that this stochastic effect was larger in larger populations (where genetic drift is overall weaker) is due to the fact that larger populations, all other things being equal, maintain more polymorphic loci, increasing the opportunity for Hill-Robertson interference. The maximum advantage of the recombination modifier occurs for deleterious mutations of intermediate effect, and the conditions corresponding to the largest advantage of sex are those where Muller's ratchet is expected to be strongest (Gordo and Campos 2008). This model gives a truly modern and complete treatment of the role of negative linkage disequilibrium in the evolution of recombination; both selection and genetic drift play key roles in how selection on a new mutation affects the fate of other loci and how recombination frees loci from the weight of linkage. However, the effects of indirect selection drop off quickly as the rate of sex in the population increases, implying that in the absence of other forces that enhance indirect selection such as spatial or temporal structure, the advantage provided by recombination in the face of purifying selection in finite populations may not be enough to explain the evolution of obligate sex (Roze 2014).

The rate of evolution for recombination modifiers (eq. 14.2) relies on the product of weak epistasis, disequilibria, current allele frequencies, and the incremental change in recombination, and is itself expected to be small. Despite this, experimental results in *Drosophila* have shown relatively large differences in sibling species' recombination rates (True et al. 1996). Whether these are likely to have arisen via this type of weak selection is unknown, although marked differences in the suppression of recombination near the centromere suggest that other mechanistic effects may play a larger role.

The Intersection of Recombination Theory with Genomic Architecture, Epistasis, and Fitness Landscapes

I now turn to a consideration of what recent theory and models of genomic architecture (including the form and extent of epistasis and both the global and local properties of fitness landscapes) can tell us about the generation of linkage disequilibrium and the evolution of sex. As previously discussed, empirical studies measuring epistasis have found mixed results. However, in a multilocus model considering a broad range of epistatic effects, Kouyos et al. (2006) found that epistatic interactions of a given strength could generate very different types of linkage disequilibrium. Epistatic interactions had the greatest impact when selection was weak, and so the evolution of recombination under mutation-selection balance might be driven by a small number of interactions under weak selection, rather than by the average epistasis of all interactions. It is this latter quantity that is generally measured in empirical studies, leaving open the question as to whether epistasis is a major force in generating the linkage disequilibrium that can drive the evolution of recombination.

How are epistatic interactions generated biologically and when might we expect to see the negative epistasis predicted by theory to favor recombination and sex? One ecological explanation for the generation of negative epistasis is density-dependent regulation of population size under limiting resources via truncation selection (Crow and Kimura 1979; Kondrashov 1988; de Visser and Elena 2007). If individuals carrying more than some threshold number of deleterious mutations are completely inviable while individuals that fall under the threshold survive, selection imposes an extreme form of negative epistasis. Metabolic control theory also predicts negative epistasis under some conditions (de Visser and Elena 2007). Szathmáry (1993) showed that selection for maximum flux along an enzymatic pathway caused deleterious mutations to show positive epistasis, while selection for optimum flux caused negative epistasis. If maximum flux is important with plentiful resources, but optimum flux is important when resources are scarce, negative epistasis should be observed under highly competitive (low resource) situations (de Visser and Elena 2007).

Kondrashov's (1984, 1988) mutational deterministic hypothesis posits that negative epistasis amongst deleterious mutations is necessary to avoid excessive mutational load. If the per generation genomic deleterious mutation rate is greater than one ($U > 1$), the more efficient removal of deleterious alleles in the presence of recombination leads to an advantage of sex that can be more than twofold, overcoming the twofold cost of producing males. He argued that selection acting on many deleterious mutations independently leads to a mutational load incompatible with survival (in populations of moderate size) unless there is synergistic epistasis between deleterious alleles (negative epistasis for their deleterious effects [Kondrashov 1995]). Finding large values of U may thus be indirect evidence of negative epistasis for deleterious mutations. For this to be a general explanation for the evolution and maintenance of sex and recombination, both $U > 1$ and synergistic epistasis for deleterious mutations would need to be common. In organisms where genomic mutation rates have been estimated, there is mixed support for genomic mutation rates greater than one, with evidence of U near or above one in at least some eukaryotes (de Visser and Elena 2007; Kouyos et al. 2007; Hartfield and Keightley 2012).

The mutational deterministic hypothesis focuses on the overall genomic rate of deleterious mutations and the resulting load experienced by populations. However, the distribution of mutational effects is also important. If both mildly deleterious alleles and strongly deleterious alleles occur, purging of strongly deleterious alleles by selection decreases effective population size, thereby increasing the strength of genetic drift, the rate of accumulation of mildly deleterious alleles, and the rate of Muller's ratchet (Gordo and Charlesworth 2001).

We need also to consider the interaction between sex and genetic architecture. Negative epistasis can be caused by genetic robustness—phenotypic stability or invariance in the face of repeated mutation or other perturbations (de Visser and Elena 2007). In this way, genetic robustness may itself favor recombination. This robustness, however, comes at a cost; it allows mutations to accumulate within the genome, eventually leading to a dramatic decrease in fitness. A. Gardner and Kalinka (2006) showed that intermediate levels of recombination allowed lineages to escape this cost. Recombination breaks down the association between the target gene, which is subject to deleterious mutations, and the robustness gene, decoupling the short-term benefit of robustness (increased fitness in the presence of a mutation) from the long-term cost (increased frequency of mutations at mutation-selection balance). This decoupling causes the benefit to be reaped only by the robust individuals, while the cost is paid by the entire population in the form of mutation load.

Recombination itself may cause negative epistasis, increasing the opportunity for its own evolution and persistence. Since recombination increases the variability in genetic backgrounds experienced by any particular locus from one generation to the next, it may select for greater genetic robustness in sexually reproducing organisms, which in turn generates negative epistasis. As discussed earlier, models using artificial gene networks showed that recombination increased negative epistasis by increasing robustness (Azevedo et al. 2006); however, whether natural systems are under this type of selective pressure is unclear. Rapid advances in the understanding of gene networks and the use of genomic data may answer this question.

More complex forms of epistasis are important in considering the potential advantages and disadvantages of recombination in the face of adaptive evolution. An important category of epistasis is sign epistasis, where the sign of an allele's fitness effect (whether it is beneficial or deleterious) varies across genetic backgrounds (Weinreich et al. 2005). This type of allelic interaction creates a "rugged" fitness landscape, with local minima and maxima, and constrains the possible pathways that can be taken by adaptive evolution (Crona et al. 2013). Consideration of the effects of sign epistasis and complex adaptive landscapes has led to contradictory results. Using an empirically derived fitness landscape showing sign epistasis between individually deleterious mutations, de Visser et al. (2009) used simulations of asexual and sexual populations to show a general disadvantage to recombination. They found a slight advantage of sex during early stages of adaptation, likely due to the formation of allelic combinations needed to reach local optima via the breakup of negative linkage disequilibrium, a Fisher-Muller effect. However, recombination generally prevented populations from escaping local maxima by breaking down "escape" genotypes.

In contrast to that study, which assumed uniform recombination across a nonstructured genome, Watson et al. (2011) considered a model with explicitly modular genomes. Their approach viewed genes as necessarily modular units within the larger genome, with tight linkage and high epistasis between sites within a locus, and free recombination and low or no epistasis between sites on different loci. This work borrows from ideas in evolutionary computation theory and the genetic algorithm literature, where genetic algorithms with sexual recombination outperform mutation-only algorithms owing to their ability to select on and recombine larger building blocks rather than small changes. In this optimality model, they considered not only how quickly sexual and asexual populations converged on the fittest genotype, but also whether the populations could escape local optima. Their individual-based simulations found the same "speed advantage" to recombination that is seen in the classic Fisher-Muller model, and also found that asexual populations became trapped in local optima, while the sexual populations were able to access the globally optimal genotype. By considering a fitness landscape that has different selectively accessible routes between alleles via mutation, we see that recombination frees alleles from their genetic backgrounds; in the absence of recombination, genotypes rather than alleles compete for fixation. Without recombination between loci, competition between alleles at one locus is interfered with by competition between alleles at other loci, so that clonal interference (as defined by Gerrish and Lenski 1998) not only slows the rate of evolution but also limits the net increase in fitness achievable via adaptive evolution.

Further computer simulation work that considered so-called "tunably rugged" fitness landscapes found a transitory advantage to recombination, which reverses at longer timescales when recombining populations are more likely to become trapped at local fitness peaks (Nowak et al. 2014). These studies deal with population dynamics at the limit where selection is strong compared with recombination, i.e., at the limit of tight linkage. An open question is how populations under relatively weak selection respond to these types of fitness landscapes. With relatively weak selection, recombination would be expected to play a greater role in breaking up linkage disequilibrium, allowing sexual populations to follow trajectories defined by "allele frequency space" rather than "genotype sequence space" (Watson et al. 2011), and thus escape the trap of local fitness maxima.

Finally, the time dependence of the advantage found by Nowak et al. (2014) necessitates consideration of fitness landscapes that themselves change with time. Under a changing fitness landscape, the transitory advantage of recombination can continue indefinitely, as long as the timescale of fitness landscape change is shorter than the timescale of advantage for asexual populations. A similar result was found for a model considering the fate of modifiers of recombination (amongst other evolutionary forces) under changing environments. Here, the evolutionary dynamics of recombination modifiers were shown to be sensitive to the particular details of environmental fluctuations (Carja et al. 2014). The rate of recombination evolved toward a nonzero value that decreased with increasing environmental variability, again pointing out the key importance of timescale. However, this model assumed an infinite population size, and so the effects of drift in generating negative linkage disequilibrium amongst selected loci were not considered. An obvious link exists between these findings and the Red Queen family of models where selection varies across time and/or space, and we once again see the importance of heterogeneity in expanding the conditions under which recombination is favored (proposition 6, table 14.1).

In the Absence of Sex: What the Study of Asexuality and Clonal Reproduction Can Tell Us about the Evolution of Sex and Recombination

The relative rarity of ancient asexual lineages within eukaryotes has been seen as evidence of the importance of sex and recombination. How, then, are those few ancient asexuals managing in the absence of sex? While it is not easy to prove a negative, so that a lack of evidence for sex is not quite the same thing as evidence for asexuality, ancient asexual lineages are thought to exist within protists, plants, fungi, and animals (Judson and Normark 1996). While there are large and diverse clades with no ancient asexual species, such as in the angiosperms where all fully asexual species are thought to be closely related to sexual species (Whitton et al. 2008), the fact that long-term asexuality can be found across the tree of life is nevertheless intriguing. Within these asexual lineages, there are groups both ancient and species rich, such as the bdelloid rotifers, the darwinulid ostracods, and various groups within the oribatid mites, implying that they are managing without sex quite well indeed (Mark Welch and Meselson 2000; Schön and Martens 2003; Maraun et al. 2004; Schaefer et al. 2006).

What can these "evolutionary scandals" (to paraphrase Maynard Smith 1986) tell us about the theory of the evolution of sex? Recent genomic work in the bdelloid rotifers implies that they have come up with alternative ways to slow Muller's ratchet and to generate the genetic variation necessary to escape from local fitness optima. Flot et al. (2013) give evidence of ongoing horizontal gene transfer, likely mediated via double-stranded DNA breaks caused by repeated cycles of desiccation. Double-strand breaks also promote gene conversion during repair; concerted evolution mediated by gene conversion is proposed to slow Muller's ratchet, allowing restoration of the fittest genotype. As discussed above, a eukaryote with a relatively large genome in a finite population should be constantly accumulating deleterious mutations. In the model of Connallon and Clark (2010), gene conversion acts to increase the expected size of the zero-mutation class, $E(n_0)$ (eq. 14.1). In the work by Watson et al. (2011) discussed previously, the benefit of recombination in allowing populations to reach higher fitness maxima depended on the ability of the population to generate allelic diversity or to exploit standing genetic variation. Horizontal gene transfer produces genotypes that could not be produced by a single asexual lineage, allowing asexual lineages to escape local fitness maxima in much the same way that recombination frees sexual lineages from the same fate.

Another example where genomic architecture may facilitate long-term asexuality can be found in the ciliate *Tetrahymena*, where some asexual lineages may be millions to tens of millions of years old (Doerder 2014; Zufall 2016). Amicronucleate lineages of *Tetrahymena* that have lost the germline micronucleus, which allows for sexual reproduction, still retain a somatic macronuclear genome containing approximately forty-five copies of each chromosome. The process of "phenotypic assortment," which produces asexual progeny that differ in the number of copies of segregating alleles, generates genetic variation and allows selection to purge deleterious mutations, again providing a means for slowing Muller's ratchet in the absence of recombination (Zufall 2016). Thus, in lineages with long-term asexuality, genomic processes such as gene conversion, horizontal gene transfer, and phenotypic assortment may provide alternatives to sexual recombination for breaking up negative disequilibrium and allowing an escape from long-term population fitness decline (proposition 7, table 14.1).

Finally, a vast number of organisms engage in both asexual and sexual reproduction, with populations switching serially between the two modes, or having individuals follow one or the other route to reproduction within a generation. In models of these types of systems, increased asexual reproduction and decreased sexual reproduction can lead to higher mutation load and lower mean fitness (Muirhead and Lande 1997; Pálsson 2001). However, increased asexual reproduction can sometimes lead to higher mean fitness under relatively high genomic mutation rates (Marriage and Orive 2012). A key aspect is the relative numbers of recessive deleterious mutations held in heterozygous or homozygous form; thus the interaction of ploidy and segregation with recombination is a vital consideration. When trade-offs exist between the proportions of the population undergoing asexual and sexual reproduction, including selfing, asexual reproduction leads to the maintenance of recessive or partially recessive deleterious mutations in heterozygous form and can shield a proportion of the population from deleterious mutations arising meiotically (Marriage and Orive 2012). Sexual reproduction in diploids can bring together deleterious mutations whose individual fitness effects would otherwise be unseen by selection; asexual reproduction in diploids acts to freeze this within-locus variation.

A recent review of empirical studies considering species with rare or cryptic sex (Hartfield 2016) found inconsistent evidence for the type of within-individual allelic sequence divergence expected under long-term asexual reproduction (the "Meselson effect" [Mark Welch and Meselson 2000; Butlin 2002]). It is not yet clear whether this less than expected amount of allelic divergence is caused by gene conversion or by other genome-wide forces such as the effects of linked selection. However, it is clear that a more nuanced understanding of how genetic diversity is shaped in the absence of sexual recombination will in turn greatly aid our understanding of how genetic diversity is shaped in its presence.

Overview and Summary

The theory of the evolution of sex is one of the richest and most quantitatively sophisticated bodies of theory within evolutionary biology. There are any number of ways to classify the various types of models within this field, and a rich history of these types of classifications exists (e.g., Maynard Smith 1978a; G. Bell 1982, 1985; Michod and Levin 1988; Kondrashov 1993; Feldman et al. 1996; Otto and Lenormand 2002). I have chosen to focus on the theory of recombination and to contrast two main theoretical approaches: optimality theory and modifier theory. These two approaches differ fundamentally in the role of selection: optimization of a specific criterion, often under implicitly group selection arguments, versus changes in recombination modifier frequency via direct individual selection. Tests of both of these bodies of theory, carried out in various experimental systems, give support for the potential role played by both types of selection, and utilization of genomic techniques and fitness assays points to the importance of various forces identified in the corresponding theories. For example, comparison of fitness after experimental adaptation in sexual and asexual populations of yeast showed an increase in the rate of adaptation (the optimality criterion under consideration) in sexual populations (M. J. McDonald et al. 2016). Whole-genome sequencing of whole-population samples showed clear evidence that recombination alleviates clonal interference in the sexual populations, yielding a group-level advantage to sex. Experiments considering the maintenance of sex within populations of facultatively sexual rotifers have shown higher rates of sex evolving in spatially or temporally heterogeneous environments (Becks and Agrawal 2010, 2012). The fitness distributions of both sexual and asexual progeny give evidence that, at least under temporally changing environments, it is the long-term advantage of recombination in generating more variable progeny that results in an increase in sex (proposition 5, table 14.1).

In addition to differing with regard to the action of natural selection, there exists a clear contrast in how these two bodies of theory originated and developed, and in what aspects of the evolution of sex they sought to explain; my formalization of the constitutive theory of the evolution of sex makes this distinction clear. Historically, optimality theory and modifier theory differ in where they lie along an axis of verbal to quantitative models (Phillips, chap. 4). The initial development of the optimality theory of recombination was largely verbal (e.g., Fisher 1930; Muller 1932), although later work framed these ideas in an explicitly mathematical way. In contrast, modifier theory was from the very start a highly mathematical body of models, with an explicit relative frequency (and therefore relative fitness) approach to tracking the fate of modifiers of recombination. In addition, the equilibrium versus nonequilibrium nature of optimality versus modifier theory can also be seen to reflect a different frame of reference for the question "Why recombination?" Optimality theory contrasts populations with and without sexual recombination and describes key differences between them; modifier theory asks what processes actively shape the evolution of sexual recombination. In general, recombination theory has moved decisively to a more quantitative, process-focused body of theory, mirroring other bodies of evolutionary theory in this regard.

At the core of modern theory for the evolution of sex is the buildup and breakdown of linkage disequilibrium and the role that negative linkage disequilibrium plays in favoring the evolution of recombination. This negative linkage disequilibrium, whereby positive and negative fitness alleles find themselves bound on the same genetic background, can be generated both stochastically and deterministically. A great deal of the more recent theory for how evolution can favor recombination focuses on determining the relative roles of genetic drift due to finite population size and of epistasis generated by selection in creating negative disequilibrium, all in an explicitly mathematical framework.

Recent theory has also considered the intersection of recombination theory and aspects of genomic architecture, such as the form and extent of epistasis, as well the global and local properties of fitness landscapes. As genomic data become more readily available, both for organisms undergoing recombination and for those lineages that have apparently evolved alternative ways to both generate genetic diversity and avoid the buildup of deleterious mutations in the absence of recombination, we are seeing ways in which existing theory is supported. But there are also indications that a simple explanation is unlikely for this "queen of problems." For example, the use of computer simulations and genetic algorithms highlights the importance of considering both the details of genomic modularity and the timescale under which evolution is considered. The presence of sex across the evolutionary tree of life, in organisms with populations both large and small, and in genomes with widely disparate architectures, argues for a view that the reality of sex, much like the theory needed to fully explain its evolution and persistence, is multifaceted and complex.

Acknowledgments

The author thanks J. Blumenstiel and the EEB "Weird Sex" reading group, especially F. S. Chang, for helpful comments and discussion. The final version of this chapter was greatly improved by comments from S. Scheiner and several anonymous reviewers. Finally, the author acknowledges support from NSF grant DEB 1354754 during the writing of this chapter.

FIFTEEN

Speciation

SCOTT V. EDWARDS, ROBIN HOPKINS,
AND JAMES MALLET

> Although clearly a major focus of evolutionary biology, the study of speciation
> has never emerged as a coherent discipline.
>
> —(Harrison 1991, *282*)

Speciation is the process by which ancestral species diverge into two or
more descendant lineages. As such, speciation has occurred at each of the
branching events that have generated the 8–14 million species thought to
exist across the domains of life (Hawksworth et al. 1995; Mora et al. 2011).
The theory of speciation has played an important role in the modern development of evolutionary thinking and indeed could be said to have been
at the forefront of evolutionary theory since the publication of Darwin's
(1859) *On the Origin of Species*.

The Diversity of Reproductive Barriers between Species

In this chapter we review speciation theory and describe how it has provided insights into the fundamental ecological, genetic, and geographic
processes causing speciation. Our focus is exclusively on speciation in sexually reproducing organisms, which is commonly conceived of as the evolution of barriers to gene exchange (Mayr 1963; Coyne and Orr 2010). Isolating barriers are often divided into those acting before or after fertilization:
Prezygotic barriers occur prior to the formation of a zygote and include
ecological, temporal, geographic, behavioral, and mechanical barriers (col-
lectively known as pre-mating barriers), as well as post-mating prezygotic
barriers such as gamete interactions and female reproductive tract interactions. Postzygotic barriers include zygotic mortality, hybrid sterility, and
hybrid inviability. Barriers resulting from interactions with the biotic or

Table 15.1.   The Theory of Speciation

Domain: The process of speciation through the evolution of reproductive isolation.

Propositions:

1. Barriers to gene flow can occur at both pre- and postzygotic life stages and can be driven by intrinsic and extrinsic factors.
2. Reproductive barriers can evolve owing to divergent evolution between allopatric lineages experiencing little or no gene flow.
3. Intrinsic postzygotic reproductive isolation can result from negative epistatic interactions between loci evolving independently in diverging lineages.
4. Selection to decrease costly hybridization favors the accumulation of prezygotic reproductive isolation between sympatric lineages (reinforcement).
5. Local adaptation to divergent environments can drive pre- and postzygotic isolation between lineages (ecological speciation).
6. Reproductive isolation can evolve between sympatric lineages experiencing gene flow.
7. Sexual selection can drive divergence in mating systems, contributing to reproductive isolation.
8. Genomic changes such as chromosomal rearrangements and polyploidy can contribute to divergence and reproductive isolation

abiotic environment are known as extrinsic barriers, while those that act independently of the environment are known as intrinsic barriers. This categorization of speciation barriers as extrinsic or intrinsic (table 15.1) does not imply that they all cause species divergence, since most speciation events likely involve a combination of such forces. Instead the ideas classify particular barriers to reproduction that together explain a lack of current gene flow between diverging lineages. Although they used different terms, Darwin and the early Darwinians also discussed isolating barriers and theorized how they might arise. However, the major work exploring the evolution of reproductive barriers began during the modern synthesis and has been extended since.

What Is Speciation? What Are Species?

Both empirical and theoretical speciation research ultimately rest on cogent definitions of species, a topic that is fraught with disagreement (Nathan and Cracraft, chap. 6). Nonetheless, we believe that discussions of speciation can proceed productively even in the absence of universal agreement on the definition of species. Species are divergent metapopulation lineages (de Queiroz 2007). However, this fundamental lineage concept of species fails to distinguish species from nonspecies, but instead argues that species as taxa can be recognized by a variety of criteria, including the biological species concept (Mayr 1942), the phylogenetic species concept

(Cracraft 1983), and others. However, if speciation is to be distinguished from within-species evolution, we need a criterion of species rather than a generalized concept of gradual divergence. The conflicting criteria for species (e.g., monophyly, diagnosability, reproductive isolation, evolutionary independence) are manifestations of fundamental genetic and ecological processes that are subject to the vagaries of Earth history, divergence times and rates of divergence via natural selection, and drift or other processes (de Queiroz 2007).

In this chapter, by species we mean divergent forms that can overlap spatially and maintain differences at multiple regions of their genomes. These genomic differences might be maintained via intrinsic incompatibilities or extrinsic, ecological selection. For geographically or temporally separated forms the definition becomes somewhat arbitrary, but we do not believe that this hinders understanding models of speciation. Speciation theory can focus on how species diverge to achieve the ability to overlap spatially or in other words to achieve reproductive isolation—if not across the whole genome then at least at key genes that allow further phenotypic and genetic divergence.

The Goals of Speciation Theory

Speciation theory can help both to explain findings from empirical studies and to provide testable predictions to guide future empirical studies. Speciation theory is built on a large foundation of verbal models based on empirical observations and experimental evidence. Most of these verbal ideas are based on an understanding of how geography (allopatry versus sympatry) and timing (pre- versus postzygotic) influence the evolution of reproductive isolation (proposition 1, table 15.1)—an understanding that does not necessarily depend on mathematics (Turelli et al. 2001). Built on these verbal arguments are many hundreds of mathematical and numerical studies detailing which specific conditions are conducive to the evolution of barriers to reproduction. The sheer number of complex mathematical models can make a broad understanding of general speciation theory challenging (Kirkpatrick and Ravigné 2002). It can also provide an impenetrable shield to empiricists attempting to find broader meaning in their results.

There are many communication difficulties between empiricists and theoreticians in the realm of speciation. For example, empiricists are often interested in and tend to investigate the evolutionary forces acting on existing trait variation. How much reproductive isolation does the trait confer? Is the trait under natural selection or sexual selection? Is there migration between populations or gene flow between species? By contrast, when theoreticians study speciation they generally investigate the evolutionary conditions that allow divergence to occur (Turelli et al. 2001) and are less interested in which traits might contribute to reproductive isolation. How strong does nonrandom mating need to be? How strong does natural selection or sexual selection need to be? Can there be gene flow and if so how much?

Another example of a gap between theory and empirical studies concerns how biologists measure reproductive isolation. Empiricists will often conduct crosses or compare fitness of hybrids and parental types to estimate the degree of isolation between species (Sobel and Chen 2014). By contrast, theoreticians often assume a simple measure of genetic differentiation as an index of a history of isolation, such as when species differ in their allelic identities within a set of polymorphic loci. Gametic association of alleles at different loci between species, also known as linkage disequilibria, provide another common measure of the progress of speciation (see the seminal theoretical papers on speciation of Felsenstein 1981; Kirkpatrick and Ravigné 2002). In models leading to the evolution of reproductive barriers, the reduction of gene flow causes elevated linkage disequilibrium. Without reproductive isolation, in a population characterized by random mating and recombination, linkage disequilibria will decline to zero exponentially with time. As reproductive isolation evolves, gametic associations accumulate between alleles at multiple loci that differentiate species. Strong linkage disequilibrium is also implied by our definition of species as divergent forms that can overlap spatially and maintain differences at multiple regions of their genomes. In contrast, empiricists rarely discuss reproductive isolation in terms of linkage disequilibrium, although other measures of genetic differentiation are sometimes used (Coyne and Orr 1997).

The proliferation of highly specific mathematical theories is likely a result of the complexity of evolutionary factors influencing speciation. Selection, epistasis, gene-by-environment interactions, nonrandom mating, spatial structure, and gene flow are all difficult to model singly, and combining them all into a broad theory of speciation is exceedingly difficult (Kirkpatrick and Ravigné 2002; Gavrilets 2014). Furthermore, it is widely accepted that, in most cases, speciation results from the evolution of many reproductive barriers. Mathematical models do not yet account for the diversity of evolutionary forces affecting these multiple barriers. The solution to this complexity has been to develop models that investigate a particular scenario or set of circumstances for a certain type of barrier to reproduction. Hundreds of these models, both analytical and using numerical simulations, have been published, each exploring specific combinations of evolutionary forces with specific assumptions. In the absence of a general mathematically based theory about speciation, we often fall back on verbal theories to guide our understanding of the evolution of reproductive isolation.

Evolution of Reproductive Isolation in Allopatry

Mayr firmly established the importance of allopatry in speciation, a paradigm that is not strongly challenged today (proposition 2, table 15.1). We currently have methods, albeit indirect and subject to bias, to estimate the proportion of speciation events on a phylogeny that may have occurred in allopatry or involved some sort of sympatry (Barraclough and Vogler 2000; Fitzpatrick and Turelli 2006). Some of these methods implicitly or explicitly assume that allopatry is the expected speciation scenario in the absence of additional evidence—the null model. But it is not obvious that the null model of allopatry should be accepted even if the simple alternative hypothesis of sympatry is rejected. Given the increasing prevalence in the literature of scenarios of speciation involving gene flow at some time or another, lack of evidence for secondary contact or ongoing genetic or behavioral interactions between diverging species is not a solid basis for accepting that species have evolved in allopatry.

The increasing interest in sympatric speciation, reinforcement, porous species boundaries, and other scenarios involving interactions between individuals and populations raises questions: For populations that have diverged in allopatry, is secondary contact necessary to complete the speciation process? What proportion of speciation events have occurred solely in allopatry without some gene flow? The challenges in answering these questions lie largely in the difficulties of reconstructing historical ranges of species accurately on recent or ancient timescales. Although new tools such as niche modeling and geographic information systems permit informed guesses as to the dynamics of species ranges in the past, and fossils occasionally provide insight into whether diverging species experienced sympatry in the past, it is rare that these multiple sources of data allow us to reconstruct the geography of speciation unambiguously.

Theories of Intrinsic Postzygotic Barriers to Reproduction

How postzygotic reproductive isolation could evolve was one of the first conundrums of speciation theory. How do two lineages diverge such that their hybrids are inferior but neither of the two lineages passes through a period of low fitness? This mystery plagued Darwin as he was working through his theories on natural selection and the evolution of species. Darwin (1859) argued that natural selection could not favor hybrid sterility or hybrid inferiority. He argued that rather than being intended and the essence of speciation, as creationists such as Buffon (1753) had postulated earlier, hybrid sterility and inviability were accidental by-products of evolutionary divergence. New insights into this problem were gained during the modern synthesis as a new understanding of genes and population genetics emerged. These theories centered on understanding how genic intrinsic postzygotic reproductive isolation can evolve.

Dobzhansky-Muller Incompatibilities (DMIs)

It was proposed that epistatic interactions between alleles at different genes could result in intrinsic reproductive isolation (proposition 3, table 15.1). Although this idea was initially discussed by W. Bateson (1909), the genetic hypothesis was first clearly developed by Dobzhansky (1937a) and Muller (1942), and so loci that show this pattern are often today referred to as Dobzhansky-Muller incompatibilities (DMIs). In the basic verbal theory, two lineages that diverge in allopatry (or with virtually no gene flow between them) can accumulate neutral or selected substitutions at different genes. Within each lineage, genes evolve without the opposition of intermediate steps or allelic combinations by selection. Then, crosses between these two diverged lineages bring together divergent alleles in combinations never before seen by natural selection. These new combinations of alleles have negative interactions that make hybrids unfit, sterile, or inviable (fig. 15.1).

Based on the negative epistatic mechanism of incompatibility, theory predicts that the numbers of DMIs will accumulate faster than linearly as the number of substitutions differentiating the two species increases. The numbers of DMIs are expected to "snowball" and accumulate with the square of the number of substitutions separating two diverging lineages; if three or more loci generate an incompatibility, the snowballing effect will be even faster (H. A. Orr 1995; H. A. Orr and Turelli 2001). Because each new substitution in one lineage has the potential to interact with any of the previous substitutions, the number of possible interactions between substitutions is much greater than the raw number of substitutions. The beauty of the overall idea is that DMIs will inevitably accumulate between isolated populations as divergence occurs between species, even if neutral

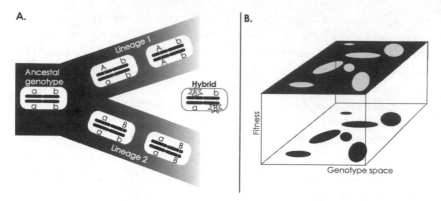

Figure 15.1. Negative epistatic interactions between loci can cause hybrid inviability or sterility. (A) Schematic of the evolution of a Dobzhansky-Muller-incompatibility. Two loci independently evolve in diverging lineages such that one lineage has a mutation in one locus ($a \rightarrow A$) while the other lineage has a mutation in a second locus ($b \rightarrow B$). In the hybrid the two derived alleles may interact to reduce hybrid fitness. (B) A schematic of a holey landscape as described by Gavrilets (1997) in which fitness is either zero or one. The genotypic space is scattered with regions of incompatibilities causing low fitness.

drift within species is the only driving force. In its simplest form, evolution within species is neutral and the field of gene combinations is a "holey landscape": a flat surface representing equal fitness, while some gene combinations are in deep "holes" of low fitness (Gavrilets 1997).

Recent Studies on Incompatibilities

Research in the past two decades has clarified the relationships between DMIs and speciation. The early 1990s was a time of renewed awareness of DMIs and the first empirical characterizations of putative incompatibilities (H. A. Orr 1993; H. A. Orr and Turelli 1996; Presgraves and Orr 1998; Presgraves et al. 2003; Phadnis and Orr 2009). This work frequently focused specifically on hybrid breakdown and attempted to link the hybrid breakdown observed in the laboratory, and sometimes confirmed in nature, with the speciation process. Because they cause hybrid inviability or sterility, DMIs became termed "speciation genes"—a "perhaps unfortunate term" (H. A. Orr 2005). It has subsequently been argued that many such speciation genes might not have been the original drivers of speciation because they could have evolved long after speciation was complete (Via and West 2008; Mallet 2010; K. L. Shaw and Mullen 2011).

Separation of populations in allopatry was classically considered a key requirement for the evolution of DMIs (Orzack and Sober 2001; Coyne and Orr 2004). It is commonly acknowledged that all species segregate deleterious recessive mutations (some causing lethality), and through the same logic recently it has been observed that deleterious incompatibilities between loci can also segregate within populations (Corbett-Detig et al. 2013). Thus, DMIs could play an important role in the distribution of fitness effects not only between species but within species as well. Given that empirical examples of DMIs often reveal positive selection (Tang and Presgraves 2009; Sweigart and Flagel 2015; Tang and Presgraves 2015), the upper surface of the holey landscape (fig. 15.1B) is unlikely to be completely flat for these loci, and this positive selection may counter some level of gene flow. Complete allopatry may not be required.

The verbal theory of the origin of DMIs has been validated not only mathematically (H. A. Orr and Turelli 2001), but also via empirical investigations of gene loci in *Drosophila* that epistatically generate hybrid incompatibility (Presgraves et al. 2003; Brideau et al. 2006; Masly and Presgraves 2007; Tang and Presgraves 2009). Experimental evolutionary studies in yeast have perhaps yielded the clearest picture yet of the relationship between divergent evolution in allopatry and the genic basis of DMIs (Dettman et al. 2007; J. B. Anderson et al. 2010; Stukenbrock 2013), as well as compelling examples of ecological speciation. Although the relevance to natural systems can always be debated, these studies yield sensible and intelligible connections between the genes exhibiting DMIs in hybrids and their role in divergent ecological settings in ways that genetic studies on long-diverged species in nature have not.

The snowball theory of the accumulation of DMIs over time has been tested in animals (Matute et al. 2010) and plants (Moyle and Nakazato 2010), and in a meta-analysis (Gourbière and Mallet 2010), all of which generally confirm the nonlinear accumulation of incompatibilities predicted by the theory. The logic of snowball DMI evolution is compelling and successfully explains Haldane's rule (see below). Two problems may lead to difficulties with tests using comparative data: First, stochastic accumulation of a few incompatibilities initially, each of which may have quite large and variable fitness effects on hybrids (Turelli and Moyle 2007), can result in very noisy changes in hybrid fitness, making the expected curvature hard to detect (Gourbière and Mallet 2010; Turelli et al. 2014). Second, the extent to which other, nonepistatic incompatibilities such as underdominance of chromosome rearrangements, or the effects of additive loci undergoing quantitative genetic adaptation to divergent niches, contribute to hybrid inviability and sterility between species is still unclear (Gourbière and Mallet 2010). In many systems, it is likely that ecological divergence and mate choice contribute more to reproductive isolation between recently diverged species than incidental DMIs (Schemske 2010; Turelli et al. 2014). Overall the prediction and testing of the snowball effect represent major refinements in our understanding of the genetics of incompatibilities, but the extent to which DMIs cause speciation in nature remains unclear.

Haldane's Rule

Perhaps the most famous examples of DMI evolution obey Haldane's rule, which states that if hybrids of one sex suffer most from inviability or sterility, it is most likely the heterogametic sex, the sex with different sex chromosomes (Haldane 1922). The heterogametic sex is the male (XY) in *Drosophila* and mammals, or the female (ZW) in birds and butterflies. Haldane's original observation sparked decades of empirical work investigating the generality of the rule as well as theoretical work attempting to understand the causal mechanism of the pattern. To date, four general theoretical explanations for Haldane's rule have been put forward: dominance theory, faster-male theory, faster-X theory, and meiotic drive (Schilthuizen et al. 2011). Each of these explanations consists of a cluster of models within the constitutive theory of speciation. The genetic scenarios described in these theories are extensions of scenarios in DMI evolution, albeit applied to the heterogametic sex.

The dominance theory posits that homogametic hybrids, for example, male birds (ZZ) or female monkeys (XX), will suffer only from DMIs that are dominant. Those loci may be linked or unlinked to the X or Z sex chromosomes and do not cause a major bias between the sexes. F1 hybrids will additionally suffer from DMIs affected by recessive alleles on the X or Z chromosome that are expressed only in the heterogametic sex, although the interacting partner loci may be dominant and on autosomes. In the flour-beetle *Tribolium*, autosomal loci differing between geographic populations as well as environmental effects influence the expression of Haldane's rule, as well as sex-linked recessives; this suggests that local adaptation to different environments may be a driver of the DMIs involved in Haldane's rule (Wade et al. 1994; Wade et al. 1999).

In addition, faster-male theory proposes that more rapid selective processes occurring in males (e.g., sexual selection on seminal fluids involved in sperm competition) lead to an accumulation of greater hybrid sterility or inviability in males than in females. A third idea, faster-X theory, proposes that the sometimes heterogametic X or Z chromosome evolves faster because advantageous recessive mutations will be exposed to selection in the heterogametic sex more rapidly than in the homogametic sex. Even after incorporating the frequent observation of faster evolution for genes with male-biased expression that presumably underlie male traits subject to sexual selection, recessiveness on the X chromosome was still required to explain Haldane's rule (Turelli and Orr 1995). Both faster male evolution and a more rapid accumulation of DMIs on the X or Z chromosome have been observed in natural systems (e.g., Masly and Presgraves 2007). However, for obvious reasons, faster-male theory cannot explain Haldane's rule in birds and butterflies, whereas dominance theory certainly can. It is unclear how great a role faster-X theory plays in general, but in *Drosophila* there is some evidence for more rapid accumulation of hybrid incompatibility-causing alleles on the X than on the autosomes (e.g., Masly and Presgraves 2007).

Other proposed mechanisms leading to Haldane's rule include accumulation of suppressors of meiotic drive on the sex chromosomes. Early statements of this model (Frank 1991; Hurst and Pomiankowski 1991) envisioned suppressors of meiotic drive accumulating on the sex chromosomes, causing rapid co-evolutionary arms races that increase divergence and incompatibilities between species. Empirical results consistent with this model have appeared, mostly in *Drosophila*, but a number of caveats apply, not the least of which is the inability of the theory to explain the ac cumulation of meiotic drive suppressors on sex chromosomes as opposed to autosomes (McDermott and Noor 2010). In conclusion, meiotic drive and a bias toward faster sex chromosome evolution, together with faster male evolution in male-heterogametic species, all play a role in Haldane's rule incompatibilities in *Drosophila*, but linkage of a recessive allele to an asymmetrically inherited genomic component (the dominance theory), usually a sex chromosome, is required for the very widespread obedience to the rule (Turelli and Moyle 2007).

Reinforcement and the Evolution of Prezygotic Reproductive Isolation

Alfred Russel Wallace (1889) was the first to lay out a coherent argument that reproductive isolation could evolve by natural selection, even though, in earlier correspondence, he had seemingly accepted Darwin's (1868) arguments against it. His idea, sometimes termed the Wallace effect (N. A.

Johnson 2008), was elaborated and refined during the modern synthesis, and when applied to the evolution of assortative mating, is today termed reinforcement. The general verbal theory asserts that two lineages, after diverging in allopatry, produce hybrids that may suffer from postmating incompatibilities when the two meet in secondary contact. The cost of producing hybrids or of mating with the other lineage will create a selection pressure to increase prezygotic reproductive isolation. Reinforcement is therefore the evolution of prezygotic reproductive isolation via selection to reduce the costly effects of hybridization (proposition 4, table 15.1; fig. 15.2).

The history of reinforcement research is marked by controversy (N. A. Johnson 2008). During the modern synthesis, Dobzhansky (1937a, 1940) championed the theory of reinforcement. He and others performed empirical studies documenting patterns of reproductive isolation consistent

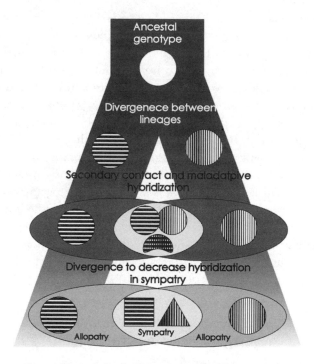

Figure 15.2. A schematic of the process of reinforcement. Two lineages diverge from a common ancestor and, in secondary contact, produce maladaptive, sterile, or inviable hybrids. This creates selection to decrease hybridization and leads to further divergence between the lineages in sympatric populations.

with reinforcement and performed experimental evolution to demonstrate the feasibility of the process. On the other hand, although Ernst Mayr (1942, 1963, 1970) believed that mating behavior evolved to aid species recognition, he argued that evolution of divergent mating behavior was normally achieved in allopatry, rather than by a process of reinforcement. Theoretical results accumulated that appeared to support the feasibility of reinforcement (Maynard Smith 1966; Dickinson and Antonovics 1973). However, by the 1980s additional theory suggested that gene flow and recombination would often prevent a buildup of linkage disequilibrium between divergently selected loci that caused postzygotic isolation and loci that affected mating behavior (Felsenstein 1981; Spencer et al. 1986; Butlin 1987). This sort of recombination may lead one species to go extinct because, for example, it acquires alleles that cause it to be attracted to the other species and to produce offspring with reduced fertility or viability.

Despite theoreticians' doubts, empiricists continued to report examples of reproductive isolation that likely evolved owing to reinforcement. Testing for reinforcement was popular in the 1980s and was reinvigorated by landmark studies in *Drosophila* and other groups, confirming key predictions of the reinforcement model, including a greater propensity to mating assortatively in areas of sympatry with sister species and a decrease in fitness of interspecific hybrids (Noor 1995). The identification of specific genes contributing to reinforcement in natural populations has also reinvigorated this model (Hopkins and Rausher 2011), and the measurement of reinforcing selection in a natural system has demonstrated the role that it can play in speciation (Hopkins and Rausher 2012).

Accompanying the abundance of empirical examples of reinforcement has been a deluge of theoretical studies investigating the feasibility of reinforcement as a function of the strength of selection, the extent of gene flow, and the genetic architecture of traits (Liou and Price 1994; Servedio and Kirkpatrick 1997; Barton and de Cara 2009). Clearly reinforcement suffers acutely from the problem faced by most speciation theory—too many specific models and little general theoretical framework. Nevertheless, there is currently a general understanding that reinforcement can occur within a wide range of parameters (N. A. Johnson 2008). As discussed below, selection can be strong enough to maintain high linkage disequilibrium in the face of gene flow and recombination. Reinforcement has thus emerged as an increasingly viable hypothesis for species divergence in natural systems, particularly when researchers seek it in appropriate study systems and use a carefully controlled experimental design.

One of the reasons why reinforcement may be more successful in nature than the models of Felsenstein (1981) and others would predict is that genetic divergence, for instance in ecological adaptation, can often have an indirect or pleiotropic effect that can lead to assortative mating. For example, divergence in ecological niche may entail habitat choice, but because individuals tend to mate with others in the same location, this divergence will also affect mating behavior (Diehl and Bush 1989). Any ecologically divergent trait that also affects mating behavior can thus improve the chances of speciation (Servedio et al. 2011).

The direct experimental evidence for reinforcement, however, does not answer questions about the generality of reinforcement; for these, comparative studies are required. Perhaps the most comprehensive dataset that exists is that compiled by Coyne and Orr (1997) for *Drosophila* species. Taking genetic divergence between species pairs as a surrogate for time since separation, the study examined levels of premating and postmating variation as measured in captivity. The study compared sympatric and allopatric species pairs and showed that in both, postmating isolation evolved at roughly the same rate. Premating isolation in currently allopatric species evolved at a similar rate to postmating isolation, but in sympatry premating isolation seems to have evolved much faster than postmating barriers. The greater rate of assortative mating evolution in pairs of sympatric species argues that reinforcement was likely pervasive.

Recently, Coyne and Orr's data were updated and reanalyzed in the hopes of finding some patterns in the causes of premating isolation. Curiously, in recently evolved pairs of sympatric species, there was no evidence that the strength of postmating incompatibilities, host plant differences (a surrogate for ecological postmating isolation), or X-chromosome size, which is expected to be correlated with increased rapidity of DMI evolution via Haldane's rule, had any effects on mating isolation (Turelli et al. 2014). And yet around half of all recently evolved species overlap spatially, show strong mating isolation, and are probably separated in some way ecologically. These results suggest that underestimates of the strength of postmating isolation in sympatric species are caused by inaccuracies of laboratory assays and shifts in host plant use. Although geography is undoubtedly involved, "the pervasiveness of the reinforcement pattern and the commonness of range overlap for close relatives indicate that speciation in *Drosophila* is often not purely allopatric" (Turelli et al. 2014, 1176).

Reproductive Isolation through Local
Adaptation to Divergent Environments

Ecological Speciation

In the past twenty years, the role of ecology in speciation has garnered renewed attention, with diverse scenarios coalescing under the title of "ecological speciation" (proposition 5, table 15.1). During the modern synthesis, ecological and habitat differences were found to be widespread among closely related species in nature, and ecological divergence formed a key component of the process of speciation (Turesson 1922; Clausen et al. 1939; Mayr 1963; Schluter 2000; Rundle and Nosil 2005; Nosil 2012). The notion that ecological adaptation drives divergence and eventually speciation has a strong and consistent history in the plant literature (Lowry 2012), although research in animals has been increasing. The more recent increase in appreciation of ecology in speciation was driven in part by new statistical tools that allowed the measurement of natural selection in the wild and the fitness of hybrid individuals in the habitats of the parental species as well as intermediate habitats (Hoekstra et al. 2001; Kingsolver et al. 2001; Ramsey et al. 2003; Wang et al. 2013). Ecological speciation has also become more popular because of the ability to employ data from both the field and experiments in the lab to test hypotheses about mechanisms (Lowry et al. 2008a). Finally, the recent focus on ecological speciation (Nosil 2012) has been driven largely by spectacular findings from natural, nonmodel systems, pointing to mechanisms whereby adaptation to novel environments or niches ultimately results in the evolution of pre- or postzygotic isolating mechanisms in nature (Rundle et al. 2000; Hawthorne and Via 2001; Jiggins et al. 2001; H. D. Bradshaw and Schemske 2003; Lowry et al. 2008b). These systems demonstrate ecological causes of divergent selection; evidence for ecological differences with pleiotropic effects on assortative mating; and linkages between ecological selection and genes driving reproductive isolation (Rundle and Nosil 2005).

The theoretical foundations for ecological speciation can be traced back to Darwin's principle of divergence. In general, empirical studies (e.g., host shifts of Rhagoletis fruit flies from hawthorn to apple, G. L. Bush 1969; Feder et al. 1988; McPheron et al. 1988; Berlocher 2000; Feder et al. 2003) have played a more prominent role in the development of the field of ecological speciation than has mathematical theory (M. R. Orr and Smith 1998; Schemske 2010; Nosil 2012). However, the two approaches are complementary, with hybrid sterility studies emphasizing genetic mechanisms of incompatibility and ecological genomics emphasizing the generation of incompatibilities via ecological adaptation. Although some argue that the focus on ecological speciation is hardly new (Harrison 2012), it is arguably an advance over the more restricted search for speciation genes if only because it explicitly incorporates ecological and behavioral drivers of divergence and their consequences for genomic divergence and incompatibility.

The Evolution of Reproductive Isolation in Sympatry

Speciation in sympatry, or in the presence of gene flow, has been far more controversial than speciation in allopatry (proposition 6, table 15.1). In his monumental work *Animal Species and Evolution*, Mayr's (1963) emphasis on allopatric speciation led many in the field to discount the possibility of sympatric speciation. Mayr's insistence, longevity, and strong influence on the field led to the idea that allopatric speciation should be a strong null hypothesis that should generally be accepted, unless a clear set of tests favor sympatric speciation or speciation in the face of substantial gene flow (Coyne and Orr 2004). Many complex patterns of phenotypic variation, including ring species (Moritz et al. 1992; Irwin et al. 2001; Alcaide et al. 2014), appear compatible with allopatric speciation, or at least with parapatric speciation accompanied by limited gene flow. But even Mayr (2002) eventually admitted the possibility and occasional occurrence in some groups (especially freshwater fishes) of sympatric speciation. As stated previously, it is inappropriate to accept allopatric speciation as a null model merely because it cannot be rejected. Additionally, recent findings of cryptic introgression between phenotypically distinct species in many groups, such as hominins (Green et al. 2010), birds (Rheindt and Edwards 2011), and butterflies (S. H. Martin et al. 2013), suggest that more complex speciation scenarios than simple isolation in divergence may often be warranted.

Others, particularly those working with host-specific parasitic or phytophagous insects, have long argued that there are simply too many species to be explained entirely by the slow grind of Earth's geography. Additionally, the evolution of ecological niches can create mating barriers between derived and ancestral populations (G. L. Bush 1975b, 1975a; White 1978; P. W. Price 1980). Pleiotropy between locally adapted traits and reproductive isolating barriers will be abundantly present in the adaptation of, for example, a parasitic insect to a new host. This debate in the speciation literature has sparked theoretical investigations of how reproductive isolation evolves with gene flow between diverging lineages.

Models of Sympatric Speciation

We can trace the origin of quantitative models of sympatric speciation to Maynard Smith's (1966) model involving two niches, the establishment of a stable polymorphism for the ecological (or postmating isolation) trait, and the subsequent evolution of premating reproductive isolation. This classic paper was not so much an argument for sympatric speciation as an outline of the conditions under which it might happen. The most favorable situation proposed by Maynard Smith was what he called pleiotropism: a single-locus ecological trait causing habitat selection that also has a pleiotropic effect on mating behavior. Maynard Smith raised and immediately dismissed the idea in three lines as very unlikely. Yet in modern and perhaps less clear terminology "pleiotropism" is none other than a "magic trait," an ecologically adapted trait that also acts as a barrier to reproduction. Today this seems rather more likely than hitherto (Servedio et al. 2011), and it was the basis of earlier verbal arguments for sympatric speciation (G. L. Bush 1975b, 1975a). Felsenstein (1981), in a highly influential paper, also regarded pleiotropy as an unlikely one-allele model, giving the Maynard Smith model as an example. Felsenstein pointed out that recombination between loci conferring divergent ecological advantages in specific habitats and loci specifying assortative mating could make sympatric speciation difficult because disequilibria between the loci would be degraded. By the 1980s, many considered sympatric speciation to be virtually impossible.

Consider the following argument. Suppose m is the rate of gene flow between two incipient species, and after every gene flow event introgressed genomes recombine randomly. Magic trait models assume that $m < 0.5$ between divergently adapted genotypes, which favors speciation because divergence due to natural selection becomes easier than in panmictic populations. However, the greatest challenge is to model speciation in a single population with panmixia, where $m = 0.5$ initially between the incipient species. For this reason, models of sympatric speciation usually start with panmixia.

A recent innovation in modeling the combination of population ecology and evolutionary dynamics is known as adaptive dynamics. Traditional population genetic models have often ignored population growth, possibly to the detriment of their realism. Adaptive dynamics models of speciation suggest a relatively wide set of conditions under which sympatric speciation might take place (Dieckmann and Doebeli 1999). Adaptive dynamics theories typically propose a phenotypic model of evolutionary divergence based on demography and ecological competition. Speciation is envisaged as a branching process where one population splits into two phenotypic clusters as a result of ecological pressures (Metz 2011). Unlike population genetic models, most adaptive dynamics theory does not explicitly include genetic considerations or reproductive isolation. Rather, it explicitly models population growth and the demography of ecological competition as drivers of divergence. Darwin (1859) proposed his own principle of divergence based on Malthusian ideas of geometric population growth and ecological competition that drove divergence in very much the same vein, and the adaptive dynamics community therefore claim that they are modeling Darwin's ideas more closely than traditional population genetics models (Metz 2011). Dieckmann and Doebeli's (1999) innovation to the adaptive dynamics literature was to add a simple quantitative genetic model of an ecological phenotype and genetic assortative mating to the adaptive dynamics models. The reason that branching took place was that competition for the available resource spectrum drove the populations apart, with the populations acquiring premating isolation as a result of reinforcement.

However, the adaptive dynamics model may be unrealistic. In the Dieckmann and Doebeli (1999) model, branching took place because the model allowed mutations only of small effect so that there was better exploitation of the resource when there were two isolated panmictic species than if one was occupying the center of the resource. If variable effect mutations were allowed, a single population perfectly exploiting the Gaussian resource would be able to evolve that could not be invaded by another species (Polechová and Barton 2005). The Dieckmann and Doebeli model predicts that linkage disequilibrium between the ecological and mating loci will develop, but it requires an initial chance deviation from linkage equilibrium on which selection is able to act. This model provides an interesting case in which random drift might kick-start a higher-order selective process.

Today's many models of sympatric speciation can be viewed as extensions of a simple and very early single-population model of migration/selection balance (Haldane 1930, 1932). Haldane showed that a one-locus polymorphism will be maintained provided that divergent selection is greater than gene flow (fig. 15.3). The same will be true for more loci, although the overall selection will now act on multilocus genotypes, and a buildup of linkage disequilibrium comes into play, so that recombination as well as gene flow is important. Although we have highlighted the influential although contested contribution of Dieckmann and Doebeli (1999), many others before and since also have shown that sympatric speciation

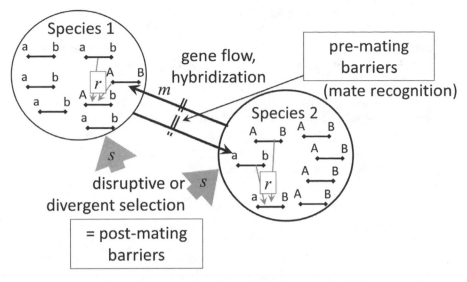

Figure 15.3. Simplified diagram of speciation with gene flow. Selection (*s*) for divergently selected loci A, B in two different ecological niches allows some differentiation to take place, maintaining a balanced polymorphism (due to migration-selection balance) and linkage disequilibrium of adapted loci between niches. This is opposed by gene flow (*m*) and recombination (*r*) that can produce less-well-adapted or even deleterious intermediates (i.e., DMIs, see fig. 15.1A) within each niche. Such polymorphisms must be selected strongly enough to avoid swamping by gene flow: only if *s* > *m* will such a polymorphism exist. A second step in most models of sympatric speciation is reinforcement: the evolution of assortative mating can then further reduce gene flow. The combination of these two steps can then allow more weakly selected divergent loci to evolve, which in turn causes an increase in cumulative selection against migrants and recombinants and further selection for reinforcement.

is possible under a wide variety of circumstances (Maynard Smith 1966; Diehl and Bush 1989; Fry 2003). It is clear that sympatric speciation can occur in theory; the only remaining questions are whether the models are realistic and how often sympatric speciation occurs in nature (Coyne and Orr 2004).

Sexual Selection and the Evolution of Reproductive Barriers

Sexual selection is differences in mating success among members of one sex. Advantages of mate choice may be direct (e.g., ensuring parental assistance) or indirect (e.g., superior fitness of offspring). There are two major flavors of indirect sexual selection, usually termed good genes models or runaway models. In good genes models, the offspring have superior fitness in general, while in runaway models, the offspring merely have superior mating success that may conflict with ecological fitness (Kirkpatrick and Ryan 1991). Speciation by reinforcement is a form of good genes sexual selection (proposition 7, table 15.1). Some models of mate choice, such as those envisioning an avoidance of "bad genes" (Iwasa et al. 1991), may capture certain aspects of reinforcement as well.

Verbal theory predicts that divergent mating preferences and sexual signals arising in allopatry could impede the merging of species upon secondary contact (Ritchie 2007; Safran et al. 2013). Sexual selection may also accelerate speciation in sympatry (Higashi et al. 1999), but unambiguous empirical examples of the process are rare or nonexistent. While sexual selection can promote speciation, it may impede speciation in sympatry (Kondrashov and Shpak 1998; Kirkpatrick and Nuismer 2004) or in peripheral isolates (Servedio and Bürger 2015).

Sexual selection as a driver of speciation has been tested in comparative studies of birds (T. Price 1998; Edwards et al. 2005; T. Price 2008), fish such as cichlids (Seehausen et al. 1999; Seehausen and van Alphen 1999; A. B. Wilson et al. 2000), and in insects, with a group of flightless Hawaiian crickets being a prominent example (Mendelson and Shaw 2005). Learning and phenotypic plasticity of mate choice can facilitate speciation in some situations (Verzijden et al. 2012), but can also retard it (Nonaka et al. 2015). More recently, little correlation between sexual dimorphism and diversification rates has been found in comparative studies (Morrow et al. 2003; Ritchie 2007). Distinguishing between sexual selection and selection on traits other than mate choice as drivers of speciation can be challenging; Safran et al. (2013) suggest that focusing on traits that have clear functions in fecundity and survival rather than in mate choice may help.

Genomic and Chromosomal Barriers to Reproduction

Plants, animals, and fungi all exhibit enormous amounts of chromosomal variability (Deakin and Ezaz 2014; Baack et al. 2015; Manicardi et al. 2015) that may play a role in speciation (proposition 8, table 15.1). Often variation occurs among populations or species, and connections between chromosomal variation, instability, and speciation have been postulated (A. C. Wilson et al. 1975; White 1978). Chromosomal rearrangements and inversions between species are potent inhibitors of hybridization, given the challenges they pose to normal pairing during meiosis (Dobzhansky 1937a; White 1973; Lande 1984). Yet the loss of fitness during meiosis can provide a strong barrier to the evolution and fixation of these very same chromosomal rearrangements in natural populations (Lande 1983; Baker and Bickham 1986). The challenges of explaining the fixation of chromosomal rearrangements has led to an extensive literature, with recent work suggesting that some types of rearrangements may not require extreme levels of genetic drift or overdominant selection to become established (Kirkpatrick and Barton 2006; Kirkpatrick 2010). The importance of genetic drift in chromosomal evolution and speciation, especially founder-effect speciation, is today generally deprecated (Barton and Charlesworth 1984; Coyne and Orr 1997, 2004). However, the possibility of evolution through Wright's (1982) shifting balance process, particularly in chromosomal evolution (Barton and Rouhani 1991), deserves to be reexamined (Coyne and Orr 1997).

Polyploidy

The abundance of examples of ploidy differences between closely related plant species has led to researchers to consider chromosomal mechanisms of speciation a major force in plant speciation (Baack et al. 2015). Because hybrids between a tetraploid and its diploid ancestor usually produce infertile triploid hybrids, polyploidy can be an instantaneous mechanism of speciation. Whole-genome duplications are inferred to have occurred in many lineages across the eukaryotic tree of life; in plants, the correspondence between polyploidy and recent speciation is particularly widespread. The possibility that changes in ploidy can drive speciation in plants is increased in cases where species can produce offspring by selfing (autopolyploidy), which in a novel polyploid would ameliorate the problem of finding a mate. Polyploidization can also occur in conjunction with hybridization (allopolyploidy). Disagreement exists as to whether allopolyploidy (Soltis and Soltis 2009) or autopolyploidy (Ramsey and Schemske 2002) is the more common mechanism. A recent survey found approximate parity between the two modes: although autopolyploids are produced more rapidly, they also seem to be more prone to extinction (Barker et al. 2016). The greater abundance of even versus odd chromosome numbers, among other evidence, suggests that approximately 15 percent of angiosperm and 31 percent of recent fern species evolved as a result of chromosome doubling (Otto and Whitton 2000; T. E. Wood et al. 2009). Macroevolutionary analyses of comparative genomics suggest an association between polyploidization and the diversification of plants (Jiao et al. 2011; Jiao et al. 2012; Tank et al. 2015). While whole genome duplication may have contributed to evolutionary flexibility and diversification deep in the tree of life, recent polyploidy events, in contrast, generally seem to slow down rates of diversification (T. E. Wood et al. 2009; Mayrose et al. 2011).

Inversions and Local Adaptation

Another type of genomic rearrangement that shows conspicuous connections with speciation is chromosomal inversion. Inversions maintain their distinctness between species by suppressing recombination, and may create fixed divergent adaptations between populations or species, even in the presence of gene flow, explaining their tendency to accumulate between species (Kirkpatrick and Barton 2006). Inversions that reduce recombination have increasingly been found to harbor genes underlying species differences and reproductive isolation (Noor et al. 2001; Lowry and Willis 2010). In recent years, genomic dissection of a number of species that exhibit chromosomal differences or striking intraspecific polymorphisms have revealed that inversions are hotspots of evolutionary novelty, sometimes harboring multiple loci that work in concert to promote reproductive isolation (Joron et al. 2011; Kunte et al. 2014; Küpper et al. 2016; Lamichhaney et al. 2016a; Tuttle et al. 2016). Polymorphic inversions represent an evolutionary alternative to speciation, since each inversion morph may allow exploitation of divergent environments in the face of potential gene flow. However, such polymorphisms could also have the potential to lead to speciation, perhaps when they become geographically isolated and additional genome-wide divergence accumulates.

Conclusion

Almost thirty years ago, Harrison (1991, 282) wrote, "Although clearly a major focus of evolutionary biology, the study of speciation has never emerged as a coherent discipline." Speciation research remains a wonderfully, if sometimes frustratingly, diverse subfield within evolutionary biology. Even today, researchers from systematists to geneticists to ecologists, publishing in a wide array of journals, purport to study speciation, often without reference to major lines of research within the field of speciation.

The focus of speciation research has changed over time, often adopting new paradigms or approaches as technical or conceptual advances have been made, or sometimes spearheaded by key publications (e.g., Maynard Smith 1966; Felsenstein 1981) or new genomic technologies. The most recent trend in speciation research is one of integration: How synthetic is a single study or research program, and how many levels in the hierarchy of mechanisms, from molecular to ecological, are spanned? Does the model system chosen for study lend itself to both ecological and genetic manipulation and investigation? These are the questions that drive the choice of species and topics in speciation research today.

Building on research agendas laid out during the modern synthesis, a new generation of experimental studies, using the latest tools of genomics and high-throughput phenotyping, has emerged in the past twenty years that focuses on the details of genetic incompatibilities or reinforcement, or that links fitness in the wild with allelic variation at specific genes. We believe that both descriptive (biogeographic/historical) and experimental research have valuable contributions to make to our understanding of speciation. Because speciation is in essence a historical event, experimental studies necessarily have to consider what is known of the particular history of the organisms being studied, and to ask how this history might constrain the details of genetic or ecological interactions observed in that study system today.

Despite the overwhelming evidence for the role of allopatry in speciation, this chapter has emphasized the importance of interactions among lineages found together in either primary or secondary sympatry. We have also emphasized the importance of verbal theories and the two-way flow of information between empirical and theoretical studies of speciation. Many mathematical theories of speciation understandably often envision very simple genetic mechanisms for some extrinsic traits, such as models that envision a single locus controlling mate or habitat choice (Kirkpatrick and Ravigné 2002). Such models, while necessarily simplistic, are useful in so far as they can envision a collection of processes that could result in speciation in the quantitative genetic limit of single genes controlling single traits. We agree with Turelli et al. (2001, 330) that "given the complexity of speciation, mathematical theory is subordinate to verbal theory and generalizations about data." Empirical studies of speciation are the most useful means of determining the relative strength and roles of intrinsic and extrinsic forces in speciation in nature, and arbitrate as to which mechanisms of speciation are most prevalent. In particular, the recent flush of genomic studies of speciation (Seehausen et al. 2014) has provided a detailed window into the genetic processes and genomic restructuring that accompany, or cause, speciation. Ultimately, empirical, historical, theoretical, and experimental studies will all be essential for evaluating the relative roles of selective versus random processes in speciation, whether at the level of genomes, individuals, or populations.

Speciation in the past was regarded as a key element in evolutionary theory: "Without speciation there would be no diversification of the organic world" (Mayr 1963, 621). However, Mayr's ideas conflicted somewhat with the uniformitarian Darwinian view that the processes we see operating within species, if extended, will lead naturally toward speciation. Notwithstanding Mayr's great contributions to the field, theories of speciation today tend to revert to a more Darwinian view of divergence. Speciation is seen instead as part of a continuum of diversification, with the species boundary no longer as clearly defined as it seemed to zoologists of the 1960s and 1970s (Nathan and Cracraft, chap. 6). Today we can more clearly perceive that within-species processes of natural and sexual selection, and likely some stochastic processes, can drive prezygotic isolation directly, by altering mating behavior, or by driving evolution in different directions in each lineage, leading indirectly to epistatic or other incompatibilities. Thus, today's theory of speciation would perhaps have been somewhat unexpected to earlier generations, but it is nonetheless coherent and well-formed, in contrast to Harrison's (1991) earlier view of the field. Our theory of speciation (table 15.1) thus successfully bridges theories of microevolution and of macroevolution, yet without needing to establish any clear hiatus between the two.

Acknowledgments

We are grateful for helpful editorial comments and suggestions from Trevor Price and the editors. We dedicate this chapter to the memory of our dear friend Rick Harrison. He has been a major influence on the field of speciation for many years and on our understanding of speciation in particular.

SIXTEEN

The Theory of Evolutionary Biogeography

ROSEMARY G. GILLESPIE, JUN Y. LIM,
AND ANDREW J. ROMINGER

Biogeography is the study of the distribution of species and their diversity in geographic space and through geological time. Evolutionary biology is the study of the processes that produce these patterns of species diversity and through which species then adapt and diversify. Thus, evolutionary biogeography falls at the intersection of these two fields—the interaction between the physical land or seascape and its consequences on evolutionary processes and hence the spatiotemporal distribution of species. Perhaps unsurprisingly, because the study of evolutionary biogeography encompasses such a broad array of processes and phenomena at a range of spatial and temporal scales, there have been historical differences in how evolutionary biologists and ecologists have approached examining the role of the physical environment and evolutionary contingencies in shaping contemporary assemblages.

In this chapter, we first summarize some of the major biogeographic patterns and some of the key explanatory variables that have been implicated in the patterns. We go on to examine theories that have been developed that examine different biogeographic patterns. We then develop propositions for a synthetic theory of biogeography that integrates the dynamic nature of geographic space and, in parallel, the spatial dynamics of biodiversity.

Patterns in Biogeography

No species is found everywhere. While this observation might seem trivial, it may be caused by myriad abiotic and biotic factors. Why, one might ask, are certain species or even whole assemblages congruently limited in their distribution? Are there factors that directly preclude their dispersal into other areas, or will they continue to expand their range given enough time? Why do some areas support and contain a larger number of species than others?

At the broadest spatial extents, species richness typically shows a latitudinal gradient (Pianka 1966), a pattern that appears to be dictated by evolutionary processes leading to differential species accumulation over time (Belmaker and Jetz 2015). However, there can be marked local or regional variation. For example, elevational gradients in species richness (G. C. Stevens 1992) have now been extensively documented (Sanders and Rahbek 2012). Overall, it appears that major patterns may be explained based simply on time and the size of the regional species pools (Kraft et al. 2011), and the associated effects of climate stability and time-integrated area (Fine 2015), with recent mechanistic models providing insights into the eco-evolutionary feedbacks (Pontarp et al. 2019).

Biogeographic theories in the past have aimed to integrate patterns of diversity with landscape configuration or niche limitations. The former affects entire communities that are defined by the landscape, while the latter affects individual organisms within the landscape. There are a number of core theoretical constructs that have been advanced in the past, all of which—though broadly interacting—provide understanding for some key area of evolutionary biogeography.

Vicariance Biogeography

The theory of vicariance biogeography proposes that the distributions of taxonomic groups are determined by splits (or vicariance events) in the ranges of ancestral species. The advent of vicariance biogeography, in concert with the development of cladistics and the understanding of plate tectonics in the 1970s, transformed the field of biogeography from a largely descriptive field to a rigorous science by setting the stage for concrete theories and associated testable hypotheses (Nelson and Platnick 1981). The ability to test these concepts was greatly enhanced with the advent of molecular data, which provided the needed temporal framework of colonization or fragmentation of a lineage (Riddle et al. 2008). Thus, considering the ancient breakup of Gondwana, some studies have supported an ancient history of vicariance (Ericson et al. 2002), though many others have shown the key role of dispersal rather than (or in addition to) vicariance (Upchurch 2008). The development of statistical tools has allowed the incorporation of fossils into the calibration of molecular phylogenies (Sanderson 2002; S. A. Smith and O'Meara 2012) and has greatly strengthened the inferences from ancient vicariant events, as has recently been shown in a lineage of spiders (H. M. Wood et al. 2012) and a widespread gymnosperm family (Mao et al. 2012). For more recent vicariance events, increasing sophistication of technologies and analytical tools has allowed examination of the details associated with the diversification process, including demographic changes involved, relative population sizes, and other aspects that allow insight into the interplay between geologic and climatologic history and the biology of organisms (Masta 2000; Oneal et al. 2010).

Dispersal Theory

It has long been hypothesized that species dispersal to novel environments and subsequent in situ speciation in isolation can generate biogeographically unique biota (Carlquist 1974). Although dispersal is well recognized as a key element in shaping biodiversity, dispersal studies have developed along disparate avenues: the examination of patterns of dispersal in space and time, the proximate and ultimate factors shaping dispersal patterns, and the effects of dispersal patterns on the structure of populations and communities (Nathan and Shohami 2013). With the advent of genetic data, it is now well established that species can traverse oceans and other vast distances in storms, or on vegetation rafts, icebergs, or in the case of plant seeds, in the plumage of birds. However, the rarity and presumed unpredictability of such long-distance dispersal events has, until recently, precluded further development of theory and associated testable hypotheses (Carlquist 1972; Nathan 2006; Crisp et al. 2011). Yet while a single rare long-distance dispersal event may indeed be impossible to predict, an understanding of the mechanisms involved in long-distance dispersal over extended (evolutionary) time periods can lend predictability to the process (Nathan 2006). Understanding the different modes of long-distance dispersal (e.g., wind, birds, oceanic drift, rafting) can be coupled with information about an organism's natural environment and biology to generate broad biogeographic predictions (R. G. Gillespie et al. 2012). The ecology of an organism can inform inferences of the most likely mechanism(s) for long-distance dispersal and hence—given known trajectories of long-distance dispersal vectors (e.g., direction of prevailing storms, migratory routes of birds, or ocean currents)—the most plausible avenue of arrival. Integrating such considerations with geological, paleontological, and evolutionary data can lead to predictions concerning patterns of endemism and diversification. Moreover, using probabilistic modeling of geographic range evolution, it is possible to compare biogeographical models. Such approaches have highlighted the importance of model choice in historical biogeography and allowed explicit identification of long-distance dispersal events and associated founder event speciation (Matzke 2014).

Island Biogeography

The equilibrium theory of island biogeography proposes that species diversity is a balance between immigration and extinction (MacArthur and Wilson 1967). Immigration decreases with increasing distance from a mainland source because more remote islands are more difficult to reach by mainland species than an island that is closer to the mainland source. Extinction decreases with increasing island size because larger islands can support larger populations than smaller ones and extinction risk is inversely related to population size. Subsequent work demonstrated how isolation affects extinction (in addition to immigration) through the so-called "rescue effect" (J. H. Brown and Kodric-Brown 1977). The equilibrium theory marked a turning point in biogeography (Losos and Ricklefs 2009). Moreover, it has been applied to a vast diversity of insular systems beyond islands per se, most notably to examine conservation implications of fragmentation on species diversity (Triantis and Bhagwat 2011).

Niche Theory and Species Distributions

Niche theory posits that individual species thrive only within definite ranges of environmental conditions (Hirzel and Le Lay 2008); building on this theory, species distribution modeling correlates spatiotemporal availability of broad-scale environmental features with species occurrences (Chase 2011). Niche modeling uses climate data—past, present, and future—in concert with the physiological tolerances of organisms to predict changing distributions. By placing such analyses within a phylogenetic context using continuous character analyses (Graham et al. 2004; Yesson and Culham 2006; Evans et al. 2009), it is possible to infer the evolution of climate niche variables and to identify cases of their convergent and divergent evolution. In this way, species distribution models have allowed the integration of evolution and ecology to infer limits and project changes in species distributions (Elith et al. 2006; Kozak et al. 2008; Graham et al. 2014). A criticism of niche models is that they assume that environmental tolerances define the niche space of an organism (Warren 2012). Although such an assumption may be flawed, niche models can serve as a starting point for inferring changes in distributions (Warren et al. 2014). Moreover, increasingly sophisticated approaches are being developed, including the incorporation of genetic data from empirical populations into species distribution models and the use of approximate Bayesian computation to estimate key unknown demographic parameters. Such an approach can provide insights into how demographic shifts have shaped and will shape biotas (J. L. Brown et al. 2016).

The recent surge of spatially explicit genomic-scale data has led to the growth of phylogeography and landscape genetics, both of which bring theory from population genetics to bear on the understanding of biogeographic patterns, and offer the potential to understand processes that shape early biogeographic patterns, including population divergence, species formation, and adaptation (Wang et al. 2013; Sexton et al. 2014; Rissler 2016).

Ecophylogenetics

Lying at the crossroads of ecology, biogeography, and macroevolution, ecophylogenetics incorporates historical contingencies into community ecology and evolutionary processes (Webb et al. 2002). Starting with a focus on clustering versus overdispersion patterns (Pearse et al. 2014), research has moved on to focus on multiscale and multidimensional analyses of variation in niche and trait evolution using hierarchical models and statistics (Mouquet et al. 2012). This approach has now been applied to explain different kinds of ecological phenomena, including species coexistence (HilleRisLambers et al. 2012), competitive exclusion (Mayfield and Levine 2010; Violle et al. 2011), niche conservatism (Emerson and Gillespie 2008; Losos 2008), ecological release (Cavender-Bares et al. 2009), and trophic interactions (Morlon et al. 2014).

Neutral and Unified Theories

Neutral and unified theories provide null hypotheses or baselines for developing a general theory in biogeography. Biodiversity has been studied largely independently of the milieu in which it evolved. Indeed, given the complexity of the processes involved in shaping both biodiversity and the geographic setting, it is doubly complex to understand how they might be integrated. Thus, in order to make sense of process and pattern, it is important to understand the nature of the underlying factors and determine the minimum number of variables necessary to explain processes. A major guiding principle in theory development should always be to start from the simplest basis and add complexity slowly, step by step, and only when absolutely necessary (Hubbell 2009). To this end, several minimalist and broadly generalizable theoretical constructs have been developed in recent years that have proved extraordinarily powerful in their ability to provide predictive frameworks that cut across subfields of biology. These theories include the metabolic theory of ecology (J. H. Brown et al. 2004) and the maximum entropy theory of ecology. The metabolic theory of ecology is based on the assumption that metabolism underlies all relevant processes governing the dynamics of organisms (West et al. 1999, 2001; Gillooly et al. 2002; Savage et al. 2007), populations (Damuth 1981; Marquet et al. 1990; Enquist et al. 1998), and potentially species/speciation (J. H. Brown et al. 1993; Allen et al. 2002). Because metabolism itself follows allometric predictions arising from biochemical and biophysical constraints (West et al. 1999; Gillooly et al. 2001), the metabolic theory of ecology predicts that many ecological and evolutionary phenomena should as well (J. H. Brown et al. 2004). Conversely, the maximum entropy theory of ecology predicts ecological phenomena by making no mechanistic assumptions but rather assuming that any mechanism is valid so long as the community is in a steady state (Harte 2011). It asserts that universal patterns, such as species-abundance distributions or species-area relationships, derive from the universal operation of statistical laws.

Another important theoretical construct is the neutral theory of biodiversity and biogeography (Hubbell 2001; Rosindell and Cornell 2007), which can predict species-abundance distributions and species-area relationships from simple mechanistic community dynamics that assume equal demographic rates across individuals regardless of species identity. As with the metabolic theory of ecology and the maximum entropy theory of ecology, neutral theory's predictive success and mechanistic simplicity (Hubbell 2001) have allowed researchers to explore a wide range of ecological patterns (e.g., Etienne and Olff 2004; Wootton 2005; Rominger et al. 2009; Rosindell and Phillimore 2011) and use deviations from theory to understand what additional mechanisms are needed to make a more accurate, universal theory.

One limitation of neutral and unified theories is that, in their current forms, all focus chiefly on the long-term steady state of a community, failing to incorporate spatial patterns, the dynamic nature of biodiversity, or the geographical milieu. Standard ecological models of community assembly consider the species pool as a largely unchanging entity (Mittelbach and Schemske 2015). Thus, a central goal in biodiversity research is to provide a dynamic framework that incorporates the fluctuations of the milieu with the evolution of the organisms that depend on that milieu. To this end, some intriguing developments have recently been formulated arguing that the fundamental mathematical link between the neutral theories of ecology and molecular evolution (Hubbell 2001) can be used to achieve a broader synthesis and deeper understanding of both processes (Vellend 2010).

Landscape Dynamism and Species Diversification

There is increasing recognition that landscape dynamism over broader timescales can affect the diversity dynamics of resident biota (Whittaker et al. 2008; Sandel et al. 2011; Valente et al. 2014; Valente et al. 2015; Borregaard et al. 2016; Lim and Marshall 2017). Recent work has focused primarily on insular systems, where their relative transience (compared with continental systems) is felt more readily by their biota. Building on island biogeography theory (MacArthur and Wilson 1967), the recently proposed general dynamic model incorporates island age into oceanic island biogeography (Whittaker et al. 2008). This model brings together geological dynamics with evolutionary outcomes. As hotspot islands get older, the composition shifts from biotas composed primarily of colonizing species to those dominated by new endemic species (Rominger et al. 2015). As the islands become yet older, they erode and subside and the biotas go extinct (Whittaker et al. 2008; Lim and Marshall 2017). Thus, the model provides a framework that ties together the geological ontogeny of oceanic islands with the pattern of species accumulation.

Recent approaches have started to incorporate the role of factors other than geology in shaping biodiversity. Cycles of climate change and associated changes in ocean currents, in conjunction with changes in area, isolation, and elevation, may have shaped biodiversity on some oceanic islands (Fernández-Palacios et al. 2015). By taking account of multiple factors shaping the geographical milieu, the incorporation of climatic influences is a key advance, even though it includes only the role of climate on area and isolation due to fluctuations in sea level related to glacial cycles, and not climatic conditions per se.

While these models incorporate effects of changing area and isolation on patterns of diversity, they have not yet been able to integrate the biogeography of diversification with biodiversity dynamics. Recent research shows considerable promise in this direction by modeling evolutionary change onto an explicitly dynamic landscape (Melián et al. 2015). Given that geography shapes biodiversity through processes associated with periods of isolation interspersed with periods of gene flow, a general framework should build on these natural cycles of fusion and fission of habitable

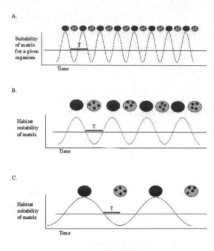

Figure 16.1. The effects of periodicity of geological and climatological fluctuations on evolution and diversification showing the periodicity over time and space over which a contiguous habitat is suitable for a given population to flourish. T is the time needed for speciation, which can be reduced by selection; one migrant per generation is sufficient to prevent divergence by drift alone (Wright 1951). The black circles represent periods when the habitat matrix is uninhabitable, leading to differentiation (above the horizontal line). (A) Ecological disturbance where T is greater than the periodicity so that there is insufficient time for differentiation and hence no change in diversity. (B) Extended disturbance such as island formation or mountain building where T is just less than the periodicity so that new species are added at each cycle. (C) Very long fragmentation such as continental breakup where T is much less than the periodicity and new species are added each cycle, but there are few cycles.

sites and how such cycles dictate the evolutionary dynamics of biodiversity. It is from these concepts, models, and theories that we build our unified theory of evolutionary biogeography.

Propositions of the Biogeographic Theory

Given our understanding of factors affecting overall biogeographic patterns, and the key theoretical developments in the past that have been

proposed to explain these patterns, here we establish a series of propositions for a theory of evolutionary biogeography. Our premise is that species diversity is the consequence of a dynamic equilibrium between dispersal (immigration or gene flow), speciation, and extinction (MacArthur
and Wilson 1963, 1967). The physical landscape affects two key variables:
(i) area, which defines the extent over which physical, ecological, and
physiological processes occur, and (ii) isolation, which may be either geographic or ecological in nature. However, the landscape modulates the interplay between dispersal, speciation, and extinction. Most important, the
evolutionary outcome of any spatial arrangement of area and isolation is
dictated by the time and periodicity of the fusion-fission cycle of a given
landscape configuration (fig. 16.1). Toward this end we have identified six
key propositions that any comprehensive theory of evolutionary biogeography must include (table 16.1).

A general theory of biogeography requires understanding of (1) how
the processes that generate biodiversity interact with the geological and climatological processes that generate the habitable sites, and (2) how the
constraints or limits to the generation of biodiversity imposed by isolation
that limit the rate of propagule arrival and area dictate how many entities
can be added to, or removed from, the system. Area and isolation represent a bounding box within which evolutionary and ecological dynamics
operate (fig. 16.2). Many of the assumptions of community ecology and
biogeography are premised on the relationship between species diversity
and area (Kadmon and Allouche 2007), which includes the concepts of
equilibrium and steady state. Much of this builds on the theory of island
biogeography (Lomolino and Brown 2009).

Table 16.1. The Theory of Evolutionary Biogeography

*Domain: The interplay between the dynamic processes of geophysical change and evolutionary change
so as to shape the distribution and diversity of species.*

Propositions:

1. All else being equal, larger areas are able to support a larger number of individuals.
2. Isolation influences the ability of organisms to colonize new or existing habitats in their
   range.
3. Time interacts with isolation to determine the evolutionary outcome of immigration and
   extinction dynamics.
4. Geological processes are dynamic and occur with characteristic frequencies.
5. Climatological processes are dynamic and exhibit periodicity.
6. Cycles of fission and fusion generated from the dynamism of geophysical and climatic
   processes over varying spatiotemporal scales shape the distribution and diversity of
   species.

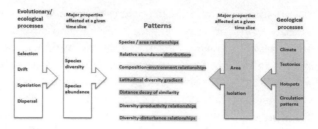

Figure 16.2. The interaction between processes and emergent properties of evolution (no shading) and geology and climatology (gray shading). Evolutionary and ecological processes can be distilled down to 4 major components: selection, drift, speciation, and dispersal (Vellend 2010). Together, these processes add species to a community. At the same time, the milieu in which organisms find themselves is itself dictated by four major geological processes: climate, tectonics, hotspots, and air and water circulation patterns. The changing dynamics of these processes dictate the area and isolation of a given locale at a given time. Patterns that have been the focus of much biogeographic research (center) usually combine some elements of both sets of processes.

Area and the Number of Individuals

The first proposition is that all else being equal, larger areas are able to support a larger number of individuals. Species richness has long been recognized to scale with area (Arrhenius 1921; Rosenzweig 1995; Storch et al. 2012; Harte et al. 2013). One of MacArthur and Wilson's (1967) key insights was that larger areas have the potential to support larger populations of species, thereby reducing extinction rates from demographic stochasticity and allowing higher species diversities to be attained (see also Hubbell 2001). Larger populations also lead to higher standing genetic diversity, which may help promote rates of evolution. Area is also broadly correlated with other aspects of the landscape that promote speciation, including habitat diversity and heterogeneity and potential allopatric barriers (Losos and Parent 2009; Kisel and Barraclough 2010; Coyne 2011).

Much has been written on the species-area relationship and the endemics-area relationship. The species-area relationship is well recognized as one of the most fundamental concepts in ecology and biogeography (H. G. Martín and Goldenfeld 2006). Variation in its shape across taxa and between regions has now been explained to an impressive degree by various scale-collapse theories (Harte 2011; Storch et al. 2012), which indicate that variation in the relationship results from variation in key underlying scaling parameters—namely, the average range size and average ratio of species to individuals in a community. However, because most studies have been based on pattern or on outcomes of processes, rather than the processes themselves, the cause for variation in these scaling parameters, and thus the ultimate explanation for variation in their relationships, remains elusive. Such variation is likely due to different evolutionary histories underlying different components of the biota, thus motivating the need for a more synthetic theory of evolutionary biogeography.

The process underlying the species-area relationship has been studied most famously in the context of immigration and extinction (MacArthur and Wilson 1967). At any instant, the number of species in an area is expected to be a balance between immigration and extinction. Immigration is affected by distance from the source; extinction is dictated by area with an additional effect of distance from the source through the rescue effect (J. H. Brown and Kodric-Brown 1977). Although the early stages of establishment have been explored (Simberloff and Wilson 1969), the model itself is static. But the nature of the equilibrium—and indeed whether an equilibrium exists—remains highly controversial. Indeed, the equilibrium is predicated on the assumption that there is a fixed mainland species pool, so immigration will decline simply because the source pool is depleted relative to the target (Harmon and Harrison 2015). Species diversities may not be set by ecological limits because biotas are open to new species.

While adaptive radiation appears to be initiated, at least partially, by the availability of niche space (R. G. Gillespie and Parent 2014), there is little evidence that ecological opportunity ever gets used up, and caution should be observed regarding "overly simplistic and outdated ideas about equilibria and carrying capacity" (Harmon and Harrison 2015, 590). For islands, then, speciation may lead to ever-increasing species diversity, limits being set only by the ontogeny of a given island (R. G. Gillespie 2013). There is some evidence to suggest this might be the case, though not always (R. G. Gillespie and Baldwin 2010; Rabosky et al. 2015). For example, similar cichlid ecomorphs (though not closely related species) can often coexist in Lake Tanganyika, apparently because the advanced age of the adaptive radiation has resulted in an accumulation of other attributes that allow them to occupy slightly different niches (Muschick et al. 2012). A similar pattern has been documented for *Anolis* lizards in the Caribbean (Losos 2009). Co-occurrence of ecologically similar species may be quite common in nature (Liebold and McPeek 2006; Wiens 2011). The corollary is that particular environments may allow many evolutionary outcomes, rather than a fixed equilibrium of defined niches. Together, studies to date suggest that there may be some form of equilibrium or steady state, but the rate at which it might be attained differs between taxa (Rominger et al. 2015). Thus, we suggest a more nuanced relationship between species richness and area: Organisms at every hierarchical level differ in the spatial framework that allows isolation, and the temporal framework in turn allows differentiation (Papadopoulou and Knowles 2015).

Landscape dynamics, at whatever temporal scale, may lead to change at a rate sufficient to prevent communities from ever fully occupying niche space, assuming this is possible at all. Overall, there is little evidence for negative relationships between diversity and abundance or diversity and niche breadth, which would be expected for saturated communities. Where studied, biotas have almost always been found to be open to new species (Harmon and Harrison 2015). While area clearly sets some kind of constraint, there appears to be no fixed limit to the number of taxa that can be accommodated at a given locale.

Isolation

The second proposition is that isolation influences the ability of an organism to colonize new or existing habitats in its range because the species pool available to communities is much reduced in more isolated areas (MacArthur and Wilson 1967). Isolation can be of two major forms: geographic and ecological. Geographic isolation tends to dominate in taxa that are dispersal limited, in which case there is a greater tendency to move between ecologically different habitats within a locale. Alternatively, when ecological isolation dominates, species will tend to retain their ancestral ecological preferences over time and space, showing a pattern of niche conservatism and with geographic separation playing a less prominent role (e.g., Wiens and Graham 2005; Wiens et al. 2010) because organisms track their habitats over space and time (Donoghue 2008). This process leads to the possibility of environmental isolation, in which unsuitable habitat isolates populations and clades in regions of suitable habitat. Environmental isolation has been used to account for differences in geographic isolation between temperate and tropical areas (Janzen 1967).

Isolation of any habitat is relative to the organism in question, in particular the extent to which it is restricted to a given habitat, the extent to which it can cross the intervening matrix of inhospitable habitat, the distances involved, and the dispersal capability of the organism at a given time. As habitats become isolated, the frequency of dispersal is reduced to the point that gene flow is insufficient to prevent speciation but sufficient to allow repeated anagenesis (Rosindell and Phillimore 2011). With increased isolation, gene flow becomes sufficiently low that most species arise through cladogenetic speciation (Edwards et al., chap. 15). In these most isolated habitats, the available ecological space can be filled through in situ evolution of new species (MacArthur and Wilson 1967), often through adaptive radiation (Schluter 2000). The isolation necessary for the rate of in situ cladogenesis to exceed immigration and anagenesis has been termed the "radiation zone" (MacArthur and Wilson 1967). The physical separation required for this effect to be manifest is clearly dependent on dispersal abilities; for example the radiation zone for mammals is much closer to a source (geographically) than it is for many insects (R. G. Gillespie and Baldwin 2010). The interaction between speciation and colonization is complex and not additive. In more isolated locations, different niches will tend to remain relatively open for long periods and consequently may be filled by evolution as readily as or more readily than by dispersal.

Interactions with Time

The third proposition is that time interacts with isolation to determine the evolutionary outcome of immigration dynamics. The importance of time is assumed in many biogeographic models, as it clearly plays a fundamental role in patterns of species accumulation through the net effects of immigration, speciation, and extinction. For example, studies on defaunated islands show that species accumulate over time, with those close to a mainland source reaching some kind of steady state in species richness (Simberloff 1974). On more isolated islands, accumulation takes longer, for example with the youngest island in the Hawaiian chain still accumulating species (R. G. Gillespie and Baldwin 2010).

Genetic isolation among populations, which increases with distance, also depends upon time. Genetic isolation is the inverse of connectivity—as populations become less connected by gene flow, they become more isolated and are more likely to diverge genetically by selection or random drift. Simple models suggest that the amount of isolation needed for divergence to proceed by drift alone corresponds to much less dispersal than one migrant every four generations (Wright 1951; Crow and Kimura 1970). With more isolation, divergence through drift is possible after a given amount of time, an effect accelerated by selection. With less isolation and time (i.e., more connectivity), divergence will not be possible.

Geological Processes

The fourth proposition is that geological processes are dynamic and occur with characteristic frequencies. The geographical processes that lead to contrasting periods of isolation and interchange between biotic communities never stop. However, they differ enormously in frequency (days through years to millennia, and epochs to eons) and amplitude (magnitude of the barrier that they impose on the biota). Plate tectonics and supercontinent cycles provide perhaps the longest periods between episodes of ocean opening and closing (J. T. Wilson 1966). Post-Archean Earth has gone through repeated periods of assembly and breakup of supercontinents. Examination of the frequency of these cycles in the context of the fossil record shows that they are strongly associated with a controlling connection between supercontinents, climate, and biogenesis (Worsley et al. 1984; Nance et al. 1988; Worsley et al. 1991). While our knowledge of the details of the supercontinent cycle has progressed considerably in recent years, the basic premise remains the same (Nance and Murphy 2013). It appears that there have been five pre-Pangaean supercontinents (Nance et al. 1986) at ca. 0.6, 1.1, 1.8–1.6, 2.0, and 2.6 Ga (fig. 16.3), each having a major effect on biogenesis.

Other events also are associated with the frequency of tectonic events. Divergent boundaries that occur along spreading centers as plates are moving apart give rise to either midoceanic ridges or midcontinental rift zones. For example, the Africa Rift Lakes are created as the African Plate pulls apart into the Somali and Nubian Plates; it will ultimately break apart with the formation of a new ocean basin. This process has led to the periodic formation of isolated lakes in the middle of the continent, which has provided the setting for adaptive radiations such as that of cichlid fish (Salzburger et al. 2014). Convergent boundaries also occur where plates come together such as subduction zones where an ocean plate meets a continental plate. Episodes of supercontinentality have also been associated with epeirogenic uplift, accretionary orogeny, low sea level and loss of shallow marine habitat, and cold climates (Nance and Murphy 2013). All these processes have cycles of fusion and fission.

Associated with tectonic movements, island hotspots also show cycles of emergence and decline. Most islands within the deep ocean basins are hotspot volcanoes and may lie on midocean ridges (e.g., Iceland), near ridges (the Azores, the Galápagos), or far from them in the plate interiors (e.g., the Hawaiian and Marquesa Islands in the Pacific Ocean, Réunion and Mauritius in the Indian Ocean). Such remote volcanic island chains may have formed owing to the movement of a plate over a stationary

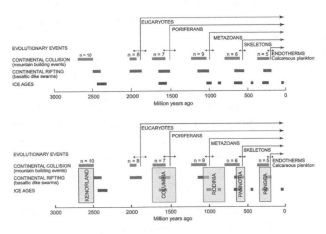

Figure 16.3. A comparison of episodic events in Earth history linked
to the supercontinent cycle (after Nance and Murphy 2013).

"hotspot" in the mantle (J. T. Wilson 1963); as the overriding lithosphere moves the volcanoes away from the hotspot, the island sinks.

This proposition holds for plate tectonics, plate boundaries, global plate reorganization, normal magmatism, melting anomalies, volcanic chains, and mantle geochemistry (D. L. Anderson 2005). In each case, landmasses go through cycles of appearance and recession, together with fusion and fission, though the geographic areas involved, the lifespan of the landmass, and the frequency of separation and joining vary enormously. However, it is the scale of these events that dictates how organisms evolve in a biogeographic context.

Climatological Processes

The fifth proposition is that climatological processes are dynamic and exhibit periodicity. That periodicity can be regular or irregular. Paleoclimates change continually as continents move and are redistributed along with changes in atmospheric chemistry, solar output, and the Earth's orbit (Zachos et al. 2001). The same cycles can affect other earth processes, most obviously sea level fluctuations. The frequency of these cycles is complicated by multiple patterns of millennial and submillennial variability that are superimposed on long glacial cycles (McManus et al. 1999). Many theories have been developed to explain these fluctuations, with promising insights being provided through the development of theory that builds on stochastic, dynamic systems (Crucifix 2012).

Climate fluctuations tend to have a shorter period than supercontinents, with 40 kyr and 100 kyr cycles, owing to the collective effects of eccentricity, axial tilt, and precession of the Earth's orbit. In the Paleozoic (542 to 251 mya) there were shorter-term base-level changes of ~0.5–3.0 myr duration and others that cycled every ~1.7 myr (Haq and Schutter 2008). In the Mesozoic, periods of warming seem to be related to ~2.4-myr eccentricity modulation cycles, while in the Cenozoic periods of cooling are correlated with astronomical cycles, suggesting that different forces play the dominant role during an icehouse versus a greenhouse world (Boulila et al. 2011). The well-known climatic fluctuations of the Pliocene and the Pleistocene—known for the formation of ice sheets and associated glacial cycles—had an average period of about 40 kyr (Ruddiman et al. 1986) until about 800 kyr ago. Climate cycles affect additional events, such as volcanism, resulting in complex feedbacks between systems (Kutterolf et al. 2013). Environments are constantly going through cycles that make them alternately suitable and unsuitable for habitation by different biotic assemblages.

Varying Spatiotemporal Scales

The sixth proposition is that cycles of fission and fusion generated from the dynamism of geophysical and climatic processes over varying spatiotemporal scales shape the distribution and diversity of species. The separation and connection of suitable habitats through the processes of tectonics or other geological and climatological events provide conditions for colonization, speciation or extinction, hybridization or introgression (fig. 16.2). Area and isolation are not involved in such processes themselves; rather, they impose limits on those processes. Cycles of fusion and fission can take place at many different scales. At the longest timescales, the supercontinent Pangaea was formed in the Carboniferous, with Laurasia (which represents the present-day Northern Hemisphere landmasses), moving toward the equator to join the large Southern Hemisphere landmass Gondwana (Briggs 1987). The subsequent fission of both Gondwana and Laurasia imposed isolation on previously connected biotas, which led to some of the most well-characterized biotic disjunctions. Here, the isolation created by geological events is fundamental. At the same time, repeated fusion also plays a key role. For example, the Great American Interchange associated with the formation of the Isthmus of Panama ca. 3 Ma allowed the exchange of biotas of North and South America, each having evolved in isolation (Leigh et al. 2014). The result was extensive mixing, coupled with extinction of nonrandom elements of the biota.

Geological and climatological processes over shorter timescales and over smaller spatial scales are also keys to the diversification process. Over periods of 10–100 kya, climatological events have played a role as species "pumps," in which periods of warming or drying served to alternately isolate and reunite biotas, as illustrated by Indonesian ants in which diversification has been attributed to repeated Pliocene changes in sea level (Quek et al. 2007), Mediterranean beetles for which island connectivity cycles have driven population divergence (Papadopoulou and Knowles 2015), and the Amazonian vertebrates whose Pleistocene diversification is due to episodic isolation associated with shifts in geology or climate (Haffer 1997). Dynamics of geology and climate also can constrain diversification, as suggested for reduced genetic diversity in coral reef fish in French Polynesia associated with a reduction of habitat with sea level change in the Holocene (Fauvelot et al. 2003). Even in the oceans, which are characterized by the potential for extensive mixing, opposing processes of isolation and exchange appear to have been responsible for much diversification (Bowen et al. 2013).

At more recent timescales and localized areas, the same mechanisms of fusion and fission resulting from the combined role of geological and climatological events have played a key role in the generation of adaptive variation and functional novelty within populations of cichlid fish in African lakes (Loh et al. 2013). The repeated isolation and subsequent mixing of populations in new combinations may provide an evolutionary crucible that can facilitate and potentially accelerate diversification (Carson 1990), with admixture among successively introduced populations playing a potentially pivotal role by providing the genetic variation to allow adaptive evolution (Rius and Darling 2014).

Conclusion

The abiotic environment has a supremely dynamic nature with changes dictating the size, isolation, and overall hospitability of a given site for supporting life. Most of the forces that affect the environment have cycles, leading to repeated emergence and subsidence together with separation and merging of favorable sites. Geological cycles can lead to fragmentation and merging of landmasses over very long (e.g., plate tectonics) or short (e.g., lava flows fragmenting forests with subsequent regeneration) timescales. Likewise, climatological cycles can lead to warming and cooling, also affecting existence and connectance of habitable sites.

Organismal diversity evolves across this dynamic geographical milieu.

Evolutionary biogeography is the product of evolutionary change that can occur within the context of geological or climatological events that allow habitable landmasses to appear, disappear, merge, or fragment. If the period between isolation and merging of an existing landmass is too short (or if selection is lacking), there is insufficient time for species to diverge (fig. 16.1a). At a certain frequency, a cycle of fragmentation and fusion can provide sufficient time for speciation (fig. 16.1b). Extended intervals between cycles create unique biotas, though not adding to species diversity (fig. 16.1c). Thus, from an evolutionary perspective, it is the rate of adaptation and differentiation of organisms relative to the periodic availability of suitable habitat that dictates the biogeographic outcome.

Most biogeographic theories to date have focused on the effects of area and isolation in dictating patterns of species diversity. In other words, they consider a landscape that is effectively static. Incorporation of changes in the environment is limited to niche theory that explicitly models the response of organisms to abiotic shifts, though with increasingly sophisticated approaches for examining variation in niche and trait evolution using hierarchical models and statistics. The theory presented here (table 16.1) combines the key importance of changing area and isolation—and hence likelihoods of anagenesis, cladogenesis, or extinction—as part of niche modeling and phylogenetic reconstruction. Still needed are specific models built on this theory that can be used to understand and predict particular biogeographic patterns.

Most biogeographic studies have been based on patterns or outcomes of processes, rather than the processes themselves. A synthetic theory of biogeography requires incorporation of the key role, not only of the habitat of an organism, but of time it has in that habitat. Thus, the theory requires assessment not only of area and isolation, but also of how these change so as to facilitate divergence, speciation, and extinction. It requires integration of the abiotic processes that dictate not only habitable area (as can be achieved through niche modeling), but also the configuration and longevity of that habitable area for a given organism. Neutral theory, together with the metabolic theory of ecology and the maximum entropy theory of ecology, with their predictive success and mechanistic simplicity, provides a potential avenue forward, although none can, as yet, incorporate the dynamic nature of biodiversity or the geographical milieu. A synthetic theory that can incorporate the effect of the changing landscape on evolutionary processes of adaptation, species formation, and extinction is key to understanding biodiversity, past, present, and future.

In formalizing the constitutive theory of evolutionary biogeography, the major advance made is in recognizing that biogeography lies at the intersection of two evolutionary processes, each with its own largely independent dynamics: (1) The matrix within which organisms exist undergoes cycles of expansion and contraction, appearance and disappearance. In the same way (2) organisms increase and decrease in numbers as a result of the matrix and adapt or go extinct according to the arrangement and area of the matrix, its age and longevity. Both dynamics can be understood in terms of various domain-specific theoretical paradigms. However, major advance in the theory of biogeography will clearly require developing connections across domains.

Acknowledgments

We are most grateful to Sam Scheiner and David Mindell for inviting us to contribute this chapter and for all their suggestions and insights. The manuscript was much improved by their comments as well as those from two anonymous reviewers. The research was supported by the National Science Foundation DEB 1241253.

SEVENTEEN

Macroevolutionary Theory

DAVID JABLONSKI

Although we have a detailed mathematical theory of microevolutionary change, the salient features of macroevolutionary change are not explained by an extrapolation of this theory.

—(Hansen 2012, *220*)

The theory of macroevolution is a theory of scale and hierarchy. Evolution above the species level involves the same fundamental components as evolutionary theory as a whole—the generation and sorting of variation (Jablonski 2000)—but macroevolutionary theory becomes important when ostensibly general evolutionary theories afford exclusive agency to genes of minute and random effects, to the local scale, and to the genic or organismic level. Here I explore propositions relating to large spatial and temporal scales and the hierarchical organization and dynamics of genealogical units (table 7.1). Thus, the origin and fates of major evolutionary novelties, the long-term evolutionary role of rare events ranging from the internal redeployment of gene regulatory networks to externally driven mass extinctions, and the potential for emergent properties or dynamics at different hierarchical levels are key macroevolutionary issues.

A multilevel, multiscale view of evolution is seen in many works, including Darwin's *The Origin of Species* (1859; see Gould 2002; Futuyma 2015). History, scale, and hierarchy are now entrenched in the evolutionist's toolkit to an unprecedented degree, and paleontology and developmental biology are more fully incorporated into evolutionary theory and analysis than ever before. Whether these changes represent an overturning, an expansion, or a minor polishing of the neo-Darwinian theory of fifty years ago depends entirely on whose version of neo-Darwinism is used;

Table 17.1.   The Theory of Macroevolution

Domain: Evolution above the species level.

Propositions:

1. Evolution occurs within a hierarchy of genealogical units.
2. Entities at each level can have emergent properties, but the dynamics of the levels are mechanistically linked via upward and downward causation.
3. Traits at one level, e.g., organismic phenotypes, can hitchhike on the differential origination or extinction of units at high levels, e.g., clades.
4. Macroevolutionary currencies (taxonomic richness, functional diversity, phenotypic disparity) are loosely correlated, but the relationship can sometimes be strongly nonlinear or nonsignificant.
5. Contingency (historical events) plays a significant part in the evolution of phenotypic and taxonomic richness of clades.
6. Phenotypic variation is nonrandom owing to the underlying structure of developmental processes.
7. Phenotypes can accommodate localized change owing to the modular nature of development and the epigenetic signals and responses inherent in the developmental systems.
8. Existing developmental pathways can be altered or redeployed to produce new phenotypes, allowing for more efficient generation of variation than the great majority of purely *de novo* mutations.
9. Species originate along a variety of phenotypic trajectories, with all combinations of evolutionary tempo and mode.
10. Evolutionary trends may occur at any level in the taxonomic hierarchy, producing nested patterns that need not coincide.
11. Speciation and extinction rates are governed in part by intrinsic organismic and species-level traits.
12. Interaction among clades can be positive or negative and multi-way, reciprocal or one-sided.
13. Mass extinctions can remove or marginalize incumbent taxa and promote diversification, but unevenly among surviving clades.

compare, for example, Gould (2002), Jablonski (2007), Pigliucci and Müller (2010a), Futuyma (2015), and Laland et al. (2015). Here, I start with some basic components such as scale, hierarchy, and contingency and then develop key concepts in the origin and sorting of variation in a macroevolutionary framework, including potential relationships among different aspects of biological diversity ("macroevolutionary currencies") and how rare events such as mass extinctions can be incorporated into theory. Owing to the history of macroevolutionary study, the focus is almost entirely on multicellular animals. I conclude with a brief consideration of whether the macroevolutionary process itself has evolved and a final comment on some of the most promising ways for generating a fuller macroevolutionary theory.

Scale and Hierarchy

Scale enters the domain of macroevolution because evolutionary phenomena viewed on long timescales need not flow smoothly and predictably from those observed over the short term, and phenomena at the provincial, continental, or global scale do not always flow from those observed locally. Empirical examples, important because they were not expected from dominant theories and models of the time, range from the morphological stasis or nondirectional random walks common in the fossil record at the 1–10 myr timescale, rather than the sustained evolutionary transformations formerly expected in light of the evolutionary responsiveness of local populations on annual or decadal timescales (Hunt 2007b), to evidence that mass extinction events can qualitatively change survivorship patterns and thus redirect evolutionary trajectories in ways inconsistent with dynamics in calmer intervals (Jablonski 2005b). Such predictive failures do not necessarily mean that novel processes are required to operate at those scales, but at the very least indicate that a macroevolutionary theory cannot consist of simple extrapolation from short-term, local models and observations.

In contrast to scalar measures, many biological hierarchies involve nested entities—individuals in the philosophical sense (Nathan and Cracraft, chap. 6)—with distinctive properties at each level, such that events at each level can propagate upward to larger, more inclusive entities and downward to their constitutive components (proposition 2, table 17.1). However, one attribute of a nested hierarchy is asymmetry of effects: dynamics at lower levels need not be manifest at higher levels, whereas dynamics at higher levels always propagate downward (e.g., Valentine and May 1996; Salthe 2013). Thus, a parasitic DNA sequence might never proliferate to the point of reducing the fitness of the host organism, and many selectively driven changes in organismal phenotype may have little effect on the extinction probability of their species or clade relative to a sister group. But the preferential loss of certain species owing to their narrow geographic ranges will necessarily remove, by downward causation, the associated array of clade-specific organismic traits, along with clade-specific parasitic DNA sequences in their cells (proposition 3, table 17.1). And independent, deterministic selective forces within species can produce effective randomness at the clade level (e.g., McShea and Brandon 2010).

The basic framework of macroevolutionary theory is a genealogical hierarchy, comprising genes, organisms, demes (genetically defined conspecific populations), species, and clades (reviewed from different viewpoints by Valentine and May 1996; Gould 2002; Jablonski 2007; and Tëmkin and Eldredge 2015; see also Kearney, chap. 7). Macroevolution is often analyzed using another hierarchy, that of formal taxonomy, frequently focused at the genus level, in part to reduce species-level sampling biases, but also as a rough proxy for ecological and functional diversity. Although taxonomic ranks are notoriously subjective, evidence is accumulating that genera, while imperfect, correspond sufficiently to genealogical units that they can be used for many purposes (e.g., Jablonski and Finarelli 2009; Soul and Friedman 2015), and an analysis of genetic distances finds that the lower taxonomic ranks are more comparable across orders, classes, and phyla than generally assumed (Holman 2007). Just as important, progress in modeling dynamics in the taxonomic hierarchy has led to insights from the frequency distribution of lower taxa within higher ones, including support for the claim that such distributions reflect natural evolutionary processes and not simply random agglomerations of low-ranked taxa (Holman 1985; Foote 2012; Maruvka et al. 2013; Humphreys and Barraclough 2014). More work is needed on integrating phenotypically based taxa and molecular phylogenies, including taxonomic ranking protocols that maximize the utility of such taxa for macroevolutionary analysis—particularly given that virtually all fossil taxa will forever lack sequence data but provide an essential window into the timing, location, and dynamics of past phenotypes.

Macroevolutionary Currencies

Biodiversity has many dimensions, but three macroevolutionary currencies that have received special attention are taxonomic richness, morphological disparity, and functional variety (proposition 4, table 17.1). These variables tend to be broadly correlated, and the use of higher taxa as rough proxies for disparity and functional variety has been validated repeatedly (Erwin 2007; Jablonski 2007; Chao et al. 2014), although such relationships tend to break down at finer timescales and among geographic regions (Jablonski 2008a; Edie et al. 2018). Further, higher taxa tend to correspond to functional groups or adaptive zones for animals, but major plant clades often split along reproductive lines with multiple convergences in phenotype and function (Donoghue 2008). Because the times and places where the different currencies are least correlated or most strongly nonlinear in their association are of much interest, evolutionary models must go beyond the proxy assumption and treat the different currencies independently. Study of the relationships among those currencies, and how they directly or indirectly affect one another, is clearly a growth area for macro-evolutionary theory.

Contingency

The hierarchical framework is essential, but mechanistic models are difficult because macroevolutionary outcomes also depend heavily on history (proposition 5, table 17.1): the initial conditions—the raw material provided by biological entities at any hierarchical level—and the environmental context (Beatty 2006; Erwin 2015b; Turner 2015; Mindell and Scheiner, chap. 1). The challenge is to develop theories and models that illuminate the relative contribution of historical events to macroevolutionary patterns, as they intertwine with and influence the factors discussed below. The most straightforward approach would simply be to run replicated, controlled experiments from the same starting point, plainly impossible for metazoans over geologic timescales, but feasible for long-term laboratory populations, which frequently exhibit contingency effects (Blount 2016).

Macroevolutionary Lags and Contingency

One indirect indicator of contingency is the perpetual difficulty in pinpointing key innovations, i.e., characters or character states that trigger taxonomic diversification. Although some newly evolved traits appear to be closely, and even repeatedly, associated with diversifications, many cherished novelties, from pharyngeal jaws in teleosts to multicellularity in eukaryotes, are associated with prolific diversification in some clades but not others, or have proven to originate long before the diversifications once causally attributed to them. Further, some striking traits seem never to promote diversification, as in the low species numbers of flamingos and anteaters despite their impressive divergences from ancestral phenotypes. Still other diversifications are not associated with any recognizable evolutionary novelty. The evolution of, for example, elaborately cross-regulated gene networks, internal chemosymbionts, or endothermic metabolism should pay immediate dividends by promoting ecological dominance or prolific diversification; each feature is arguably integral to the success of one or more major groups, but the fossil record indicates prolonged delays.

The temporal gap between the origin of a clade and its diversification or rise to ecological dominance (very different issues) has been termed a macroevolutionary lag (Jablonski and Bottjer 1990). Such lags are neglected tools for probing the relation between character evolution and clade dynamics and provide a vehicle for the dissection of evolutionary contingency and its causes and consequences. At least three general mechanisms have been proposed. The first is artifacts, whether of the geometry of exponential diversification, or of the vagaries of paleontological or phylogenetic sampling (Jablonski and Bottjer 1990; Foote 2010; Patzkowsky 2016). The second is intrinsic factors, such as a need to accrue additional traits downstream of a putative "key innovation" before diversification can occur, although character states that alter speciation rates or extinction rates, for example via geographic ranges and genetic population structures, might also trigger diversifications long after more dramatic organismic traits are in place. The third is extrinsic factors such as ecological opportunities, whether by arrival on an empty island or by elimination of major competitors through a mass extinction, providing belated "key opportunities" (Moore and Donoghue 2007). Macroevolutionary lags and other poor correspondences between character evolution and taxonomic diversification may indicate that changes in both extrinsic conditions and intrinsic traits are almost always required for prolific diversifications or radiations (Bouchenak-Khelladi et al. 2015).

Convergence and Contingency

In contrast, catalogs of convergences, another pervasive evolutionary phenomenon (McInerney, chap. 5), have been used to downplay the role of contingency (e.g., Conway Morris 2003; Vermeij 2006). However, convergences also derive in part from the contingent, limited developmental and thus evolutionary capabilities of lineages confronted by similar environmental challenges (e.g., Wimsatt 2001; McGhee 2011). Further, convergences are almost always inexact, so that later modifications and elaborations are unlikely to be equivalent functionally or in terms of potential evolutionary directions. The squid camera-eye cannot be mistaken anatomically for that of a mammal, for example. Convergences are yet more incomplete at the clade level: even the celebrated convergence of Australian marsupials and placental mammals includes unique forms on each side—kangaroos and koalas, bats and elephant seals—and alternative solutions to a given evolutionary challenge also abound (McGhee 2011). Finally, as Sterelny (2005) argues, subjective equivalencies can mask profound evolutionary differences, as with human agriculture versus the "agriculture" practiced by leafcutter and other ants (classed as convergence by Conway Morris 2003).

A more effective meta-analytical approach to evolutionary contingency cross-tabulates apparent convergences against phylogenetic distance. If contingency is important, increasing phylogenetic distance should erode inherited similarities in phenotype, evolutionary-developmental capabilities and limitations, and biotic and abiotic pressures, so that the frequency of convergence should decline accordingly, as found by Ord and Summers (see also Losos 2010, 2015). These results show, if nothing else, another way in which phylogeny informs models for any of the macroevolutionary currencies.

The Origin of Variation

Mindell and Scheiner (chap. 1) place the origin of variation outside the immediate domain of evolutionary biology, but a theory of macroevolution must account for the distinctly nonrandom production of phenotypic variation at large spatial and temporal scales (proposition 6, table 17.1). This nonrandom variation occurs at multiple hierarchical levels in all currencies and is manifest throughout the history of life. These patterns require attention because the starting point for most theories of variation—sometimes acknowledged as an operational simplification—has generally been random mutational inputs. The next assumption is generally that traits are universally underlain by many genes of small additive effect that mutate independently, with the probability of an increase in fitness inversely related to the magnitude of their phenotypic effects (Fisher 1930; Tenaillon 2014). Those genes interact to affect multiple traits, and the extent and apparent randomness of those effects, relative to the attributes of the phenotype, increase the probability that a mutation is deleterious. Such a model is powerful for short-term population studies (Frank and Fox, chap. 9), but when we consider large-scale evolutionary change, these assumptions must be relaxed or modified—not toward older models of macromutation and evolutionary saltation, although such events as genome duplications and acquisition of endosymbionts might represent modern incarnations of such discontinuities, but to incorporate growing knowledge of the relation between development and evolution (Love, chap. 8). Here I will touch only on issues that enter into the macroevolutionary framework but have yet to be fully integrated into theory or provide a foundation for modeling patterns.

Control Hierarchy

A starting proposition for a macroevolutionary theory of variation is the now-commonplace observation that development is governed by semi-hierarchical networks of genes, meaning a few-to-many structure that also contains cross-level feedbacks (e.g., S. B. Carroll et al. 2005; Peter and Davidson 2015; Love, chap. 8). This structure is widely held to affect the probability of different transitions, so that the probability distribution of the raw material for evolution in a genotype or phenotype space is not isotropic but uneven, skewed, or channeled. This probabilistic approach to developmental constraint, with some evolutionary directions absolutely unavailable and others accessible to varying degrees (Maynard Smith et al. 1985; Schwenk and Wagner 2004; Klingenberg 2005; Gerber 2014), is related to the concept of evolvability (Brigandt 2015, and below) and may enable stronger mechanistic connections between development and differences in clade behavior in morphospace (Salazar-Ciudad and Jernvall 2010; Gerber 2014).

Modularity

Also widely recognized is that development, and therefore its evolution, is modular, i.e., organized into semi-independent compartments, such that changes in gene expression in one module are more likely to affect gene regulatory networks, and ultimately the phenotype, of that module than of other modules (proposition 7, table 17.1). Thus "universal pleiotropy" and epistasis—the rule that each gene affects many traits and traits are determined by many genes of equal and small effect (Fisher 1930; Wright 1968)—is not as pervasive or chaotic as often assumed (G. P. Wagner and Zhang 2011). Nonetheless, the discreteness and long-term stability of developmental modules, and the relation between modules of molecular circuits and sets of phenotypic characters that covary in a modular fashion, are still subject to much debate and research (e.g., Goswami 2006; Klingenberg 2014; G. P. Wagner 2014). The fact that mosaic evolution is entrenched in our textbooks reflects the extensive role of modularity in macroevolution—as does the bland statement that every taxon is an amalgam of derived and primitive characters. Modularity and its converse, integration, must influence phenotypic transition probabilities at any point in time, and as they must also evolve, the maintenance, fusion, and parcellation of modules are likely targets of selection that can have macroevolutionary consequences by disallowing or promoting future directions for change.

Tinkering

Much phenotypic change can be effected by "tinkering" with development (proposition 8, table 17.1), i.e., small modifications in the timing, location, or combinations of developmental events (S. B. Carroll et al. 2005; Lieberman and Hall 2007; Shubin et al. 2009). Such tinkering is of course facilitated by developmental modularity and can generate a range of phenotypic outputs, from the imperceptible to the dramatic. Thus, new structures need not evolve from scratch but can arise by significantly modifying or redeploying ancient gene regulatory networks; highly polygenic structures can shift phenotypically—in certain directions—by alteration of preexisting gene regulatory networks, and not only by mutations in protein-coding genes.

Epigenetics

Gene regulatory networks, and other levels of control that include chromatin and noncoding RNAs, can be responsive—within limits—to extrinsic signals, ranging from chemical or physical interactions among modules in developing embryos to environmental factors such as temperature. These epigenetic responses promote the incorporation of altered modules into an integrated, functional phenotype (Hallgrímsson and Hall 2011) and create the potential for adaptive phenotypic plasticity (West-Eberhard 2003; Gilbert and Epel 2009; Kirschner and Gerhart 2010; Scheiner, chap. 13).

From Populations to Clades

Theory is beginning to incorporate these insights to explore how the relation between development and evolution should transform classic population-genetic models so that they can be used in a macroevolutionary context. A more realistic genetic architecture that has a range of effect and interaction sizes and that can itself evolve and be related to the generation of phenotypic variation is an important component (Hansen 2006, 2012; Rice 2012; Rajon and Plotkin 2013; Badyaev and Walsh 2014), and the evolution of the genotype-to-phenotype map and its modularity can be approached in population- and quantitative-genetic terms (G. P. Wagner 2010; Pavlicev and Wagner 2012). However, we are far from integrating such models with, for example, the behavior of clades in morphospace. The venerable framework of the fitness landscape may prove to be useful. However, a model landscape whose topography is determined by the fitness of genotypes becomes increasingly more rugged as it becomes more realistic for large spatial and temporal scales. History—and the inhomogeneous production of phenotypic variation around starting points—has little effect on the behavior of populations on smooth landscapes with a single fitness peak: they will reliably converge on that peak, albeit at different rates (Lachapelle et al. 2015). In contrast, a rugged landscape, with multiple peaks and valleys as Wright (1931) envisioned, hinders access to all but one or a few local peaks owing to maladaptive gene combinations between the peaks, so that the contingencies of starting position and the dynamics of the landscape itself become important.

Such rugged fitness landscapes are almost certainly the rule, and are often conceptualized in phenotypic terms as "adaptive landscapes" (G. G. Simpson 1944b; S. J. Arnold et al. 2001; M. A. Bell 2012; Hansen 2012). When phenotypes are under selection to satisfy multiple requirements (e.g., feeding, growth, reproduction), trade-offs reduce absolute fitness but allow different trait combinations to be roughly equivalent (Niklas 2009; Marshall 2014). Here too, chance and history become increasingly important with increasing ruggedness, as lineages tend to be confined to the peak nearest their starting location. The developmental factors discussed here, however, offer potential mechanisms for crossing what would otherwise be fitness valleys by phenotypic changes coordinated among parts; in the most extreme view, developmental coordination among traits makes the fitness valleys disappear. The corollary to this potential, however, is that the patterns of taxonomic occupation within a morphospace (e.g., the uneven distribution of species or higher taxa through the space) cannot be taken as a map of the adaptive landscape, because that density is a function not only of fitness but also of accessibility given the starting points of clades, the properties of their developmental systems, and factors governing speciation and extinction rates that need not be closely tied to fitness differences at the organismic level (McGhee 2011; Huang et al. 2015).

Species within Clades

The origin of variation within and among species, and how it is translated into net change at the clade level, hinges on two basic issues: first, how the developmental system of organisms governs (or not) the direction of species movement through phenotypic space, and second, the tempo and mode of such changes.

Direction

For species within clades, one question is how aspects of a phenotype, or its underlying gene regulatory networks and developmental modules, are related to the probability density function of phenotypic change around a given starting point. One simple approach would be to take the cloud of phenotypic variances and covariances existing around that starting point as the probability density function (e.g., Schluter 1996; Shubin and Wake 1996; Hunt 2007a). Not all studies of such within-versus-among species variation can distinguish the raw inputs of developmental systems from the selective processes that filter that variation. Nevertheless, these observations suggest a potential first-order hypothesis for macroevolutionary analysis applicable to both extant and fossil systems: new species within a clade are most likely to originate in the existing directions of variation within each parent species. This hypothesis can in turn be related to an influential explanation for the association of phenotypic change with speciation, the view that speciation stabilizes otherwise-transient phenotypic variation by severing or attenuating gene flow to a local population (Futuyma 1987, 2015; Edwards et al., chap. 15). Such a process may also account for correlations between speciation and both clade-level phenotypic evolution (e.g., Rabosky et al. 2013) and rates of molecular divergence (Ezard et al. 2013; Bromham et al. 2015), although the causal direction of these correlations remains controversial.

A model relating the direction of speciation in phenotype space to within-population variation has a rich set of implications for clade-level dynamics. The inhomogeneous nature of potential variation around any species in phenotype space means that proximity of other points in that space is a poor indicator of how readily they can be reached even if favored by selection (Gerber 2014). Integrating this appreciation of evolutionary accessibility with analyses of clades in phenotype space will enhance the integration of development with macroevolution. For example, the more similar the probability density functions around all the species in a clade—a little-evaluated possibility—the more likely the clade as a whole will generate a directional trend via successive speciation events.

At the clade level, evolutionary changes must often impose further directional restrictions and biases: all aquatic amniotes retain many ancestral traits betraying their terrestrial ancestry, for example. Thus, the probabilistic view of evolutionary constraint adopted here cannot be static but must incorporate how prior steps channel future ones, which cuts to the essential role of history in shaping clade trajectories. Unfortunately, a clade's deployment in phenotype space in the absence of developmental data can falsify a constraint hypothesis but is insufficient to prove one. Some portions of phenotype space are physically untenable, but those vacancies are theoretically less interesting than the feasible but unoccupied portions of the space. The fossil record is packed with phenotypes absent from the modern biota, from giant ground sloths to uncoiled nautiloids and echinoids with proboscis-like test extensions, each demonstrating the evolutionary accessibility of currently vacant portions of phenotype space. One could argue that extrinsic forces are responsible for each of the aforementioned gaps in modern clades, but those missing phenotypes may still reflect intrinsic constraint (with past forms no longer accessible from younger starting points owing to developmental changes) or preemption owing to biotic factors such as competitors, or even the waiting time expected between recursions to relatively improbable trait combinations. A similar approach to missing phenotypes, phylogenetic rather than temporal, considers features or functions accessible to some clades but apparently not to others (Vermeij 2015); for example, photosymbiosis and chemosymbiosis are widespread among invertebrate phyla but absent in echinoderms. Such an approach can sharpen hypotheses on impediments such as developmental factors or energetic trade-offs that might underlie the uneven distribution of certain adaptations.

Evolvability

A considerable body of theory exists on the potential of clades to differ in evolvability, i.e., their ability to respond to directional selection (e.g., Hansen 2006; Kirschner and Gerhart 2010; G. P. Wagner 2010; G. P. Wagner and Draghi 2010; Sterelny 2011; G. P. Wagner 2014), but rigorously guiding that theory with data has been difficult. With respect to macroevolution, evolvability must be concerned not just with the production of heritable variation but with sustained responses to selection, and thus the lineage's ability to accommodate and field variation over many rounds of selection and change. At this scale, evolvability might be quantified operationally by the net amount of phenotype space traversed or encompassed by a clade relative to another clade, standardized by species richness and time. Ultimately, however, it will need to reflect back to the nature of the genotype-phenotype map as mediated by development. For macroevolution, a common-garden design—i.e., analysis over a specified time interval within a single biogeographic province—is needed to hold some external variables constant, but may be insufficient. For example, co-occurring sister clades of neotropical fishes have each traversed roughly the same amount of phenotype space since separation, with one ricocheting within narrow bounds—so that total change greatly exceeds net change—and the other diffusing freely from its starting point (Sidlauskas 2008), though the roles of development and ecology in this apparent contrast in evolvability cannot yet be separated.

Tempo and Mode

Regardless of the biases in the direction of evolutionary change within species, the dynamics of phenotypic change among species has drawn much attention (proposition 9, table 17.1). When the axes are defined in terms of tempo (continuously gradual versus static with punctuations) and mode (branching versus nonbranching), the famous punctuated equilibrium / phyletic gradualism end-members (Eldredge and Gould 1972) simply become two cells in a matrix of models for evolutionary change: punctuated cladogenesis and gradual anagenesis (Jablonski 2007). All of the end-member patterns occur in the fossil record, plus intermediates and switching among tempos and mode over time, with sustained directional change being the least frequent (Hunt 2007a; Hunt et al. 2015). Although patterns consistent with random walks are recorded, the temporal scaling and rates of change are generally intermediate between random walks and stasis and closer to the latter (Hunt 2013). Evolutionary stasis has been defined in many ways, but the key feature is statistically negligible net phenotypic change at the species level, i.e., a lack of directionality rather than of evolutionary lability, and species in the fossil record can show high total rates of evolution while accumulating little net change. The challenge is to characterize and account for the distribution of species-level tempo and mode across the tree of life (Jablonski 2000, 2007). To succeed, we need a deeper understanding of the relative contribution of intrinsic factors at different levels (biased or constrained variation at the organismic level, factors governing genetic population structures at the species level, and so on) and extrinsic factors (the sorting of variation, as dictated by, for example, the tempo and mode of environmental change).

Various authors have proposed that different genetic population structures or environmental tolerances predispose clades to different evolutionary tempos and modes, as do different habitat types, such as surface ocean versus shallow seafloor (Jablonski 2008c). The many hypotheses, not all mutually exclusive, for the mechanism behind species-level evolutionary stasis (e.g., M. A. Bell 2012; S. J. Arnold 2014; Hunt and Rabosky 2014; Futuyma 2015) require tests that separate the alternatives to assess their relative frequencies. This enterprise is aided by the application of models that can be assessed by Bayesian or information-criterion approaches, most commonly Brownian motion models, Ornstein-Uhlenbeck models, or their combination (Hunt et al. 2015). These models are valuable but are not diagnostic of specific evolutionary processes. For example, a species can fit a Brownian motion model under drift, or by tracking randomly varying environments, or more generally by the interaction of many independent forces (e.g., Pennell et al. 2015). An Ornstein-Uhlenbeck model, where the probability of change decreases with distance from a specified trait value, is often conceptualized as fluctuation around an optimum, but can result from other intrinsic and extrinsic factors (Futuyma 2015). And of course shifts from one stable position to another can be mediated by many mechanisms, from drift (an early favorite), to environmental tracking by an isolated subpopulation, to correlated responses to selection on one character that drags others with it (e.g., Hansen 2012).

Evolutionary Novelties

The term novelty has a tormented history, but some useful distinctions are summarized by Gunter Wagner (2014, 2015). The first distinction is between functional and phenotypic novelty; some have suggested the restriction of "innovation" to the former, and "novelty" to the latter (Müller and Wagner 1991; Love 2006; G. P. Wagner 2014, 2015); for different usages, see Erwin (2012). The origin of new functional capacities, such as flight or endothermy, is often related to morphologic change, of course, but the magnitude of functional divergence from an ancestor may be only loosely related to the magnitude of phenotypic divergence (e.g., Wainwright 2007, but see Jablonski 2008a). A second distinction lies between two types of evolutionary novelty (e.g., Müller 2003; Moczek 2008; G. P. Wagner 2014, 2015). Type I entails the origin of a structure or body part lacking a structural homolog in the ancestral clade, such as the vertebrate head. Type II entails the radical transformation of an existing body part, such as forelimbs to wings or flippers. "Radical" here indicates an evolutionary and developmental commitment to these modifications such that a reversion to the ancestral form is highly unlikely. The fin to limb transition was an impressive morphological change, but it involved modifications to a preexisting appendage. Nonetheless, when vertebrates return to a purely aquatic existence, they never produce true fins (with fin rays) but modify the tetrapod limb plan in various ways. As in most biological definitions, the boundaries here are not sharp. Beetle horns are novel morphologic structures that involve the *distalless* gene regulatory network that generates limbs, a "deep homology" (Shubin et al. 2009) that constitutes phenotypic novelty, but it commandeers an existing developmental pathway. Still unclear are the macroevolutionary consequences of these different types of novelty, e.g., for future range or directions of evolutionary change. Type I novelties might open more evolutionary trajectories to a clade than do Type II novelties (as the vertebrate head surely did, relative to, say, wings), but that expectation has not been tested.

By virtually any definition, evolutionary novelty exploits the modularity and hierarchical control of gene regulatory networks. Changes in the location and timing of developmental events can promote more dramatic and coordinated phenotypic changes than expected from a Fisherian model, and such changes are more likely to be adaptive than the Fisher expectation because they draw on existing ontogenetic pathways (e.g., Gould 1977, 2002). Heterotopy, changes in the location of developmental events, has increasingly been appreciated for its evolutionary potential (Zelditch and Fink 1996; Baum and Donoghue 2002). Heterochrony, changes in the timing of developmental events, has been dismissed as simply drawing on existing variation (e.g., Zelditch and Fink 1996), but two points ameliorate that view. First, heterochrony in clades having multiphase life cycles can abruptly yield significant ecological changes, i.e., functional innovation. Heterochrony in such lineages can also alter dispersal abilities (as a by-product or even as the direct target of selection) that in turn influence genetic population structures and geographic range sizes, thereby affecting origination and extinction rates—excellent examples of upward causation in action. If heterochronies are initiated by relatively simple genetic changes, we might expect such transitions to occur repeatedly, as is in fact the case for permanently aquatic, paedomorphic salamanders (C. K. Johnson and Voss 2013).

The modular nature of development permits a much greater array of developmental shifts involving specific structures or regions of the body. Such local, as opposed to global or whole-organism, heterochronies have occurred frequently, again with potential impact on ecology and gene flow. The appendage heterochronies that produced the skeletal structures supporting wings in pterosaurs, birds, and bats are good examples, among many others. Thus, while Raff (1996) and others are correct that most evolutionary transformations are not underlain by heterochrony, as was sometimes implied during the renewed wave of enthusiasm for the concept, heterochrony is one of the mechanisms linking development to macroevolution (Hanken 2015).

The larger challenge is to develop a theory linking developmental changes to the differential behavior of clades and traits over time. Contrasts in rates and patterns of phenotypic or taxonomic evolution are often associated with clade-level differences in development, but establishing the causal direction is difficult. Are the macroevolutionary dynamics driven by the developmental differences, or do the developmental differences arise later, stabilizing favored phenotypes? Intriguing signposts for theory do exist. For example, the exceptional beak diversity achieved by finches relative to other songbird clades might stem from the simplicity of the gene regulatory networks governing their beaks' three-dimensional form, so that simple changes in a handful of genes could generate a wealth of beak forms (Mallarino et al. 2011; Fritz et al. 2014). Integrated experiments and model simulations can push theory beyond the generalizations presented here to more powerful, predictive models. For example, the strength of developmental integration, or its converse modularity, and the eccentricity of phenotypic covariation within populations around a multivariate centroid, appear to influence the differential evolution of skull disparity between mammalian clades (Goswami 2006; Haber 2012; Goswami et al. 2014). The next step is to test for generality of such patterns, for example in other modules within the tetrapod body.

The differential occupation of phenotype space among clades can be a vehicle for addressing many macroevolutionary issues, particularly if it can be coupled with the developmental underpinnings of form. Thus, if beetles can build horns by a repositioned *distalless* module, why are tetrapods unable to coopt that module to make dorsal wings, as have sprung from the human imagination many times, from winged horses to dragons? The absence of such wings in the real world—and the fact that tetrapods have always evolved powered flight via modified forelimbs instead—has often exemplified developmental constraint (e.g., Erwin 2007; Losos 2011). Thus, we need a theory that both accounts for the origin of beetle horns and disallows certain other appendages that might use the same core developmental pathway. The answer may lie in epigenetics and the limits to how novel structures are integrated into the developing body, in fundamental differences between protostome and deuterostome development, or in the selective value of incipient structures. This somewhat absurd example of a missing form underscores ingredients for a macroevolutionary theory of variation (table 17.1). Experimental manipulation of developmentally important genes has begun to probe these limits to development, and we are at the threshold of new advances in this area.

Requiring further theoretical exploration are the nonrandom temporal and spatial patterns in the origin of evolutionary novelties (Jablonski 2010b). Temporally, the Cambrian explosion of animal body plans, and the first appearance of most phylum-level taxa, are unmatched in the preceding 4 billion years or the ensuing half billion (Erwin and Valentine 2013; Erwin 2015b). Mass extinctions also promote evolutionary pulses, but differ qualitatively from the Cambrian explosion, serving mainly to remove ecological dominants and allowing existing but marginal clades to diversify taxonomically, functionally, and morphologically. Higher taxa representing new morphologies associated with substantial functional changes (evolutionary innovations sensu G. P. Wagner 2014) preferentially first appeared in shallow-water environments and spread across the continental shelf over millions of years. In contrast, the first appearances of genera and families tend to conform to their clade-specific bathymetric diversity gradients (Jablonski 2005a; Harper 2010; Kiessling et al. 2010). Biogeographically, many significant evolutionary transitions have been traced to the tropics, and paleontological work has confirmed this general tendency for marine invertebrate orders (Jablonski 1993, 2005b; Jablonski et al. 2006; P. R. Martin et al. 2007; Kiessling et al. 2010; for phenotypic novelty per se Vermeij 2012; Jansson et al. 2013; Jablonski et al. 2017), although the latitudinal dynamic is less clear on a per-taxon basis. These empirical patterns show how important ecology, development, and their intersection are to macroevolution, but we have only a weak theoretical foundation for weighing the relative roles of these factors in some of the most striking regularities in the history of life.

Diversifications and Diversity-Disparity Relationships

A long-standing focus for macroevolutionary research has been diversification, i.e., the net proliferation of a monophyletic group of species or higher taxa, and the related phenomenon of adaptive radiation, which is generally defined as a rapid and extensive gain in functional or phenotypic diversity (e.g., Gavrilets and Losos 2009). Much research on diversifications attempts to link them to the acquisition of specific phenotypic or functional triggers (Bouchenak-Khelladi et al. 2015). However, causal interpretation of taxonomic and phenotypic patterns observed in the fossil record is difficult. It is even more difficult when inferred from molecular phylogenies, which inevitably lack direct information on the phenotypes or numbers of extinct taxa, particularly at deep phylogenetic nodes.

Evolutionary Models and Evolutionary Process

The first macroevolutionary models for temporal dynamics were taxonomic (see Raup 1985), but models for diversification in phenotype space provided a rich new dimension to macroevolutionary analysis (Foote 1996; Erwin 2007; P. J. Wagner 2010; McGhee 2011; Chartier et al. 2014). The phenotype space approach has been augmented by a set of models for trait evolution within a phylogeny, with the three standard models being (1) a pulse early in the history of a clade with a later slowdown that may approach a steady state, (2) a steady accrual, and (3) limited diversification. As with species dynamics, these clade behaviors have been codified as the Brownian motion and Ornstein-Uhlenbeck models, along with an "early burst" model that is essentially Brownian motion with a temporally decreasing rate parameter (e.g., Harmon et al. 2010; Pennell et al. 2015). These descriptive models are widely taken to be diagnostic of specific evolutionary scenarios. For example, an early burst supposedly indicates diversity-dependent processes, such that within-clade crowding damps further diversification. However, this pattern does not rule out other extrinsic factors, such as distantly related competitors, predators, parasites, abiotic reduction in habitable area or climate shifts, or intrinsic factors such as reduced excursions in form as the "easy" transitions are exhausted. Similarly, fit to an Ornstein-Uhlenbeck model is often taken to signal stabilizing selection around one or more phenotypic optima but is again consistent with any factor that puts bounds on phenotypic diversification, from intrinsic constraints to competitive exclusion (Hansen 1997; G. J. Slater 2013, 2015).

Of course, the fit of phenotypic models to the overall behavior of clades says nothing about the evolutionary dynamics of their constituent species. Even static species can create a clade-level pattern in phenotype space that fits a Brownian motion model, for example, and gradual anagenesis among species can generate a clade that fits an Ornstein-Uhlenbeck model. Far less attention has been paid to declining or bottlenecked diversity—despite the wealth of paleontological evidence for such trajectories—primarily because molecular data must always put maximum diversity in the present day. We would seriously misinterpret the dynamics of, for example, proboscideans, horses, or hominids from molecular data alone, all of those clades being mere remnants of their former diversity and disparity.

Three Diversification Modes in Diversity-Disparity Space

Another consideration lost with exclusively neontological data is the relationship among the macroevolutionary currencies through time. For example, an early burst in both morphological disparity and taxonomic richness is fundamentally different from an early burst in disparity alone.

The macroevolutionary variables most readily quantified in the fossil record, diversity and disparity, have often been analyzed separately, but diversification dynamics can be conceptualized as a time-series of points
and compared among clades in a single bivariate space, defined by range-
standardized measures of diversity and disparity. Taxonomic diversification
is inherently exponential, and a time-homogeneous expansion or Brownian model for trait evolution yields an approximately linear increase in
morphological disparity (Slatkin 1981; Foote 1993, 1996; Ricklefs 2006).
Thus, when diversity is log-transformed and disparity is plotted arithmeti-
cally in bivariate space, diffusion in phenotype space during exponential
diversification falls on the 1:1 diagonal, with end-member types predicated
on whether taxonomic diversity lags or leads the other variable falling in
the upper left and lower right regions respectively (fig. 17.1A).

This conceptualization frames three alternative macroevolutionary dynamics, all of which occur in the fossil record when clade histories are analyzed up to their global maximum in genus- or species-level taxonomic
diversity (fig. 17.1B). Type 1 diversification, when rates or magnitudes of
phenotypic divergence are unexpectedly high, whether because of less constrained developmental processes or exceptional ecological opportunities

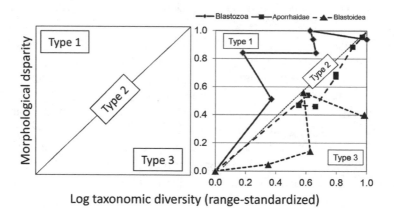

Fig. 17.1. (A) Diversity-disparity space for analyzing the relation between
taxonomic and morphological diversification. Type 1: morphology
outstrips taxonomic diversification; type 2: morphology concordant
with taxonomic diversification; type 3: morphology trails behind
taxonomic diversification. (B) Three empirical trajectories, for Cambrian-
Ordovician blastozoan echinoderms, Jurassic-Cretaceous aporrhaid
gastropods, and Ordovician-Carboniferous blastoidean echinoderms
(data from Foote 1993; Roy 1994; Foote 1996). Bootstrapped confidence
limits not shown here, but blastoids fall significantly above the 1:1 line,
blastozoans fall significantly below it, and aporrhaids never leave it.

(e.g., Valentine 1980; Jablonski 2000), is exemplified by the echinoderm class Blastozoa, an important component of the Cambrian explosion, later converging on the type 2 diagonal. Type 2 diversification, concordance among currencies and thus a roughly constant rate and magnitude in the per-taxon rate of net phenotypic evolution, occurred in aporrhaid gastropods, a significant Mesozoic marine clade. Type 3 diversification, where proliferation of genealogical units outstrips their diversification in form or function ("nonadaptive radiation," Rundell and Price 2009), occurs in the blastozoan subclade Blastoidea, which was part of the Ordovician sequel to the Cambrian explosion, with disparity lagging diversity, and never crossing into the type 1 field. These and other examples of contrasting trajectories through diversity-disparity space corroborate views that the Cambrian explosion of metazoan form was dramatic relative to taxonomic diversification whereas later diversifications were less prolific morphologically on a per-taxon basis. Distinguishing among these alternatives becomes increasingly difficult in neontological data as extinction erases the diversity signal while leaving more of the disparity signal intact (e.g., Liow et al. 2010; G. J. Slater and Pennell 2014).

This scale-free approach is useful both for theoretical purposes and for comparative empirical analyses, but magnitude does matter. The evolutionary burst of body plans in the late Proterozoic and Early Cambrian greatly exceeds the evolutionary burst of mammals after the end-Cretaceous mass extinction (Valentine 1973; Erwin and Valentine 2013), which greatly exceeds the adaptive radiation of stickleback fishes in coastal lakes (Schluter 2000). Of course, exclusion of the diversity and disparity of extinct stem groups will strongly undermine analyses that rely strictly on molecular data and present-day phenotypes.

Toward Mechanistic Models

A more mechanistic approach to the interplay among taxonomic, functional, and phenotypic diversity is the Valentine-Walker model (Valentine 1980; Erwin and Valentine 2013). In this formulation, evolution operates in a landscape consisting of a mosaic of discrete bins defined by environmental conditions and resources. This landscape is entered by a clade that evolves in phenotypic steps drawn from a highly skewed size-frequency distribution (mostly small, a few large). These steps succeed only if they land on an empty cell or, for large steps, an empty clump of cells, so that rich monophyletic diversifications are increasingly unlikely as the landscape fills. However, stochastic extinction clears a steady supply of cells, mostly noncontiguous but occasionally in small clumps, so that taxonomic origination never ceases, even as net diversification slows or approaches a steady state.

This simple heuristic model accounts for a pleasing number of empirical patterns. These include the initial burst of diversification in morphological disparity and functional groups relative to taxonomic diversity in the Cambrian; the slowdown in the production of evolutionary novelty following the Cambrian explosion despite the continued rise of taxonomic diversity at lower ranks; the absence of Cambrian-like diversification following mass extinctions (because extinctions rarely fully vacate adaptive zones); the slowdown rather than leveling off of diversification in the absence of major environmental pressures. A more fully realized topological model, with a larger set of parameters, can match additional macroevolutionary phenomena (Gavrilets and Losos 2009), although its predictions are not always fit by real-world cases, e.g., diversity-disparity relationships; exploring mechanisms and reexamining assumptions behind such mismatches will be useful.

Sorting of Variation: Diversity Dynamics

The key concepts for diversity dynamics within a genealogical hierarchy have been outlined from diverse perspectives (e.g., Lewontin 1970; Hull 1980; Jablonski 2000; Gould 2002; Okasha 2006; Jablonski 2007, 2008c). Selection and other processes can operate on the heritable variation at a given focal level, but differential survival or proliferation of the units in the genealogical hierarchy can also be driven by events operating both above and below the focal level (downward and upward causation, respectively) (Vrba and Eldredge 1984; Gould 2002; Jablonski 2007; Tëmkin and Eldredge 2015). Thus, the demise, persistence, or proliferation of entities (e.g., genes, organisms, or species) within their respective larger units cannot necessarily be attributed solely to selection at those levels but may involve sorting as a by-product of processes operating at higher and lower levels (Vrba and Gould 1986). Such effects have been termed "hitchhiking" (e.g., Vrba and Gould 1986; Jablonski 2000, 2008c) by analogy to genetic hitchhiking, where selection on one or more genes alters the frequency of others that have little or no direct effect on fitness.

This hierarchical framework involves a form of multilevel selection distinct from that discussed by Goodnight (chap. 10), which emphasizes what has been termed multilevel selection 1 (Heisler and Damuth 1987; Damuth and Heisler 1988; Okasha 2006), in which the fitness of individual organisms is determined in part by their membership in groups. Macroevolution is concerned primarily with multilevel selection 2, the differential origination or persistence of genealogical units at different levels within the biological hierarchy. Strict-sense species selection occurs when selection operates on traits emergent at the species level that affect speciation and extinction, such as geographic range size or genetic population structure. Effect macroevolution is upward causation, as when organismic traits such as body size or diet influence the proliferation or survival of higher-level units, e.g., speciation or extinction rates. Differential speciation and extinction owing to intrinsic biotic properties, without reference to the focal level of the operative traits, is species sorting, often termed broad-sense species selection. This purportedly neutral term may cause more confusion than it allays, however.

Emergent Properties

Emergence is an elusive concept, but operationally a feature can be considered emergent at a given level if its evolutionary consequences do not depend on how the feature is generated at lower levels (Jablonski 2007, 2008c), an approach similar to Brandon's (1990) application of the statistical concept of "screening-off." By this logic, geographic range size can be viewed as an emergent property at the species level, because the differential survival of those genealogical units is statistically associated with broad geographic range regardless of which lower-level (organismic) traits promoted the broad range of particular species (Jablonski and Hunt 2006; Jablonski 2007). Thus, as species vary in their geographic range sizes, and variation in species range size is causally associated with species survivorship (and, more controversially, with speciation rates), and range size is also heritable at the species level (i.e., ranges of related species are more similar in size than expected by chance), this species-level trait meets the three criteria for the operation of selection above the organismic level (Jablonski 1987; Hunt et al. 2005; Waldron 2007; Borregaard et al. 2012).

In addressing the concept of emergent properties and how they influence evolutionary dynamics, empirical tractability has often been conflated with the theoretical issues. Species selection is most readily demonstrated when it overwhelms selection at the organismic level, but such opposition is an operational convenience rather than a theoretical requirement (Goodnight, chap. 10). If sorting processes operate at all levels simultaneously, albeit at different rates, and upward and downward causation is pervasive, then the long-term evolutionary behavior of a clade in phenotype space, or the waxing and waning of its species richness, will not coincide exactly with selective pressures at any one level because it is the resultant of forces operating at multiple levels. Processes at different hierarchical levels can as readily reinforce as oppose one another, as when a species that is widespread—and thus extinction-resistant by virtue of that emergent species trait—also consists of individuals with broad trophic requirements. Similarly, species sorting's macroevolutionary role is clearest when individual species exhibit stasis, but differential speciation and extinction owing to organism- or species-level traits still affect the waxing and waning of clades and their movement through phenotype space when species undergo continuous gradual transformation (Slatkin 1981; Jablonski 2008c). Cross-level conflicts occur, of course, whenever selection favors traits that drive changes in organismal or species-level properties linked to increased extinction risk, such as selection for large body size in mammals (e.g., Van Valkenburgh et al. 2004; Clauset and Erwin 2008). The most extreme cases, where increases in organismal fitness drive species into extinction, deterministically or by pushing them into states where stochastic effects make extinction inevitable, have been termed evolutionary suicide, Darwinian extinction, or self-extinction (Ferriere and Legendre 2013).

Partitioning sorting processes among hierarchical levels is a key macroevolutionary goal. Of the approaches discussed by Jablonski (2008c), two seem most promising. One postulates species- and organism-level traits and fits general linear models to evaluate the relative contributions of those traits in determining extinction or origination rates (e.g., Jablonski and Hunt 2006). The other uses a hierarchical expansion of the Price equation to partition variances between fitness and phenotype among levels (A. J. Arnold and Fristrup 1982; C. Simpson 2010, 2013; Rankin et al. 2015); this method applies to broad-sense species selection as it does not address the role of particular traits. (See Goodnight, chap. 10, for strengths and drawbacks of the Price equation in the context of multilevel selection 1.) Both of these methods found empirical support for a significant, but not exclusive, role for strict-sense and broad-sense species selection, respectively, but more work is needed to extend and refine these approaches, for example the multilevel permutation test of Hoehn et al. (2016).

Species Drift

Upward and downward causation does not require active selection. At any hierarchical level, drift can be viewed as differential replication owing to chance (Hull 1989), and such drift will affect the frequencies of lower-level entities. Drift can also propagate upward, by driving alleles to fixation, phenotypic characters to oblivion, or demes/species to extinction. Stochastic processes at higher levels have received little formal attention except as null models, but such "phylogenetic drift" (Stanley 1979) or "species drift" (Gould 2002) can change the amount and nature of variation available for selection at multiple levels. At the species level and above, random processes compose the wide variety of events encountered by clades on geologic timescales; hence Turner's (2015, 87) statement that "contingency is to species selection as drift is to selection." In fact, the small number of species contained in most clades at any one time suggests that species drift will often be a more significant factor at that level than at the level of bodies within populations (Gould 2002), but see C. Simpson and Müller (2012), who argue that the overall scarcity of sustained trends in the fossil record suggests that species drift is a minor factor. The digital clades in Gould et al.'s (1977) simulations may be too small to assess the rise and fall of orders (Stanley et al. 1981), but are about the right size, averaging about four taxa per clade, to model the behavior of most genera. Species-poor clades, these simulations show, will be not only more extinction-prone but more likely to undergo stochastic changes in composition that can, in turn, alter evolutionary dynamics.

Other Scaling Effects

Another scaling property of hierarchies is the tendency for rate constants to decrease with each ascending level, even as the potential role of drift increases. The biased replication of certain selfish genetic elements with each cell cycle is rapid relative to the generation times of most metazoan organisms, which in turn are brief relative to the speciation rates of most metazoans. This property has been used to argue against the efficacy of sorting above the organismic level, but such arguments hold only if organismal adaptation is the sole evolutionary process or outcome of interest and ignore the operation of upward and downward causation. Thus, although sorting among species may generally be too slow to construct a complex adaptation such as a wing in the course of successive organismal generations, sorting at that level may determine within and among clades the persistence and number of species bearing wings.

Trends

One reason for the enduring interest in species sorting is that it offers a mechanism for large-scale evolutionary trends in form and taxonomic richness. However, just as there are multiple models for phenotypic evolution at the species level, several alternatives exist for clade-level phenotypic evolution (Jablonski 2010a). These scenarios give us yet another perspective on the evolutionary models discussed above, underscoring both their analytical utility and their lack of mechanistic specificity. The primary distinction is between active and passive trends. Active trends (P. J. Wagner 1996, 2010) arise by directional phyletic transformation of the constituent species of a clade, by directional speciation, or by differential speciation and/or extinction in different regions of phenotype space. (In contrast, McShea's [1994, 2000] "driven trend" excludes differential speciation and extinction; see also Turner 2015.) Despite their different underlying dynamics, all such trends would fit Ornstein-Uhlenbeck models with a temporal shift in the putative optimum, or a starting point far from that optimum. Passive trends can arise by diffusion from a fixed boundary, which would fit a Brownian motion model (Stanley 1973, 1979) or by just considering the leading edge of unbounded diffusion (Gould 2002). Under this scenario, a clade starting close to an absorbing or reflecting boundary for a trait value, whether body size, organismal complexity, or geographic range size, will produce an increasing mean and variance, the classic triangular plot of many macroecological and morphological studies (Stanley 1973; McShea 1994; Gould 2002; Foote et al. 2008). Additional models are required when positive and negative interactions are included, as in the ecological or functional dynamics of clades of A. M. Bush and Novack-Gottshall (2012) and Dick and Maxwell (2015).

Trends may occur at any taxonomic level, producing nested patterns that need not coincide (proposition 10, table 17.1). Unbiased samples of large clades tend to show a variety of body-size trends for their subclades (Jablonski 1996, 1997; Klompmaker et al. 2015). The increase in mean body size in mammals is best explained as passive diffusion away from a lower bound (Stanley 1973; Clauset and Erwin 2008; G. J. Slater 2013), although a second mode at 30 kg might represent an evolutionary attractor for certain clades (Alroy 1998). However, as already noted, density maxima in phenotype space need not reflect organismal-level optima.

Intrinsic and Extrinsic Factors

The differential survival and generation of genealogical units are governed by both intrinsic and extrinsic factors (proposition 11, table 17.1). The macroevolutionary sorting of variation has long been modeled as a birth-death process among taxa of various ranks (e.g., Stanley 1979; Raup 1985;

Nee 2006; Rabosky 2014), and with the explosion of molecular phylogenetic data, related models have been applied to diversity patterns within
phylogenies for extant organisms. Phylogenies provide rich data on net diversification among sister groups and can yield insights into intrinsic and
extrinsic controls on evolutionary dynamics. However, decomposing net
diversification into its origination and extinction components is crucial for
understanding the diversification process (e.g., Foote 2010; Fritz et al. 2014).
Richness differences among clades or through time can reflect contrasting
speciation rates with extinction relatively invariant among clades (the assumption underlying the application of pure-birth rather than birth-death
models), contrasting extinction rates (P. J. Wagner and Estabrook 2014), or
slim differences between strongly covarying rates. Such information is difficult to retrieve robustly from extant species and their phylogenies because
extinction can mask true evolutionary rates or trends (e.g., Finarelli 2007;
Liow et al. 2010; Quental and Marshall 2010; Rabosky 2010). Most proposed methods for estimating taxonomic origination and extinction rates
from time-calibrated phylogenies are undermined by covariation between
rates and organismal- or clade-level traits (Maddison and FitzJohn 2015;
Rabosky and Goldberg 2015). Here too, new approaches that rigorously
fit models for evolutionary dynamics to combined molecular-phylogenetic
and paleontological data, particularly when fossil data are sparse or confined to rich but unevenly distributed time bins, will be especially valuable
and are the focus of considerable attention (e.g., Morlon et al. 2011; G. J.
Slater et al. 2012; Gavryushkina et al. 2014; Heath et al. 2014).

Mass Extinctions

Mass extinctions, meaning intense excursions above "background" extinction rates for large, phylogenetically and ecological disparate segments of
the global biota, are of theoretical interest for several reasons (Jablonski
2005b). First, they represent exceedingly rare events that, while accounting
for only a few percent of Phanerozoic extinctions of species or genera, can
have long-lasting or permanent effects—not simply on taxonomic dominance (as in the dinosaur-mammal changeover at the end of the Mesozoic), but on clade dynamics and phenotype space occupation (Jablonski
1998, 2005b; Krug and Patzkowsky 2007; Foote 2010; Huang et al. 2015),
although J. J. Sepkoski (1996) argues that the rise of the post-Paleozoic
fauna was accelerated by, not contingent on, the end-Permian mass extinction (proposition 13, table 17.1). Nonetheless, while extinctions remove
or severely deplete clades bearing clear-cut evolutionary novelties and pro-

mote prolific taxonomic diversifications among the survivors, they bring relatively modest pulses of evolutionary novelty on a per-taxon basis. Whether this damped novelty generation derives from changes in developmental lability, ecological limitation, or other factors is unclear. A general framework for addressing these changes has not been established, but the necessary components are becoming clearer (table 17.1).

Second, mass extinctions provide a succession of natural experiments in multilevel and multidimensional evolution. Many organismic and clade-level traits effective during "normal" times are evidently inconsequential during mass extinctions (Jablonski 2005a, 2008b; Hoehn et al. 2016). However, whereas broad species-level geographic range no longer contributes to clade survivorship, the buffering effect of broad clade-level range persists from "normal" to mass extinction intervals. Further, genus range-size is an emergent property, in that its survivorship-enhancing effects do not depend on whether constituent species are also widespread or are widely separated but narrow-ranging (Jablonski 2005b; Foote et al. 2016). These intensive selective events, in which many once-significant organismic and species-level features are effectively neutral, provide significant opportunities for hitchhiking of phenotypic traits on geographic range size (e.g., Jablonski 2008b; Landman et al. 2014). In addition, mass extinctions rarely empty adaptive zones completely, instead mostly thinning the number of occupants (Erwin et al. 1987; Foster and Twitchett 2014; Edie et al. 2018). This failure to fully vacate adaptive zones may explain why even the massive end-Permian extinction does not trigger a Cambrian-like explosion of new body plans (Erwin et al. 1987).

Third, recoveries from mass extinctions are more complex than generally appreciated, but most theory and analysis have emphasized the extinctions per se. Not all survivors participate in the taxonomic diversifications that follow major extinction events and may finally disappear well after the extinction itself (the "dead clade walking" phenomenon [Jablonski 2002]—widely reported but little understood). Taxonomic bottlenecks are formally equivalent to species drift: species-poor survivors will rarely provide a random sample of phenotypes present before the bottleneck and so can yield both upward and downward effects. Such macroevolutionary founder effects may play a significant role in determining large-scale evolutionary patterns (proposition 3, table 17.1). Surprisingly, the bottleneck sizes of major clades that survive mass extinctions are poor predictors of the later duration or phenotypic expansion of those clades, implying that some tightly bottlenecked groups rediversify sufficiently rapidly to avoid drifting into extinction, and others fail to take advantage of a large survivor pool (Jablonski 2002). The phenotypic consequences of these macroevolutionary founder effects have not been systematically evaluated. New models are needed to develop a richer understanding of why different large-scale patterns show unbroken continuity, continuity with setbacks, collapse followed by persistence without recovery, or unbridled diversification in the aftermath of mass extinctions.

Time Homogeneity: Has Macroevolution Evolved?

The sorting of variation has been occurring since the inception of life on Earth. We expect transient excursions in rates owing to external pressures, and lasting ones when new units of selection arise in life's major transitions. A case can be made that each of the major evolutionary transitions mentioned above created new units of selection and so represents a shift from multilevel selection 1 to multilevel selection 2 (Okasha 2006; Szathmáry 2015). Still unknown are the full scope and broader implications of long-term trends in macroevolutionary processes. Some of these trends are relatively straightforward, such as stepwise escalation of predation intensity; others may be more subtle. For example, Cambrian ecology may have promoted different clade dynamics relative to later times, as predicted by Valentine-Walker-type models and evidenced by the finding that Cambrian diversification rates were less closely tied to trait changes than in the rest of the Phanerozoic (Polly 2004; P. J. Wagner and Estabrook 2014). At a more basic level, phylum- or class-level differences in species turnover rates, or the skewness or volume of the phenotypic probability density function around those species, or the tendency to have positive versus negative effects on co-occurring clades, could result in temporal trends in clade behavior as the players change through time, even if the basic sorting mechanisms are constant.

Understanding the long-term evolution of variation-generating mechanisms is significantly more challenging. Gene regulation and the efficacy of lateral gene transfer differ profoundly between prokaryotes and eukaryotes. Macroevolution must have operated differently in the exclusively prokaryote world of the Archean Eon and must still operate differently in the prokaryote kingdoms of life relative to eukaryotes. Given the nonrandom spatial and temporal patterns in the origin of eukaryotic novelty and higher taxa, several authors have suggested that metazoan development has itself evolved in ways that have altered the range of accessible variation, for example with diminution in the evolutionary lability of high-level control genes relative to downstream gene regulatory networks (Valentine 2004; Davidson and Erwin 2010; Erwin 2015a; Peter and Davidson 2015). Genotype-to-phenotype maps were probably simpler near the origin of development and differentiation in complex multicellular clades (e.g., Salazar-Ciudad 2008; Davidson and Erwin 2010), but it is unclear what this means for phenotypic evolution. We cannot rule out that metazoan variation profiles were essentially set near the origin of Bilateria, given that so many pathways are conserved among the extant phyla. But we can also ask whether the generation of variation also changed with the stepwise duplication (and subsequent pruning) of genomes along vertebrate and plant phylogenies. On still shorter timescales, we need a clearer picture on whether the genetic architecture of traits evolves systematically over time and whether this influences phenotypic lability.

Despite the striking conservation of high-level developmental gene networks, the genotype-phenotype map of established features is clearly dynamic. Since selection does not see how the phenotype is produced, mutations that do not affect the end-product but increase developmental complexity can accumulate (Salazar-Ciudad 2008). This accumulation is the logic behind developmental systems drift, in which clearly homologous structures across clades (i.e., representing historical continuity among ancestor-descendant phenotypes) can be generated by different developmental mechanisms, such as insect body axes, tetrapod jaws, and bird beaks (True and Haag 2001; Müller 2007). What we do not know is how, or whether, this rewiring at the molecular level—which evidently occurs over millions of years—influences the rate and direction of phenotypic change. That is, we need models for developmental evolution that carry over into the behavior of clades in phenotype space, and to the size-frequency and orientation of phenotypic transitions in a Valentine-Walker model.

Conclusion: The Still-Incomplete Synthesis

In this chapter, I have attempted to outline a general theoretical framework for macroevolution (table 17.1). Much remains to be done in moving beyond this general theory to specific constitutive theories that can more fully integrate the macroevolutionary elements discussed here. We have a good start: evolutionary theory has excelled in the realm of the short-term sorting of variation. Progress has been made in understanding sorting at higher levels and over long time intervals, although full integration with the rich body of theory on short-term sorting has been elusive or rudimentary, at least in part because context, emergent properties, and rare events are so pivotal in shaping macroevolutionary trajectories. Still more challenging, but even more essential, is the integration of theory on sorting with our growing understanding of the generation of variation. Selection is a powerful force but can operate only on the variants presented to it. Significant evolutionary change often involves shifts in the timing, rate, and place of gene expression, which facilitate the origin of variation in certain directions and combinations. Thus, nonrandom variation, modularity, the potential for recruiting entire gene regulatory networks, and the possibility that all of those features might evolve over the course of a clade's history must be incorporated into models of macroevolution. Context-dependency, emergence, and rare events are as important in the generation of variation as in its sorting. However, most of the macroevolutionary work in evolutionary developmental biology is still essentially explanatory rather than predictive. A striking change in phenotype certainly is better understood by dissecting the underlying developmental change. The challenge is to build a theoretical framework that accounts for how initial conditions at the level of phenotype, or more specifically how the phenotype is generated and accommodates evolutionary change, governs access to different evolutionary paths and promotes, limits, or channels phenotypic and taxonomic diversification.

Given the large part played in macroevolution by history and chance, and the intricate potential interactions between intrinsic biotic features and extrinsic factors, macroevolutionary theory is, and will be, most powerful when it is comparative. Thus a primary goal should be the development of models that incorporate the intrinsic properties of a clade and its components—from the architecture of its gene regulatory networks to the genetic population structure and geographic range sizes of its species—as a basis for predicting or understanding the macroevolutionary differences among clades, or for a clade among time intervals. Such differences may lie in the generation of evolutionary novelty, the volume of phenotype space occupied, the direction and rate of movement through that phenotype space, the origination and extinction rates of genealogical subunits and the directions they may trend, the responses to a shared extrinsic biotic or abiotic perturbation, and many other features. Comparative biological and paleobiological approaches have successfully identified many of the key variables and provided at least part of the general framework, and the task ahead is the integration of the components of macroevolutionary theory.

Acknowledgments

I thank the editors for inviting me to contribute this chapter and for their patience during its preparation. I am grateful to J. Cracraft, M. Foote, D. W. McShea, D. P. Mindell, S. Scheiner, G. J. Slater, J. W. Valentine, and G. P. Wagner for valuable comments on the manuscript; none should be held responsible for the end result, however. My work has been supported by the National Science Foundation, NASA, and the John Simon Guggenheim Foundation.

Abascal, F., R. Zardoya, and D. Posada. 2005. ProtTest: selection of best-fit models of protein evolution. *Bioinformatics* 21:2104–5.

Abzhanov, A., W. P. Kuo, C. Hartmann, B. R. Grant, P. R. Grant, and C. J. Tabin. 2006. The calmodulin pathway and evolution of elongated beak morphology in Darwin's finches. *Nature* 442:563–66.

Abzhanov, A., M. Protas, B. R. Grant, P. R. Grant, and C. J. Tabin. 2004. Bmp4 and morphological variation of beaks in Darwin's finches. *Science* 305:1462–65.

Adams, M. B. 1968. The founding of population genetics: contributions of the Chetverikov school, 1924–1934. *Journal of the History of Biology* 1:23–39.

Adams, M. B. 1980. Sergei Chetverikov, the Kol'tsov Institute, and the evolutionary synthesis. Pp. 242–78 in E. Mayr and W. B. Provine, eds., *The Evolutionary Synthesis: Perspectives on the Unification of Biology*. Harvard University Press, Cambridge, MA.

Adams, M. B. 1990. *The Evolution of Theodosius Dobzhansky*. Princeton University Press, Princeton, NJ.

Akin, E. 1979. *The Geometry of Population Genetics*. Springer-Verlag, New York.

Alcaide, M., E. S. C. Scordato, T. D. Price, and D. E. Irwin. 2014. Genomic divergence in a ring species complex. *Nature* 511:83–85.

Allen, A. P., J. H. Brown, and J. F. Gillooly. 2002. Global biodiversity, biochemical kinetics, and the energetic-equivalence rule. *Science* 297:1545–48.

Alroy, J. 1998. Cope's rule and the dynamics of body mass evolution in North American fossil mammals. *Science* 280:731–34.

Alvarez-Ponce, D., P. Lopez, E. Bapteste, and J. O. McInerney. 2013. Gene similarity networks provide tools for understanding eukaryote origins and evolution. *Proceedings of the National Academy of Sciences* 110:E1594–1603.

Amson, E., C. de Muizon, M. Laurin, C. Argot, and V. de Buffrénil. 2014. Gradual adaptation of bone structure to aquatic lifestyle in extinct sloths from Peru. *Proceedings of the Royal Society of London B: Biological Sciences* 281:20140192.

Amundson, R. 2005. *The Changing Role of the Embryo in Evolutionary Thought: Roots of Evo-Devo*. Cambridge University Press, New York.

Andam, C. P., and J. P. Gogarten. 2013. Biased gene transfer contributes to maintain the Tree of Life. Pp. 263–74 in U. Gophna, ed., *Lateral Gene Transfer in Evolution*. Springer, New York.

Anderson, D. L. 2005. Scoring hotspots: the plume and plate paradigms. *Geological Society of America Special Papers* 388:31–54.

Anderson, J. B., J. Funt, D. A. Thompson, S. Prabhu, A. Socha, C. Sirjusingh, J. R. Dettman, L. Parreiras, D. S. Guttman, A. Regev, and L. M. Kohn. 2010. Determinants of divergent adaptation and Dobzhansky-Muller interaction in experimental yeast populations. *Current Biology* 20:1383–88.

Antonovics, J., and P. H. van Tienderen. 1991. Ontoecogenophyloconstraints? The chaos of constraint terminology. *Trends in Ecology and Evolution* 6:166–68.

Araújo, A. M. 2004. Spreading the evolutionary synthesis: Theodosius Dobzhansky and genetics in Brazil. *Genetics and Molecular Biology* 27:467–75.

Arendt, J., and D. Reznick. 2008. Convergence and parallelism reconsidered: what have we learned about the genetics of adaptation? *Trends in Ecology and Evolution* 23:26–32.

Armelagos, G. J. 2008. Biocultural anthropology at its origins: transformation of the new physical anthropology in the 1950s. Pp. 269–82 in J. Kelso, ed, *The Tao of Anthropology*. University Press of Florida, Gainesville.

Arneberg, P., A. Skorping, B. Grenfell, and A. F. Read. 1998. Host densities as determinants of abundance in parasite communities. *Proceedings of the Royal Society of London B: Biological Sciences* 265:1283–89.

Arnold, A. J., and K. Fristrup. 1982. The theory of evolution by natural selection: a hierarchical expansion. *Paleobiology* 8:113–29.

Arnold, S. J. 2014. Phenotypic evolution: the ongoing synthesis. *American Naturalist* 183:729–46.

Arnold, S. J., M. E. Pfrender, and A. G. Jones. 2001. The adaptive landscape as a conceptual bridge between micro- and macroevolution. *Genetica* 112:9–32.

Arrhenius, O. 1921. Species and area. *Journal of Ecology* 9:95–99.

Arthur, W. 2004. *Biased Embryos and Evolution*. Cambridge University Press, Cambridge, UK.

Aspi, J., A. Jäkäläniemi, J. Tuomi, P. Siikamäki, and C. Fenster. 2003. Multilevel phenotypic selection on morphological characters in a metapopulation of *Silene tatarica*. *Evolution* 57:509–17.

Atkins, K. E., A. F. Read, S. W. Walkden-Brown, N. J. Savill, and M. E. J. Woolhouse. 2013. The effectiveness of mass vaccination on Marek's disease virus (MDV) outbreaks and detection within a broiler barn: a modeling study. *Epidemics* 5:208–17.

Auffenberg, W. 1959. Editor's introduction. *BSCS Newsletter* 1:1–9.

Azevedo, R. B. R., R. Lohaus, S. Srinivasan, K. K. Dang, and C. L. Burch. 2006. Sexual reproduction selects for robustness and negative epistasis in artificial gene networks. *Nature* 440:87–90.

Baack, E., M. C. Melo, L. H. Rieseberg, and D. Ortiz-Barrientos. 2015. The origins of reproductive isolation in plants. *New Phytologist* 207:968–84.

Badyaev, A. V., and J. B. Walsh. 2014. Epigenetic processes and genetic architecture in character origination and evolution. Pp. 177–89 in A. Charmantier, D. Garant, and L. E. B. Kruuk, eds., *Quantitative Genetics in the Wild*. Oxford University Press, Oxford, UK.

Baker, R. J., and J. W. Bickham. 1986. Speciation by monobrachial centric fusions. *Proceedings of the National Academy of Sciences* 83:8245–48.

Baldwin, J. M. 1896a. A new factor in evolution. *American Naturalist* 30:441–51, 536–53.

Baldwin, J. M. 1896b. On criticisms of organic selection. *Science* 4:724–27.

Bamshad, M., and S. P. Wooding. 2003. Signatures of natural selection in the human genome. *Nature Reviews Genetics* 4:99–111.

Bapteste, E., and Y. Boucher. 2008. Lateral gene transfer challenges principles of microbial systematics. *Trends in Microbiology* 16:200–207.

Bapteste, E., Y. Boucher, J. Leigh, and W. F. Doolittle. 2004. Phylogenetic reconstruction
and lateral gene transfer. *Trends in Microbiology* 12:406–11.

Bapteste, E., P. Lopez, F. Bouchard, F. Baquero, J. O. McInerney, and R. M. Burian. 2012.
Evolutionary analyses of non-genealogical bonds produced by introgressive descent.
Proceedings of the National Academy of Sciences 109:18266–72.

Bapteste, E., M. A. O'Malley, R. G. Beiko, M. Ereshefsky, J. P. Gogarten, L. Franklin-Hall,
F.-J. Lapointe, J. Dupré, T. Dagan, Y. Boucher, and W. Martin. 2009. Prokaryotic evolution and the tree of life are two different things. *Biology Direct* 4:34.

Barker, M. S., N. Arrigo, A. E. Baniaga, Z. Li, and D. A. Levin. 2016. On the relative abundance of autopolyploids and allopolyploids. *New Phytologist* 210:391–98.

Barnagaud, J. Y., V. Devictor, F. Jiguet, and F. Archaux. 2011. When species become generalists: on-going large-scale changes in bird habitat specialization. *Global Ecology and
Biogeography* 20:630–40.

Barraclough, T. G., and A. P. Vogler. 2000. Detecting the pattern of speciation from
species-level phylogenies. *American Naturalist* 155:419–34.

Barton, N. H. 1995. A general model for the evolution of recombination. *Genetics Research* 65:123–44.

Barton, N. H., and B. Charlesworth. 1984. Genetic revolutions, founder effects, and speciation. *Annual Review of Ecology and Systematics* 15:133–64.

Barton, N. H., and M. A. de Cara. 2009. The evolution of strong reproductive isolation.
Evolution 63:1171–90.

Barton, N. H., and S. P. Otto. 2005. Evolution of recombination due to random drift.
Genetics 169:2353–70.

Barton, N. H., and S. Rouhani. 1991. The probability of fixation of a new karyotype in a
continuous population. *Evolution* 45:499–517.

Barton, N. H., and M. Turelli. 1989. Evolutionary quantitative genetics: how little do we
know? *Annual Review of Genetics* 23:337–70.

Bastolla, U., M. A. Fortuna, A. Pascual-Garcia, A. Ferrera, B. Luque, and J. Bascompte.
2009. The architecture of mutualistic networks minimizes competition and increases
biodiversity. *Nature* 458:1018–20.

Basu, M. K., L. Carmel, I. B. Rogozin, and E. V. Koonin. 2008. Evolution of protein domain promiscuity in eukaryotes. *Genome Research* 18:449–61.

Bateson, P. 2014. New thinking about biological evolution. *Biological Journal of the Linnean Society* 112:268–75.

Bateson, W. 1902. *Mendel's Principles of Heredity: A Defense.* Cambridge University Press,
Cambridge, UK.

Bateson, W. 1909. Heredity and variation in modern lights. Pp. 85–101 in A. C. Seward,
ed., *Darwin and Modern Science.* Cambridge University Press, Cambridge, UK.

Baum, D. A., and M. J. Donoghue. 2002. Transference of function, heterotopy and the
evolution of plant development. Pp. 52–69 in Q. C. B. Cronk, R. M. Bateman, and
J. A. Hawkins, eds., *Developmental Genetics and Plant Evolution.* [Systematics Association Special Volume 65.] Taylor & Francis, London.

Baum, D. A., and S. Offner. 2008. Phylogenics and tree-thinking. *American Biology Teacher*
70:222–29.

Baum, D. A., and S. D. Smith. 2013. *Tree Thinking: An Introduction to Phylogenetic Biology.*
Roberts and Company, Greenwood Village, CO.

Baum, D. A., S. D. Smith, and S. S. S. Donovan. 2005. The tree-thinking challenge. *Science* 310:979–80.

Bazzaz, F. A., and E. G. Reekie. 1985. The meaning and measurement of reproductive effort in plants. Pp. 373–87 in J. White, ed., *Studies on Plant Demography: A Festschrift for John L. Harper*. Academic Press, London.

Beatty, J. 2006. Replaying life's tape. *Journal of Philosophy* 103:336–62.

Beaulieu, J. M., D.-C. Jhwueng, C. Boettiger, and B. C. O'Meara. 2012. Modeling stabilizing selection: expanding the Ornstein-Uhlenbeck model of adaptive evolution. *Evolution* 66:2369–83.

Becks, L., and A. F. Agrawal. 2010. Higher rates of sex evolve in spatially heterogeneous environments. *Nature* 468:89–92.

Becks, L., and A. F. Agrawal. 2012. The evolution of sex is favored during adaptation to new environments. *PLoS Biology* 10:e1001317.

Bell, G. 1982. *The Masterpiece of Nature*. University of California Press, Berkeley.

Bell, G. 1985. Two theories of sex and variation. *Experientia* 41:1235–45.

Bell, M. A. 2012. Adaptive landscapes, evolution, and the fossil record. Pp. 243–56 in E. I. Svensson and R. Calsbeek, eds., *The Adaptive Landscape in Evolutionary Biology*. Oxford University Press, Oxford, UK.

Bell, M. A., D. J. Futuyma, W. F. Eanes, and J. S. Levinton. 2010. *Evolution Since Darwin: The First 150 Years*. Sinauer Associates, Sunderland, MA.

Belmaker, J., and W. Jetz. 2015. Relative roles of ecological and energetic constraints, diversification rates and region history on global species richness gradients. *Ecology Letters* 18:56371.

Benton, M. J. 2009. The Red Queen and the Court Jester: species diversity and the role of biotic and abiotic factors through time. *Science* 323:728–32.

Berlocher, S. H. 2000. Radiation and divergence in the *Rhagoletis pomonella* species group: inferences from allozymes. *Evolution* 54:543–57.

Bernstein, H., H. Byerly, F. Hopf, and R. Michod. 1985. Genetic damage, mutation, and the evolution of sex. *Science* 229:1277–81.

Bernstein, H., F. Hopf, and R. Michod. 1988. Is meiotic recombination an adaptation for repairing DNA, producing genetic variation, or both? Pp. 139–60 in R. E. Michod and B. R. Levin, eds., *The Evolution of Sex*. Sinauer Associates, Sunderland, MA.

Berrigan, D., and S. M. Scheiner. 2004. Modeling the evolution of phenotypic plasticity. Pp. 82–97 in T. J. DeWitt and S. M. Scheiner, eds., *Phenotypic Plasticity: Functional and Conceptual Approaches*. Oxford University Press, New York.

Berry, R. R., and F. H. Bronson. 1992. Life history and bioeconomy of the house mouse. *Biological Reviews* 67:519–50.

Beyer, H.-G., and H.-P. Schwefel. 2002. Evolution strategies—a comprehensive introduction. *Natural Computing* 1:3–52.

Bhaya, D., M. Davison, and R. Barrangou. 2011. CRISPR-Cas systems in bacteria and archaea: versatile small RNAs for adaptive defense and regulation. *Annual Review of Genetics* 45:273–97.

Bielby, J., G. M. Mace, O. R. P. Bininda-Emonds, M. Cardillo, J. L. Gittleman, K. E. Jones, C. D. L. Orme, and A. Purvis. 2007. The fast-slow continuum in mammalian life history: an empirical reevaluation. *American Naturalist* 169:748–57.

Bijma, P. 2010. Estimating indirect genetic effects: precision of estimates and optimum designs. *Genetics* 186:1013–28.

Bijma, P., W. M. Muir, E. D. Ellen, J. B. Wolf, and J. A. M. Van Arendonk. 2007a. Multilevel selection 2: estimating the genetic parameters determining inheritance and response to selection. *Genetics* 175:289–99.

Bijma, P., W. M. Muir, and J. A. M. Van Arendonk. 2007b. Multilevel selection 1: quantitative genetics of inheritance and response to selection. *Genetics* 175:277–88.

Bishop, C. 2006. *Pattern Recognition and Machine Learning.* Springer, New York.

Björklund, M. 2004. Constancy of the G matrix in ecological time. *Evolution* 58:1157–64.

Blanquart, F., S. Gandon, and S. L. Nuismer. 2012. The effects of migration and drift on local adaptation to a heterogeneous environment. *Journal of Evolutionary Biology* 25:1351–63.

Bloom, S., C. Ledon-Rettig, C. Infante, A. Everly, J. Hanken, and N. Nascone-Yoder. 2013. Developmental origins of a novel gut morphology in frogs. *Evolution and Development* 15:213–23.

Blount, Z. D. 2016. History's windings in a flask: microbial experiments into evolution contingency. Pp. 244–63 in G. Ramsay and C. H. Pence, eds., *Chance in Evolution.* University of Chicago Press, Chicago, IL.

Blount, Z. D., C. Z. Borland, and R. E. Lenski. 2008. Historical contingency and the evolution of a key innovation in an experimental population of *Escherichia coli. Proceedings of the National Academy of Sciences* 105:7899–7906.

Bolten, E., A. Schliep, S. Schneckener, D. Schomburg, and R. Schrader. 2001. Clustering protein sequences—structure prediction by transitive homology. *Bioinformatics* 17:935–41.

Bonner, J. T. 1952. *Morphogenesis: An Essay on Development.* Princeton University Press, Princeton, NJ.

Bordenstein, S. R., and K. R. Theis. 2015. Host biology in light of the microbiome: ten principles of holobionts and hologenomes. *PLoS Biology* 13:e1002226.

Borregaard, M. K., N. J. Gotelli, and C. Rahbek. 2012. Are range size distributions consistent with species-level heritability? *Evolution* 66:2216–26.

Borregaard, M. K., T. J. Matthews, and R. J. Whittaker. 2016. The general dynamic model: towards a unified theory of island biogeography? *Global Ecology and Biogeography* 7:805–16.

Botero, C. A., F. J. Weissing, J. Wright, and D. R. Rubenstein. 2015. Evolutionary tipping points in the capacity to adapt to environmental change. *Proceedings of the National Academy of Sciences* 112:184–89.

Bouchenak-Khelladi, Y., R. E. Onstein, Y. Xing, O. Schwery, and H. P. Linder. 2015. On the complexity of triggering evolutionary radiations. *New Phytologist* 207:313–26.

Bouckaert, R., J. Heled, D. Kühnert, T. Vaughan, C.-H. Wu, D. Xie, M. A. Suchard, A. Rambaut, and A. J. Drummond. 2014. BEAST 2: a software platform for Bayesian evolutionary analysis. *PLoS Computational Biology* 10:e1003537.

Boulila, S., B. Galbrun, K. G. Miller, S. F. Pekar, J. V. Browning, J. Laskar, and J. D. Wright. 2011. On the origin of Cenozoic and Mesozoic "third-order" eustatic sequences. *Earth-Science Reviews* 109:94–112.

Bowen, B. W., L. A. Rocha, R. J. Toonen, and S. A. Karl. 2013. The origins of tropical marine biodiversity. *Trends in Ecology and Evolution* 28:359–66.

Bowler, P. J. 1983. *The Eclipse of Darwinism.* Johns Hopkins University Press, Baltimore, MD.

Bowler, P. J. 2009. *Evolution: The History of an Idea.* University of California Press, Berkeley.

Boyce, M. S. 1984. Restitution of r- and K-selection as a model of density-dependent natural selection. *Annual Review of Ecology and Systematics* 15:427–47.

Boyd, R. 1999. Homeostasis, species, and higher taxa. Pp. 141–85 in R. Wilson, ed., *Species: New Interdisciplinary Essays*. MIT Press, Cambridge, MA.

Bradshaw, A. D. 1965. Evolutionary significance of phenotypic plasticity in plants. *Advances in Genetics* 13:115–55.

Bradshaw, H. D., and D. W. Schemske. 2003. Allele substitution at a flower locus produces a pollinator shift in monkeyflowers. *Nature* 426:176–78.

Brandon, R. N. 1990. *Adaptation and Environment*. Princeton University Press, Princeton, NJ.

Breden, F., and M. J. Wade. 1989. Selection within and between kin groups of the imported willow leaf beetle. *American Naturalist* 134:35–50.

Brideau, N. J., H. A. Flores, J. Wang, S. Maheshwari, X. Wang, and D. A. Barbash. 2006. Two Dobzhansky-Muller genes interact to cause hybrid lethality in *Drosophila*. *Science* 314:1292–95.

Brigandt, I. 2010. Beyond reduction and pluralism: toward an epistemology of explanatory integration in biology. *Erkenntnis* 73:295–311.

Brigandt, I. 2015. From developmental constraint to evolvability: how concepts figure in explanation and disciplinary identity. Pp. 305–25 in A. C. Love, ed., *Conceptual Change in Biology: Scientific and Philosophical Perspectives on Evolution and Development*. Springer, Dordrecht.

Brigandt, I., and A. C. Love. 2012. Conceptualizing evolutionary novelty: moving beyond definitional debates. *Journal of Experimental Zoology Part B: Molecular and Developmental Evolution* 318B:417–27.

Briggs, J. C. 1987. *Biogeography and Plate Tectonics*. Elsevier, New York.

Bromham, L., X. Hua, R. Lanfear, and P. F. Cowman. 2015. Exploring the relationships between mutation rates, life history, genome size, environment, and species richness in flowering plants. *American Naturalist* 185:507–24.

Brown, J. H., J. F. Gillooly, A. P. Allen, V. M. Savage, and G. B. West. 2004. Toward a metabolic theory of ecology. *Ecology* 85:1771–89.

Brown, J. H., and A. Kodric-Brown. 1977. Turnover rates in insular biogeography: effect of immigration on extinction. *Ecology* 58:445–49.

Brown, J. H., P. A. Marquet, and M. L. Taper. 1993. Evolution of body size: consequences of an energetic definition of fitness. *American Naturalist* 142:573–84.

Brown, J. L., J. J. Weber, D. F. Alvarado-Serrano, M. J. Hickerson, S. J. Franks, and A. C. Carnaval. 2016. Predicting the genetic consequences of future climate change: the power of coupling spatial demography, the coalescent, and historical landscape changes. *American Journal of Botany* 103:153–63.

Browne, J. 1995. *Charles Darwin: Voyaging. Volume 1 of a Biography*. Alfred A. Knopf, New York.

Browne, J. 2002. *Charles Darwin: The Power of Place. Volume 2 of a Biography*. Alfred A. Knopf, New York.

Büchi, L., and S. Vuilleumier. 2014. Coexistence of specialist and generalist species is shaped by dispersal and environmental factors. *American Naturalist* 183:612–24.

Büchi, L., and S. Vuilleumier. 2016. Ecological strategies in stable and disturbed environments depend on species specialisation. *Oikos* 125:1408–20.

Buffon, G.-L. L., Comte de. 1749. *Histoire naturelle, générale et particulière, avec la description du Cabinet du Roi*. Paris.

Buffon, G.-L. L., Comte de. 1753. Description de la partie du Cabinet qui a rapport à l'histoire naturelle du cheval: l'asne. Pp. 377–403 in *Histoire naturelle, générale et particulière, avec la description du Cabinet du Roi*. Paris.

Buller, A. R., and C. A. Townsend. 2013. Intrinsic evolutionary constraints on protease structure, enzyme acylation, and the identity of the catalytic triad. *Proceedings of the National Academy of Sciences* 110:E653–61.

Bulmer, M. 2004. Did Jenkin's swamping argument invalidate Darwin's theory of natural selection? *British Journal for the History of Science* 37:281–97.

Burman, J. T. 2013. Updating the Baldwin effect: the biological levels behind Piaget's new theory. *New Ideas in Psychology* 31:363–73.

Burnham, K. P., and D. R. Anderson. 2002. *Model Selection and Multimodel Inference: A Practical Information-Theoretic Approach.* 2nd ed. Springer, New York.

Burt, A., G. Bell, and P. H. Harvey. 1991. Sex differences in recombination. *Journal of Evolutionary Biology* 4:259–77.

Bush, A. M., and P. M. Novack-Gottshall. 2012. Modelling the ecological-functional diversification of marine Metazoa on geological time scales. *Biology Letters* 8:151–55.

Bush, G. L. 1969. Sympatric host race formation and speciation in frugivorous flies of the genus *Rhagoletis* (Diptera, Tephritidae). *Evolution* 23:237–51.

Bush, G. L. 1975a. Modes of animal speciation. *Annual Review of Ecology and Systematics* 6:339–64.

Bush, G. L. 1975b. Sympatric speciation in phytophagous parasitic insects. Pp. 187–206 in P. W. Price, ed., *Evolutionary Strategies of Parasitic Insects and Mites.* Plenum, New York.

Bush, R. M., C. A. Bender, K. Subbarao, N. J. Cox, and W. M. Fitch. 1999. Predicting the evolution of influenza A. *Science* 286:1921–25.

Butlin, R. 1987. Speciation by reinforcement. *Trends in Ecology and Evolution* 2:8–12.

Butlin, R. 2002. The costs and benefits of sex: new insights from old asexual lineages. *Nature Reviews Genetics* 3:311–17.

Calcott, B. 2009. Lineage explanations: explaining how biological mechanisms change. *British Journal for the Philosophy of Science* 60:51–78.

Callahan, H. S., H. Maughan, and U. K. Steiner. 2008. Phenotypic plasticity, costs of phenotypes, and costs of plasticity. *Annals of the New York Academy of Sciences* 1133:44–66.

Camper, P. 1784. Kurze Nachricht von der Zergliederung verschiedener Orang-Utans. Pp. 65–94 in *Herrn Pieter Campers kleine Schriften.* Part 2. Verlag S. L. Crusins, Leipzig.

Camper, P. 1785. Nachricht vom Sprachwerkzeuge des Orang-Utan. *Herrn Pieter Campers samtliche kleinere Schriften.* Verlag S. L. Crusins, Leipzig.

Carja, O., U. Liberman, and M. W. Feldman. 2014. Evolution in changing environments: modifiers of mutation, recombination, and migration. *Proceedings of the National Academy of Sciences* 111:17935–40.

Carlquist, S. 1972. Island biology: we've only just begun. *Bioscience* 22:221–25.

Carlquist, S. 1974. *Island Biology.* Columbia University Press, New York.

Carroll, L. 1871. *Through the Looking Glass: And What Alice Found There.* Macmillan, London.

Carroll, S. B. 2005a. *Endless Forms Most Beautiful: The New Science of Evo-Devo.* W. W. Norton, New York.

Carroll, S. B. 2005b. Evolution at two levels: on genes and form. *PLoS Biology* 3:e245.

Carroll, S. B. 2008. Evo-devo and an expanding evolutionary synthesis: a genetic theory of morphological evolution. *Cell* 134:25–36.

Carroll, S. B., J. K. Grenier, and S. D. Weatherbee. 2005. *From DNA to Diversity: Molecular Genetics and the Evolution of Animal Design.* 2nd ed. Blackwell Science, Malden, MA.

Carson, H. L. 1990. Extinction and recolonization of local populations on a growing shield volcano. *Proceedings of the National Academy of Sciences* 87:7055–57.

Carson, H. L., and A. R. Templeton. 1984. Genetic revolutions in relation to speciation phenomena: the founding of new populations. *Annual Review of Ecology and Systematics* 15:97–131.

Carvalho, D. M., S. J. Presley, and G. M. M. Santos. 2014. Niche overlap and network specialization of flower-visiting bees in an agricultural system. *Neotropical Entomology* 43:489–99.

Caswell, H. 2001. *Matrix Population Models: Construction, Analysis, and Interpretation.* Sinauer Associates, Sunderland, MA.

Cavalli-Sforza, L. L., and M. W. Feldman. 1981. *Cultural Transmission and Evolution: A Quantitative Approach.* Princeton University Press, Princeton, NJ.

Cavender-Bares, J., K. H. Kozak, P. V. Fine, and S. W. Kembel. 2009. The merging of community ecology and phylogenetic biology. *Ecology Letters* 12:693–715.

Chao, A., C.-H. Chiu, and L. Jost. 2014. Unifying species diversity, phylogenetic diversity, functional diversity, and related similarity and differentiation measures through Hill numbers. *Annual Review of Ecology, Evolution, and Systematics* 45:297–324.

Charlesworth, B. 1970. Selection in populations with overlapping generations. I. The use of Malthusian parameters in population genetics. *Theoretical Population Biology* 1:352–70.

Charlesworth, B. 1990. Optimization models, quantitative genetics, and mutation. *Evolution* 44:520–38.

Charlesworth, B. 1993. Directional selection and the evolution of sex and recombination. *Genetics Research* 61:205–24.

Charlesworth, B. 1994. *Evolution in Age-Structured Populations.* 2nd ed. Cambridge University Press, Cambridge, UK.

Charlesworth, B. 2000. Fisher, Medawar, Hamilton and the evolution of aging. *Genetics* 154:927–31.

Charlesworth, B., R. Lande, and M. Slatkin. 1982. A neo-Darwinian commentary on macroevolution. *Evolution* 36:474–98.

Charlesworth, B., M. T. Morgan, and D. Charlesworth. 1993. The effect of deleterious mutations on neutral molecular variation. *Genetics* 134:1289–1303.

Charlesworth, D. 2006. Evolution of plant breeding systems. *Current Biology* 16:726–35.

Charnov, E. L. 1982. *The Theory of Sex Allocation.* Princeton University Press, Princeton, NJ.

Charnov, E. L. 1987. On sex allocation and selfing in higher plants. *Evolutionary Ecology* 1:30–36.

Charnov, E. L., J. Maynard Smith, and J. J. Bull. 1976. Why be an hermaphrodite? *Nature* 263:125–26.

Charnov, E. L., and W. M. Schaffer. 1973. Life-history consequences of natural selection: Cole's result revisited. *American Naturalist* 107:791–91.

Chartier, M., F. Jabbour, S. Gerber, P. Mitteroecker, H. Sauquet, M. von Balthazar, Y. Staedler, P. R. Crane, and J. Schönenberger. 2014. The floral morphospace—a modern comparative approach to study angiosperm evolution. *New Phytologist* 204:841–53.

Chase, J. M. 2011. Ecological niche theory. Pp. 93–107 in S. M. Scheiner and M. R. Willig, eds., *The Theory of Ecology.* University of Chicago Press, Chicago, IL.

Chase, J. M., and M. A. Leibold. 2003. *Ecological Niches: Linking Classical and Contemporary Approaches.* University of Chicago Press, Chicago, IL.

Chesson, P., and N. Huntly. 1997. The roles of harsh and fluctuating conditions in the dynamics of ecological communities. *American Naturalist* 150:519–53.

Chesson, P., and J. J. Kuang. 2008. The interaction between predation and competition. *Nature* 456:235–38.

Chevin, L.-M., and R. Lande. 2010. When do adaptive plasticity and genetic evolution prevent extinction of a density-regulated population? *Evolution* 64:1143–50.

Chevin, L.-M., and R. Lande. 2015. Evolution of environmental cues for phenotypic plasticity. *Evolution* 69:2767–75.

Chiariello, N., and J. Roughgarden. 1984. Storage allocation in seasonal races of an annual plant: optimal versus actual allocation. *Ecology* 65:1290–1301.

Childs, D. Z., C. J. E. Metcalf, and M. Rees. 2010. Evolutionary bet-hedging in the real world: empirical evidence and challenges revealed by plants. *Proceedings of the Royal Society of London B: Biological Sciences* 277:3055–64.

Claridge, M. 2010. Species are real biological entities. Pp. 91–109 in F. J. Ayala and R. Arp, eds., *Contemporary Debates in Philosophy of Biology*. Wiley-Blackwell, West Sussex, UK.

Clausen, J., and W. M. Hiesey. 1958. *Experimental Studies on the Nature of Species. IV. Genetic Structure of Ecological Races*. Carnegie Institute of Washington, Washington, DC.

Clausen, J., D. D. Keck, and W. M. Hiesey. 1939. The concept of species based on experiment. *American Journal of Botany* 26:103–6.

Clausen, J. D., D. Keck, and W. M. Hiesey. 1940. *Experimental Studies on the Nature of Species I: Effects of Varied Environments on Western North American Plants*. Carnegie Institute of Washington, Washington, DC.

Clausen, J. D., D. Keck, and W. M. Hiesey. 1948. *Experimental Studies on the Nature of Species III: Environmental Responses of Climatic Races of Achillea*. Carnegie Institute of Washington, Washington, DC.

Clauset, A., and D. H. Erwin. 2008. The evolution and distribution of species body size. *Science* 321:399–401.

Clavel, J., R. Julliard, and V. Devictor. 2011. Worldwide decline of specialist species: toward a global functional homogenization? *Frontiers in Ecology and the Environment* 9:222–28.

Cole, L. C. 1954. The population consequences of life history phenomena. *Quarterly Review of Biology* 29:103–37.

Coleman, O., R. Hogan, N. McGoldrick, N. Rudden, and J. O. McInerney. 2015. Evolution by pervasive gene fusion in antibiotic resistance and antibiotic synthesizing genes. *Computation* 3:114–27.

Connallon, T., and A. G. Clark. 2010. Gene duplication, gene conversion and the evolution of the Y chromosome. *Genetics* 186:277–86.

Conway Morris, S. 2003. *Life's Solution*. Cambridge University Press, Cambridge, UK.

Corbett-Detig, R. B., J. Zhou, A. G. Clark, D. L. Hartl, and J. F. Ayroles. 2013. Genetic incompatibilities are widespread within species. *Nature* 504:135–37.

Coulson, T., S. Tuljapurkar, and D. Z. Childs. 2010. Using evolutionary demography to link life history theory, quantitative genetics and population ecology. *Journal of Animal Ecology* 79:1226–40.

Coyne, J. A. 2011. Speciation in a small space. *Proceedings of the National Academy of Sciences* 108:12975–76.

Coyne, J. A., N. H. Barton, and M. Turelli. 1997. Perspective: a critique of Sewall Wright's shifting balance theory of evolution. *Evolution* 51:643–71.

Coyne, J. A., and H. A. Orr. 1997. "Patterns of speciation in *Drosophila*" revisited. *Evolution* 51:295–303.

Coyne, J. A., and H. A. Orr. 2004. *Speciation*. Sinauer Associates, Sunderland, MA.

Coyne, J. A., and H. A. Orr. 2010. Speciation: a catalogue and critique of species concepts.

Appendix of Coyne and Orr 2004, 27, 447–72. Reprinted in A. Rosenberg and R. Arp, eds., *Philosophy of Biology: An Anthology*, 272–92. Wiley-Blackwell, Chichester, UK.

Cracraft, J. 1983. Species concepts and speciation analysis. Pp. 159–87 in R. F. Johnston, ed., *Current Ornithology*. Springer US, Boston, MA.

Cracraft, J. 1989. Species as entities of biological theory. Pp. 31–52 in M. Ruse, ed., *What the Philosophy of Biology Is*. Kluwer Academic, Dordrecht.

Crawford, D. J., and V. B. Smocovitis, eds. 2004. *The Scientific Papers of G. Ledyard Stebbins (1929–2000)*. ARG Gantner Verlag Ruggell, Liechtenstein.

Crespi, B. J. 1990. Measuring the effect of natural selection on phenotypic interaction systems. *American Naturalist* 135:32–47.

Crisp, M. D., S. A. Trewick, and L. G. Cook. 2011. Hypothesis testing in biogeography. *Trends in Ecology and Evolution* 26:66–72.

Crispo, E. 2007. The Baldwin effect and genetic assimilation: revisiting two mechanisms of evolutionary change mediated by phenotypic plasticity. *Evolution* 61:2469–79.

Crona, K., D. Greene, and M. Barlow. 2013. The peaks and geometry of fitness landscapes. *Journal of Theoretical Biology* 317:1–10.

Crow, J. F., and M. Kimura. 1970. *Introduction to Population Genetics Theory*. Harper and Row, New York.

Crow, J. F., and M. Kimura. 1979. Efficiency of truncation selection. *Proceedings of the National Academy of Sciences* 76:396–99.

Crucifix, M. 2012. Oscillators and relaxation phenomena in Pleistocene climate theory. *Philosophical Transactions of the Royal Society of London A: Mathematical, Physical and Engineering Sciences* 370:1140–65.

Cutter, A. D., and B. A. Payseur. 2013. Genomic signatures of selection at linked sites: unifying the disparity among species. *Nature Reviews Genetics* 14:262–74.

Dagan, T., Y. Artzy-Randrup, and W. Martin. 2008. Modular networks and cumulative impact of lateral transfer in prokaryote genome evolution. *Proceedings of the National Academy of Sciences* 105:10039–44.

Dagan, T., and W. Martin. 2007. Ancestral genome sizes specify the minimum rate of lateral gene transfer during prokaryote evolution. *Proceedings of the National Academy of Sciences* 104:870–75.

Dagan, T., and W. Martin. 2009. Getting a better picture of microbial evolution en route to a network of genomes. *Philosophical Transactions of the Royal Society of London B: Biological Sciences* 364:2187–96.

Damuth, J. 1981. Population density and body size in mammals. *Nature* 290:699–700.

Damuth, J., and I. L. Heisler. 1988. Alternative formulations of multilevel selection. *Biology and Philosophy* 3:407–30.

Danchin, É., A. Charmantier, F. A. Champagne, A. Mesoudi, B. Pujol, and S. Blanchet. 2011. Beyond DNA: integrating inclusive inheritance into an extended theory of evolution. *Nature Reviews Genetics* 12:475–86.

Darriba, D., T. Flouri, and A. Stamatakis. 2018. The state of software for evolutionary biology. *Molecular Biology and Evolution* 35:1037–46.

Darwin, C. 1837. Notebook B, "The Transmutation of Species." http://darwin-online.org.uk/content/frameset?pageseq=1&itemID=CUL-DAR121.-&viewtype=image.

Darwin, C. 1859. *On the Origin of Species by Means of Natural Selection*. John Murray, London.

Darwin, C. 1868. Letter 6033—Darwin, C. R. to Wallace, A. R., 21 Mar 1868. http://www.darwinproject.ac.uk/DCP-LETT-6033.

Darwin, C. 1871. *The Descent of Man and Selection in Relation to Sex*. John Murray, London.

Daubin, V., M. Gouy, and G. Perrière. 2002. A phylogenomic approach to bacterial phylogeny: evidence of a core of genes sharing a common history. *Genome Research* 12:1080–90.

Davidson, E. H. 2006. *The Regulatory Genome: Gene Regulatory Networks in Development and Evolution*. Academic Press, San Diego, CA.

Davidson, E. H., and D. H. Erwin. 2006. Gene regulatory networks and the evolution of animal body plans. *Science* 311:796–800.

Davidson, E. H., and D. H. Erwin. 2010. An integrated view of Precambrian eumetazoan evolution. *Cold Spring Harbor Symposia on Quantitative Biology* 74:65–80.

Davidson, E. H., and I. S. Peter. 2015. *Genomic Control Process: Development and Evolution*. Academic Press, San Diego, CA.

Dawkins, R. 1976. *The Selfish Gene*. Oxford University Press, Oxford, UK.

De Jong, G. 1993. Covariances between traits deriving from successive allocations of a resource. *Functional Ecology* 7:75–83.

De Jong, G. 1995. Phenotypic plasticity as a product of selection in a variable environment. *American Naturalist* 145:493–512.

De Jong, G., and A. J. van Noordwijk. 1992. Acquisition and allocation of resources: genetic (co)variances, selection, and life histories. *American Naturalist* 139:749–49.

De Laguérie, P., I. Olivieri, A. Atlan, and P. H. Gouyon. 1991. Analytic and simulation models predicting positive genetic correlations between traits linked by trade-offs. *Evolutionary Ecology* 5:361–69.

De Mazancourt, C., and U. Dieckmann. 2004. Trade-off geometries and frequency-dependent selection. *American Naturalist* 164:765–78.

De Meester, L., A. Gomez, and J. Simon. 2004. Evolutionary and ecological genetics of cyclical parthenogens. Pp. 122–34 in A. Moya and E. Font, eds., *Evolution from Molecules to Ecosystems*. Oxford University Press, Oxford, UK.

de Queiroz, K. 1988. Systematics and the Darwinian revolution. *Philosophy of Science* 55:238–59.

de Queiroz, K. 1998. The general lineage concept of species, species criteria, and the process of speciation: a conceptual unification and terminological recommendations. Pp. 57–75 in D. Howard and S. Berlocher, eds., *Endless Forms: Species and Speciation*. Oxford University Press, New York.

de Queiroz, K. 1999. The general lineage concept of species and the defining properties of the species category. Pp. 49–89 in R. Wilson, ed., *Species: New Interdisciplinary Essays*. MIT Press, Cambridge, MA.

de Queiroz, K. 2005. Ernst Mayr and the modern concept of species. *Proceedings of the National Academy of Sciences* 102:6600–6607.

de Queiroz, K. 2007. Species concepts and species delimitation. *Systematic Biology* 56:879–86.

de Queiroz, K. 2011. Branches in the lines of descent: Charles Darwin and the evolution of the species concept. *Biological Journal of the Linnean Society* 103:19–35.

de Queiroz, K., and J. Gauthier. 1994. Toward a phylogenetic system of biological nomenclature. *Trends in Ecology and Evolution* 9:27–31.

de Queiroz, K., and S. Poe. 2003. Failed refutations: further comments on parsimony and likelihood methods and their relationship to Popper's degree of corroboration. *Systematic Biology* 52:352–67.

De Robertis, E. M. 2008. Evo-devo: variations on ancestral themes. *Cell* 132:185–95.

de Visser, J. A. G. M., and S. F. Elena. 2007. The evolution of sex: empirical insights into the roles of epistasis and drift. *Nature Reviews Genetics* 8:139–49.

de Visser, J. A. G. M., S.-C. Park, and J. Krug. 2009. Exploring the effect of sex on empirical fitness landscapes. *American Naturalist* 174:S15–S30.

de Vladar, H. P., and N. H. Barton. 2011. The contribution of statistical physics to evolutionary biology. *Trends in Ecology and Evolution* 26:424–32.

de Vries, H. 1903. *Die Mutationstheorie. Versuche und Beobachtungen über die Enstehung der Arten in Pflanzenreich.* Veit, Leipzig.

de Winter, W. 1997. The beanbag genetics controversy: towards a synthesis of opposing views of natural selection. *Biology and Philosophy* 12:149–84.

Deakin, J. E., and T. Ezaz. 2014. Tracing the evolution of amniote chromosomes. *Chromosoma* 123:201–16.

Débarre, F., and S. Gandon. 2010. Evolution of specialization in a spatially continuous environment. *Journal of Evolutionary Biology* 23:1090–99.

Delisle, R. 2007. *Debating Humankind's Place in Nature, 1860–2000: The Nature of Paleoanthropology.* Pearson/Prentice Hall, Englewood Cliffs, NJ.

Demerec, M. 1960. Forward. *Genetics and Twentieth Century Darwinism.* Cold Spring Harbor Symposia on Quantitative Biology. Biological Laboratory, Cold Spring Harbor, NY.

Denver, D. R., K. Morris, M. Lynch, L. L. Vassilieva, and W. K. Thomas. 2000. High direct estimate of the mutation rate in the mitochondrial genome of *Caenorhabditis elegans.* *Science* 289:2342–44.

Desdevises, Y., S. Morand, and P. Legendre. 2002. Evolution and determinants of host specificity in the genus *Lamellodiscus* (Monogenea). *Biological Journal of the Linnean Society* 77:431–43.

Dettman, J. R., C. Sirjusingh, L. M. Kohn, and J. B. Anderson. 2007. Incipient speciation by divergent adaptation and antagonistic epistasis in yeast. *Nature* 447:585–88.

Devictor, V., J. Clavel, R. Julliard, S. Lavergne, D. Mouillot, W. Thuiller, P. Venail, S. Villeger, and N. Mouquet. 2010. Defining and measuring ecological specialization. *Journal of Applied Ecology* 47:15–25.

Devictor, V., R. Julliard, J. Clavel, F. Jiguet, A. Lee, and D. Couvet. 2008. Functional biotic homogenization of bird communities in disturbed landscapes. *Global Ecology and Biogeography* 17:252–61.

DeWitt, T. J., A. Sih, and D. S. Wilson. 1998. Costs and limits of phenotypic plasticity. *Trends in Ecology and Evolution* 13:77–81.

Dick, D. G., and E. E. Maxwell. 2015. The evolution and extinction of the ichthyosaurs from the perspective of quantitative ecospace modelling. *Biology Letters* 11:20150339.

Dickinson, H., and J. Antonovics. 1973. Theoretical considerations of sympatric divergence. *American Naturalist* 107:256–74.

Dieckmann, U., and M. Doebeli. 1999. On the origin of species by sympatric speciation. *Nature* 400:354–57.

Diehl, S. R., and G. L. Bush. 1989. The role of habitat preference in adaptation and speciation. Pp. 345–65 in D. Otte and J. A. Endler, eds., *Speciation and Its Consequences.* Sinauer Associates, Sunderland, MA.

Diniz-Filho, J. A. F., and L. M. Bini. 2008. Macroecology, global change and the shadow of forgotten ancestors. *Global Ecology and Biogeography* 17:11–17.

Dobzhansky, T. 1937a. Genetic nature of species differences. *American Naturalist* 71:404–20.

Dobzhansky, T. 1937b. *Genetics and the Origin of Species.* Columbia University Press, New York.

Dobzhansky, T. 1940. Speciation as a stage in evolutionary divergence. *American Naturalist* 74:312–21.

Dobzhansky, T. 1941. The race concept in biology. *Scientific Monthly* 52:161–65.

Dobzhansky, T., and M. F. A. Montagu. 1947. Natural selection and the mental capacities of mankind. *Science* 105:587–90.

Doebeli, M., and U. Dieckmann. 2005. Adaptive dynamics as a mathematical tool for studying the ecology of speciation processes. *Journal of Evolutionary Biology* 18:1194–1200.

Doebeli, M., and I. Ispolatov. 2014. Chaos and unpredictability in evolution. *Evolution* 68:1365–73.

Doerder, F. P. 2014. Abandoning sex: multiple origins of asexuality in the ciliate Tetrahymena. *BMC Evolutionary Biology* 14:1–13.

Donoghue, M. J. 2008. A phylogenetic perspective on the distribution of plant diversity. *Proceedings of the National Academy of Sciences* 105:11549–55.

Donohue, K. 2003. The influence of neighbor relatedness on multilevel selection in the Great Lakes sea rocket. *American Naturalist* 162:77–92.

Donohue, K. 2004. Density-dependent multilevel selection in the Great Lakes sea rocket. *Ecology* 85:180–91.

Doolittle, W. F. 1999. Phylogenetic classification and the universal tree. *Science* 284:2124–28.

Doolittle, W. F. 2000. Uprooting the tree of life. *Scientific American* 282:90–95.

Doolittle, W. F. 2013. Is junk DNA bunk? A critique of ENCODE. *Proceedings of the National Academy of Sciences* 110:5294–5300.

Doolittle, W. F., and S. A. Inkpen. 2018. Processes and patterns of interaction as units of selection: an introduction to ITSNTS thinking. *Proceedings of the National Academy of Sciences, USA* 115:4006–14.

Dougherty, E. C. 1955. Comparative evolution and the origin of sexuality. *Systematic Zoology* 4:145–90.

Douglas, A. E. 2014. Symbiosis as a general principle in eukaryotic evolution. *Cold Spring Harbor Perspectives in Biology* 6:a016113.

Downes, S. M. 1992. The importance of models in theorizing: a deflationary semantic view. *PSA: Proceedings of the Biennial Meeting of the Philosophy of Science Association* 1:142–53.

Downes, S. M. 2011. Scientific models. *Philosophy Compass* 6:757–64.

Dronamraju, K. 2011. *Haldane, Mayr, and Beanbag Genetics.* Oxford University Press, Oxford, UK.

Drummond, A. J., M. A. Suchard, D. Xie, and A. Rambaut. 2012. Bayesian phylogenetics with BEAUti and the BEAST 1.7. *Molecular Biology and Evolution* 29:1969–73.

Duhem, P. 1954. *The Aim and Structure of Physical Theory.* Princeton University Press, Princeton, NJ.

Dupré, J. 1981. Natural kinds and biological taxa. *Philosophical Review* 90:66–90.

Dupré, J. 1994. The philosophical basis of biological classification. *Studies in History and Philosophy of Science Part A* 25:271–79.

Dupré, J. 1995. *The Disorder of Things: Metaphysical Foundations of the Disunity of Science.* Harvard University Press, Cambridge, MA.

Dupré, J. 1999. On the impossibility of a monistic account of species. Pp. 3–20 in R. Wilson, ed., *Species: New Interdisciplinary Essays.* MIT Press, Cambridge, MA.

East, E. M. 1918. The role of reproduction in evolution. *American Naturalist* 52:273–89.

Easterling, M. R., S. P. Ellner, and P. M. Dixon. 2000. Size-specific sensitivity: applying a new structured population model. *Ecology* 81:694–94.

Edens-Meier, R., and P. Bernhardt, eds. 2014. *Darwin's Orchids: Then and Now.* University of Chicago Press, Chicago, IL.

Edie, S. M., D. Jablonski, and J. W. Valentine. 2018. Contrasting responses of functional diversity to major losses in taxonomic diversity. *Proceedings of the National Academy of Sciences* (115:732–37).

Edwards, A. W. F. 1998. Natural selection and the sex ratio: Fisher's sources. *American Naturalist* 6:564–69.

Edwards, A. W. F. 2000. Carl Düsing (1884) on *The Regulation of the Sex-Ratio. Theoretical Population Biology* 58:255–57.

Edwards, S., K. A. Tolley, B. Vanhooydonck, G. J. Measey, and A. Herrel. 2013. Is dietary niche breadth linked to morphology and performance in Sandveld lizards *Nucras* (Sauria: Lacertidae)? *Biological Journal of the Linnean Society* 110:674–88.

Edwards, S. V. 2009. Is a new and general theory of molecular systematics emerging? *Evolution* 63:1–19.

Edwards, S. V., S. B. Kingan, J. D. Calkins, C. N. Balakrishnan, W. B. Jennings, W. J. Swanson, and M. D. Sorenson. 2005. Speciation in birds: genes, geography, and sexual selection. *Proceedings of the National Academy of Sciences* 102:6550–57.

Einstein, A. 1934. On the method of theoretical physics. *Philosophy of Science* 1:163–69.

Einum, S., and I. A. Fleming. 2004a. Does within-population variation in egg size reduce intraspecific competition in Atlantic Salmon, Salmo salar? *Functional Ecology* 18:110–15.

Einum, S., and I. A. Fleming. 2004b. Environmental unpredictability and offspring size: conservative versus diversified bet-hedging. *Evolutionary Ecology Research* 6:443–55.

Eldakar, O. T., D. S. Wilson, M. J. Dlugos, and J. W. Pepper. 2010. The role of multilevel selection in the evolution of sexual conflict in the water strider *Aquarius remigis. Evolution* 64:3183–89.

Eldredge, N., and J. Cracraft. 1980. *Phylogenetic Patterns and the Evolutionary Process.* Columbia University Press, New York.

Eldredge, N., and S. J. Gould. 1972. Punctuated equilibria: an alternative to phyletic gradualism. Pp. 82–115 in T. Schopf, ed., *Models in Paleontology.* Freeman, Cooper, San Francisco, CA.

Elias, M., Z. Gompert, C. Jiggins, and K. Willmott. 2008. Mutualistic interactions drive ecological niche convergence in a diverse butterfly community. *PLoS Biology* 6:e300.

Elith, J., C. H. Graham, R. P. Anderson, M. Dudík, S. Ferrier, A. Guisan, R. J. Hijmans, F. Huettmann, J. R. Leathwick, A. Lehmann, J. Li, L. G. Lohmann, B. A. Loiselle, G. Glenn Manion, C. Craig Moritz, M. Miguel Nakamura, Y. Nakazawa, J. M. M. Overton, A. T. Peterson, S. J. Phillips, K. Richardson, R. Scachetti-Pereira, R. E. Schapire, J. Soberón, S. Williams, M. S. Wisz, and N. E. Zimmermann. 2006. Novel methods improve prediction of species' distributions from occurrence data. *Ecography* 29: 129–51.

Emerson, B. C., and R. G. Gillespie. 2008. Phylogenetic analysis of community assembly and structure over space and time. *Trends in Ecology and Evolution* 23:619–30.

ENCODE Project Consortium. 2012. An integrated encyclopedia of DNA elements in the human genome. *Nature* 489:57–74.

Endersby, J. 2016. *Orchid: A Cultural History.* University of Chicago Press, Chicago, IL.

Endler, J. A. 1980. Natural selection on color patterns in Poecilia reticulata. *Evolution* 34:76–91.

Endler, J. A. 1986. *Natural Selection in the Wild*. Princeton University Press, Princeton, NJ.

Engen, S., R. Lande, and B.-E. Saether. 2013. A quantitative genetic model of r- and K-selection in a fluctuating population. *American Naturalist* 181:725–36.

Enquist, B. J., J. H. Brown, and G. B. West. 1998. Allometric scaling of plant energetics and population density. *Nature* 395:163–65.

Enright, A. J., and C. A. Ouzounis. 2001. Functional associations of proteins in entire genomes by means of exhaustive detection of gene fusions. *Genome Biology* 2:34.1–34.7.

Enright, A. J., S. Van Dongen, and C. A. Ouzounis. 2002. An efficient algorithm for large-scale detection of protein families. *Nucleic Acids Research* 30:1575–84.

Epstein, B., and P. Forber. 2013. The perils of tweaking: how to use macrodata to set parameters in complex simulation models. *Synthese* 190:203–18.

Ereshefsky, M. 1992. Eliminative pluralism. *Philosophy of Science* 59:671–90.

Ereshefsky, M. 1999. Species and the Linnaean hierarchy. Pp. 285–305 in R. Wilson, ed., *Species: New Interdisciplinary Essays*. MIT Press, Cambridge, MA.

Ereshefsky, M. 2010. Microbiology and the species problem. *Biology and Philosophy* 25: 553–68.

Ereshefsky, M. 2011. Mystery of mysteries: Darwin and the species problem. *Cladistics* 27: 67–79.

Ericson, P. G., L. Christidis, A. Cooper, M. Irestedt, J. Jackson, U. S. Johansson, and J. A. Norman. 2002. A Gondwanan origin of passerine birds supported by DNA sequences of the endemic New Zealand wrens. *Proceedings of the Royal Society of London B: Biological Sciences* 269:235–41.

Erwin, D. H. 2007. Disparity: morphological pattern and developmental context. *Palaeontology* 50:57–73.

Erwin, D. H. 2012. Novelties that change carrying capacity. *Journal of Experimental Zoology Part B: Molecular and Developmental Evolution* 318:460–65.

Erwin, D. H. 2015a. Novelty and innovation in the history of life. *Current Biology* 25:R930–R940.

Erwin, D. H. 2015b. Was the Ediacaran-Cambrian radiation a unique evolutionary event? *Paleobiology* 41:1–15.

Erwin, D. H., and J. W. Valentine. 2013. *The Cambrian Explosion*. Ben Roberts, Greenwood Village, CO.

Erwin, D. H., J. W. Valentine, and J. J. Sepkoski. 1987. A comparative study of diversification events: the early Paleozoic versus the Mesozoic. *Evolution* 41:1177–86.

Espinosa-Soto, C., and A. Wagner. 2010. Specialization can drive the evolution of modularity. *PLoS Computational Biology* 6:e1000719.

Estes, S., P. C. Phillips, D. R. Denver, W. K. Thomas, and M. Lynch. 2004. Mutation accumulation in populations of varying size: the distribution of mutational effects for fitness correlates in *Caenorhabditis elegans*. *Genetics* 166:1269–79.

Etienne, R. S., and H. Olff. 2004. A novel genealogical approach to neutral biodiversity theory. *Ecology Letters* 7:170–75.

Evans, M. E., S. A. Smith, R. S. Flynn, and M. J. Donoghue. 2009. Climate, niche evolution, and diversification of the "Bird-Cage" evening primroses (Oenothera, Sections Anogra and Kleinia). *American Naturalist* 173:225–40.

Evenson, W. E. 1983. Experimental studies of reproductive energy allocation in plants. Pp. 249–74 in C. E. Jones and R. J. Little, eds., *Handbook of Experimental Pollination Biology*. Van Nostrand Reinhold, New York.

Ewens, W. J. 1989. An interpretation and proof of the fundamental theorem of natural selection. *Theoretical Population Biology* 36:167–80.

Ewens, W. J. 1990. Population genetics theory—the past and the future. Pp. 177–227 in S. Lessard, ed., *Mathematical and Statistical Developments of Evolutionary Theory*. Kluwer Academic, Dordrecht.

Ewens, W. J. 1992. An optimizing principle of natural selection in evolutionary population genetics. *Theoretical Population Biology* 42:333–46.

Excoffier, L., and M. Slatkin. 1995. Maximum-likelihood estimation of molecular haplotype frequencies in a diploid population. *Molecular Biology and Evolution* 12:921–27.

Eyre-Walker, A., and P. D. Keightley. 2007. The distribution of fitness effects of new mutations. *Nature Reviews Genetics* 8:610–18.

Ezard, T. H. G., T. Aze, P. N. Pearson, and A. Purvis. 2011. Interplay between changing climate and species' ecology drives macroevolutionary dynamics. *Science* 332:349–51.

Ezard, T. H. G., G. H. Thomas, and A. Purvis. 2013. Inclusion of a near-complete fossil record reveals speciation-related molecular evolution. *Methods in Ecology and Evolution* 4:745–53.

Falconer, D. S., and T. F. C. Mackay. 1996. *Introduction to Quantitative Genetics*. 4th ed. Longman, Essex, UK.

Farris, J. S. 1977. Phylogenetic analysis under Dollo's Law. *Systematic Biology* 26:77–88.

Farris, J. S. 1983. The logical basis of phylogenetic analysis. Pp. 7–36 in N. I. Platnick and V. A. Funk, eds., *Advances in Cladistics*, vol. 2. Columbia University Press, New York.

Fauvelot, C., G. Bernardi, and S. Planes. 2003. Reductions in the mitochondrial DNA diversity of coral reef fish provide evidence of population bottlenecks resulting from Holocene sea-level change. *Evolution* 57:1571–83.

Feder, J. L., C. A. Chilcote, and G. L. Bush. 1988. Genetic differentiation between sympatric host races of the apple maggot fly *Rhagoletis pomonella*. *Nature* 336:61–64.

Feder, J. L., J. B. Roethele, K. Filchak, J. Niedbalski, and J. Romero-Severson. 2003. Evidence for inversion polymorphism related to host race formation in the apple maggot fly, *Rhagoletis pomonella*. *Genetics* 163:939–53.

Feldman, M. W., F. B. Christiansen, and L. D. Brooks. 1980. Evolution of recombination in a constant environment. *Proceedings of the National Academy of Sciences* 77:4838–41.

Feldman, M. W., S. P. Otto, and F. B. Christiansen. 1996. Population genetic perspectives on the evolution of recombination. *Annual Review of Genetics* 30:261–95.

Felsenstein, J. 1974. The evolutionary advantage of recombination. *Genetics* 78:737–56.

Felsenstein, J. 1978. Cases in which parsimony or compatibility methods will be positively misleading. *Systematic Biology* 27:401–10.

Felsenstein, J. 1981. Skepticism towards Santa Rosalia, or why are there so few kinds of animals? *Evolution* 35:124–38.

Felsenstein, J. 1985. Phylogenies and the comparative method. *American Naturalist* 125:1–15.

Felsenstein, J. 1988. Sex and the evolution of recombination. Pp. 74–86 in R. E. Michod and B. R. Levin, eds., *The Evolution of Sex: An Examination of Current Ideas*. Sinauer Associates, Sunderland, MA.

Fernández-Palacios, J. M., K. F. Rijsdijk, S. J. Norder, R. Otto, L. Nascimento, S. Fernández-Lugo, E. Tjørve, and R. J. Whittaker. 2015. Towards a glacial-sensitive model of island biogeography. *Global Ecology and Biogeography* 25:817–30.

Ferriere, R., and S. Legendre. 2013. Eco-evolutionary feedbacks, adaptive dynamics and evolutionary rescue theory. *Philosophical Transactions of the Royal Society of London B: Biological Sciences* 368:20120081.

Fichman, M. 2004. *An Elusive Victorian: The Evolution of Alfred Russel Wallace*. University of Chicago Press, Chicago, IL.

Finarelli, J. 2007. Mechanisms behind active trends in body size evolution of the Canidae (Carnivora: Mammalia). *American Naturalist* 170:876–85.

Fine, P. V. 2015. Ecological and evolutionary drivers of geographic variation in species diversity. *Annual Review of Ecology, Evolution, and Systematics* 46:369–92.

Firestein, S. 2016. *Failure: Why Science Is so Successful*. Oxford University Press, New York.

Fischer, B., G. S. v. Doorn, U. Dieckmann, and B. Taborsky. 2014. The evolution of age-dependent plasticity. *American Naturalist* 183:108–25.

Fisher, R. A. 1918. The correlation between relatives on the supposition of Mendelian inheritance. *Transactions of the Royal Society of Edinburgh* 52:399–433.

Fisher, R. A. 1922. On the dominance ratio. *Proceedings of the Royal Society of Edinburgh* 42:321–41.

Fisher, R. A. 1930. *The Genetical Theory of Natural Selection*. Oxford University Press, Oxford, UK.

Fitch, W. M. 1970. Distinguishing homologous from analogous proteins. *Systematic Zoology* 19:99–113.

Fitch, W. M. 2000. Homology: a personal view on some of the problems. *Trends in Genetics* 16:227–31.

Fitzpatrick, B. M., and M. Turelli. 2006. The geography of mammalian speciation: mixed signals from phylogenies and range maps. *Evolution* 60:601–15.

Flot, J.-F., B. Hespeels, X. Li, B. Noel, I. Arkhipova, E. G. J. Danchin, A. Hejnol, B. Henrissat, R. Koszul, J.-M. Aury, V. Barbe, R.-M. Barthelemy, J. Bast, G. A. Bazykin, O. Chabrol, A. Couloux, M. Da Rocha, C. Da Silva, E. Gladyshev, P. Gouret, O. Hallatschek, B. Hecox-Lea, K. Labadie, B. Lejeune, O. Piskurek, J. Poulain, F. Rodriguez, J. F. Ryan, O. A. Vakhrusheva, E. Wajnberg, B. Wirth, I. Yushenova, M. Kellis, A. S. Kondrashov, D. B. Mark Welch, P. Pontarotti, J. Weissenbach, P. Wincker, O. Jaillon, and K. Van Doninck. 2013. Genomic evidence for ameiotic evolution in the bdelloid rotifer *Adineta vaga*. *Nature* 500:453–57.

Fontaine, C., P. R. Guimarães, S. Kéfi, N. Loeuille, J. Memmott, W. H. van der Putten, F. J. F. van Veen, and E. Thébault. 2011. The ecological and evolutionary implications of merging different types of networks. *Ecology Letters* 14:1170–81.

Foote, M. 1993. Discordance and concordance between morphological and taxonomic diversity. *Paleobiology* 19:185–204.

Foote, M. 1996. Models of morphological diversification. Pp. 62–86 in D. Jablonski, D. H. Erwin, and J. H. Lipps, eds., *Evolutionary Paleobiology*. University of Chicago Press, Chicago, IL.

Foote, M. 2010. The geologic history of biodiversity. Pp. 479–510 in M. A. Bell, D. J. Futuyma, W. F. Eanes, and J. S. Levinton, eds., *Evolution since Darwin: The First 150 Years*. Sinauer Associates, Sunderland, MA.

Foote, M. 2012. Evolutionary dynamics of taxonomic structure. *Biology Letters* 8:135–38.

Foote, M., J. S. Crampton, A. G. Beu, and R. A. Cooper. 2008. On the bidirectional relationship between geographic range and taxonomic duration. *Paleobiology* 34:421–33.

Foote, M., K. A. Ritterbush, and A. I. Miller. 2016. Geographic ranges of genera and their constituent species: structure, evolutionary dynamics, and extinction resistance. *Paleobiology* 42:269–88.

Forber, P. 2011. Reconceiving eliminative inference. *Philosophy of Science* 78:185–208.

Forister, M. L., L. A. Dyer, M. S. Singer, J. O. Stireman, and J. T. Lill. 2012. Revisiting the evolution of ecological specialization, with emphasis on insect-plant interactions. *Ecology* 93:981–91.

Fortin, F.-A., F.-M. De Rainville, M.-A. Gardner, P. Marc, and C. Gagné. 2012. DEAP: evolutionary algorithms made easy. *Journal of Machine Learning Research* 13:2171–75.

Foster, W. J., and R. J. Twitchett. 2014. Functional diversity of marine ecosystems after the Late Permian mass extinction event. *Nature Geoscience* 7:233–38.

Fox, G. A. 1989. Consequences of flowering-time variation in a desert annual: adaptation and history. *Ecology* 70:1294–1306.

Fox, G. A. 1992. Annual plant life histories and the paradigm of resource allocation. *Evolutionary Ecology* 6:482–99.

Frank, S. A. 1986. Hierarchical selection theory and sex ratios I. General solutions for structured populations. *Theoretical Population Biology* 29:312–42.

Frank, S. A. 1991. Divergence of meiotic drive-suppression systems as an explanation for sex-biased hybrid sterility and inviability. *Evolution* 45:262–67.

Frank, S. A. 1997. The Price equation, Fisher's fundamental theorem, kin selection, and causal analysis. *Evolution* 51:1712–29.

Frank, S. A. 1998. *Foundations of Social Evolution*. Princeton University Press, Princeton, NJ.

Frank, S. A. 2009. Natural selection maximizes Fisher information. *Journal of Evolutionary Biology* 22:231–44.

Frank, S. A. 2012a. Natural selection IV: the Price equation. *Journal of Evolutionary Biology* 25:1002–19.

Frank, S. A. 2012b. Natural selection V: how to read the fundamental equations of evolutionary change in terms of information theory. *Journal of Evolutionary Biology* 25:2377–96.

Frank, S. A. 2012c. Wright's adaptive landscape versus Fisher's fundamental theorem. Pp. 41–57 in E. Svensson and R. Calsbeek, eds., *The Adaptive Landscape in Evolutionary Biology*. Oxford University Press, New York.

Frank, S. A. 2013a. Natural selection VI: partitioning the information in fitness and characters by path analysis. *Journal of Evolutionary Biology* 26:457–71.

Frank, S. A. 2013b. Natural selection VII: history and interpretation of kin selection theory. *Journal of Evolutionary Biology* 26:1151–84.

Frank, S. A. 2014. The inductive theory of natural selection: summary and synthesis. arXiv:1412.1285.

Frank, S. A., and M. Slatkin. 1992. Fisher's fundamental theorem of natural selection. *Trends in Ecology and Evolution* 7:92–95.

Frickey, T., and A. Lupas. 2004. CLANS: a Java application for visualizing protein families based on pairwise similarity. *Bioinformatics* 20:3702–4.

Frigg, R., and S. Hartmann. 2012. Models in science. In E. N. Zalta, ed., *The Stanford Encyclopedia of Philosophy*, http://plato.stanford.edu/entries/models-science/.

Fritz, J. A., J. Brancale, M. Tokita, K. J. Burns, M. B. Hawkins, A. Abzhanov, and M. P. Brenner. 2014. Shared developmental programme strongly constrains beak shape diversity in songbirds. *Nature Communications* 5:3700.

Fry, J. D. 2003. Multilocus models of sympatric speciation: Bush versus Rice versus Felsenstein. *Evolution* 57:1735–46.

Futuyma, D. J. 1987. On the role of species in anagenesis. *American Naturalist* 130: 465–73.

Futuyma, D. J. 2013. *Evolution*. 3rd ed. Sinauer and Associates, Sunderland, MA.

Futuyma, D. J. 2015. Can modern evolutionary theory explain macroevolution? Pp. 29–85 in E. Serrelli and N. Gontier, eds., *Macroevolution*. Springer International, Switzerland.

Futuyma, D. J. 2017. Evolutionary biology today, and the call for an extended synthesis. *Interface Focus* 7:20160145.

Futuyma, D. J., and G. Moreno. 1988. The evolution of ecological specialization. *Annual Review of Ecology and Systematics* 19:207–33.

Gaba, S., and D. Ebert. 2009. Time-shift experiments as a tool to study antagonistic co-evolution. *Trends in Ecology and Evolution* 24:226–32.

Gadgil, M., and W. H. Bossert. 1970. Life historical consequences of natural selection. *American Naturalist* 104:1–24.

Galison, P. 1997. *Image and Logic: A Material Culture of Microphysics*. University of Chicago Press, Chicago, IL.

Galloway, L. F. 2005. Maternal effects provide phenotypic adaptation to local environmental conditions. *New Phytologist* 166:93–100.

Galton, F. 1886. Regression towards mediocrity in hereditary stature. *Journal of the Anthropological Institute of Great Britain and Ireland* 15:246–63.

Gardner, A. 2008. The Price equation. *Current Biology* 18:R198–R202.

Gardner, A. 2015. The genetical theory of multilevel selection. *Journal of Evolutionary Biology* 28:305–19.

Gardner, A., and A. Grafen. 2009. Capturing the superorganism: a formal theory of group adaptation. *Journal of Evolutionary Biology* 22:659–71.

Gardner, A., and A. T. Kalinka. 2006. Recombination and the evolution of mutational robustness. *Journal of Theoretical Biology* 241:707–15.

Gardner, M.-A., C. Gagné, and M. Parizeau. 2015. Controlling code growth by dynamically shaping the genotype size distribution. *Genetic Programming and Evolvable Machines* 16:1–44.

Gatesy, J., C. Hayashi, D. Motriuk, J. Woods, and R. Lewis. 2001. Extreme diversity, conservation, and convergence of spider silk fibroin sequences. *Science* 291:2603–5.

Gauthier, J. A., M. Kearney, J. A. Maisano, O. Rieppel, and A. D. B. Behlke. 2012. Assembling the squamate tree of life: perspectives from the phenotype and the fossil record. *Bulletin of the Peabody Museum of Natural History* 53:3–308.

Gavrilets, S. 1986. An approach to modeling the evolution of populations with consideration of genotype-environment interaction. *Soviet Genetics* 22:28–36.

Gavrilets, S. 1997. Evolution and speciation on holey adaptive landscapes. *Trends in Ecology and Evolution* 12:307–12.

Gavrilets, S. 2004. *Fitness Landscapes and the Origin of Species*. Princeton University Press, Princeton, NJ.

Gavrilets, S. 2014. Models of speciation: where are we now? *Journal of Heredity* 105:743–55.

Gavrilets, S., and A. Hastings. 1996. Founder effect speciation: a theoretical reassessment. *American Naturalist* 147:466–91.

Gavrilets, S., and J. B. Losos. 2009. Adaptive radiation: contrasting theory with data. *Science* 323:732–37.

Gavryushkina, A., D. Welch, T. Stadler, and A. J. Drummond. 2014. Bayesian inference of sampled ancestor trees for epidemiology and fossil calibration. *PLoS Computational Biology* 10:e1003919.

Gayon, J. 1998. *Darwinism's Struggle for Survival: Heredity and the Hypothesis of Natural Selection*. Cambridge University Press, Cambridge, UK.

Gerber, S. 2014. Not all roads can be taken: development induces anisotropic accessibility in morphospace. *Evolution and Development* 16:373–81.

Gerhart, J. 2000. Inversion of the chordate body axis: are there alternatives? *Proceedings of the National Academy of Sciences* 97:4445–48.

Geritz, S. A. H., É. Kisdi, G. Meszéna, and J. A. J. Metz. 1998. Evolutionarily singular strategies and the adaptive growth and branching of the evolutionary tree. *Evolutionary Ecology* 12:35–57.

Geritz, S. A. H., E. van der Meijden, and J. A. Metz. 1999. Evolutionary dynamics of seed size and seedling competitive ability. *Theoretical Population Biology* 55:324–43.

Gerrish, P. J., and R. E. Lenski. 1998. The fate of competing beneficial mutations in an asexual population. *Genetica* 102:127–44.

Gharib, W. H., and M. Robinson-Rechavi. 2013. The branch-site test of positive selection is surprisingly robust but lacks power under synonymous substitution saturation and variation in GC. *Molecular Biology and Evolution* 30:1675–86.

Ghiselin, M. T. 1974a. *The Economy of Nature and the Evolution of Sex*. University of California Press, Berkeley.

Ghiselin, M. T. 1974b. A radical solution to the species problem. *Systematic Biology* 23:536–44.

Ghiselin, M. T. 1989. Individuality, history and laws of nature in biology. Pp. 53–66 in M. Ruse, ed., *What the Philosophy of Biology Is*. Kluwer Academic, Dordrecht.

Ghiselin, M. T. 1997. *Metaphysics and the Origin of Species*. SUNY Press, Albany, NY.

Giere, R. N. 1988. *Explaining Science: A Cognitive Approach*. University of Chicago Press, Chicago, IL.

Giere, R. N. 2004. How models are used to represent reality. *Philosophy of Science* 71: 742–52.

Gilbert, S. F. 2013. *Developmental Biology*. 10th ed. Sinauer and Associates, Sunderland, MA.

Gilbert, S. F. 2014. A holobiont birth narrative: the epigenetic transmission of the human microbiome. *Frontiers in Genetics* 5:282.

Gilbert, S. F., and D. Epel. 2009. *Ecological Developmental Biology*. Sinauer Associates, Sunderland, MA.

Gilbert, S. F., J. Sapp, and A. I. Tauber. 2012. A symbiotic view of life: we have never been individuals. *Quarterly Review of Biology* 87:325–41.

Gillespie, J. H. 1974. Natural selection for within-generation variance in offspring number. *Genetics* 76:601–6.

Gillespie, R. G. 2013. Adaptive radiation: convergence and non-equilibrium. *Current Biology* 23:R71–R74.

Gillespie, R. G. and B. G. Baldwin. 2010. Island biogeography of remote archipelagoes. Pp. 358–87 in J. B. Losos and R. E. Ricklefs, eds., *The Theory of Island Biogeography Revisited*. Princeton University Press, Princeton, NJ.

Gillespie, R. G., B. G. Baldwin, J. M. Waters, C. I. Fraser, R. Nikula, and G. K. Roderick. 2012. Long-distance dispersal: a framework for hypothesis testing. *Trends in Ecology and Evolution* 27:47–56.

Gillespie, R. G., and C. E. Parent. 2014. Adaptive radiation. In J. B. Losos, ed., *Oxford Bibliographies in Evolutionary Biology*, Oxford Bibliographies, DOI: 10.1093/obo/9780199941728-0004.

Gillooly, J. F., J. H. Brown, G. B. West, V. M. Savage, and E. L. Charnov. 2001. Effects of size and temperature on metabolic rate. *Science* 293:2248–51.

Gillooly, J. F., E. L. Charnov, G. B. West, V. M. Savage, and J. H. Brown. 2002. Effects of size and temperature on developmental time. *Nature* 417:70–73.

Glansdorff, N., Y. Xu, and B. Labedan. 2008. The last universal common ancestor: emergence, constitution and genetic legacy of an elusive forerunner. *Biology Direct* 3:29.

Glass, B., ed. 1980. *The Roving Naturalist: Travel Letters of Theodosius Dobzhansky.* American Philosophical Society, Philadelphia, PA.

Glassford, W. J., W. C. Johnson, N. R. Dall, Y. Liu, W. Boll, M. Noll, and M. Rebeiz. 2015. Co-option of an ancestral *Hox*-regulated network underlies a recently evolved morphological novelty. *Developmental Cell* 34:520–31.

Godfray, H. C. J., and G. A. Parker. 1992. Sibling competition, parent-offspring conflict and clutch size. *Animal Behaviour* 43:473–90.

Godfrey-Smith, P. 2009. *Darwinian Populations and Natural Selection.* Oxford University Press, New York.

Godfrey-Smith, P. 2014. *Philosophy of Biology.* Princeton University Press, Princeton, NJ.

Goldschmidt, R. B. 1940. *The Material Basis of Evolution.* Yale University Press, New Haven, CT.

Gompel, N., B. Prud'homme, P. J. Wittkopp, V. A. Kassner, and S. B. Carroll. 2005. Chance caught on the wing: *cis*-regulatory evolution and the origin of pigment patterns in *Drosophila. Nature* 433:481–87.

Gontier, N., ed. 2015. *Reticulate Evolution.* Springer International, New York.

Goodnight, C. J. 1985. The influence of environmental variation on group and individual selection in a cress. *Evolution* 39:545–58.

Goodnight, C. J. 1990a. Experimental studies of community evolution I: the response to selection at the community level. *Evolution* 44:1614–24.

Goodnight, C. J. 1990b. Experimental studies of community evolution II: the ecological basis of the response to community selection. *Evolution* 44:1625–36.

Goodnight, C. J. 2013a. Defining the individual. Pp. 37–54 in F. Bouchard and P. Hueneman, eds., *From Groups to Individuals: Evolution and Emerging Individuality.* MIT Press, Cambridge, MA.

Goodnight, C. J. 2013b. On multilevel selection and kin selection: contextual analysis meets direct fitness. *Evolution* 67:1539–48.

Goodnight, C. J., E. Rauch, H. Sayama, M. A. M. De Aguiar, M. Baranger, and Y. Bar-yam. 2008. Evolution in spatial predator-prey models and the "prudent predator": the inadequacy of steady-state organism fitness and the concept of individual and group selection. *Complexity* 13:23–44.

Goodnight, C. J., J. M. Schwartz, and L. Stevens. 1992. Contextual analysis of models of group selection, soft selection, hard selection, and the evolution of altruism. *American Naturalist* 140:743–61.

Goodnight, C. J., and L. Stevens. 1997. Experimental studies of group selection: what do they tell us about group selection in nature? *American Naturalist* 150:S59–S79.

Gordo, I., and P. R. A. Campos. 2008. Sex and deleterious mutations. *Genetics* 179:621–26.

Gordo, I., and B. Charlesworth. 2001. The speed of Muller's ratchet with background selection, and the degeneration of Y chromosomes. *Genetics Research* 78:149–61.

Goswami, A. 2006. Cranial modularity shifts during mammalian evolution. *American Naturalist* 168:270–80.

Goswami, A., J. B. Smaers, C. Soligo, and P. D. Polly. 2014. The macroevolutionary consequences of phenotypic integration: from development to deep time. *Philosophical Transactions of the Royal Society of London B: Biological Sciences* 369:20130254.

Goudge, T. 1961. Darwin's heirs. *University of Toronto Quarterly* 30:246–50.

Gould, S. J. 1977. *Ontogeny and Phylogeny.* Harvard University Press, Cambridge, MA.

Gould, S. J. 1980. G. G. Simpson, paleontology, and the modern synthesis. Pp. 153–72 in E. Mayr and W. B. Provine, eds., *The Evolutionary Synthesis: Perspectives on the Unification of Biology*. Harvard University Press, Cambridge, MA.

Gould, S. J. 1983. Irrelevance, submission, and partnership: the changing role of paleontology in Darwin's three centennials, and a modest proposal for macroevolution. Pp. 347–66 in D. S. Bendall, ed., *Evolution from Molecules to Men*. Cambridge University Press, Cambridge, UK.

Gould, S. J. 2002. *The Structure of Evolutionary Theory*. Harvard University Press, Cambridge, MA.

Gould, S. J., and N. Eldredge. 1977. Punctuated equilibria: the tempo and mode of evolution reconsidered. *Paleobiology* 3:115–51.

Gould, S. J., and R. C. Lewontin. 1979. The spandrels of San Marco and the Panglossian paradigm: a critique of the adaptationist programme. *Proceedings of the Royal Society of London B: Biological Sciences* 205:581–98.

Gould, S. J., D. M. Raup, J. J. Sepkoski, J. M. S. Thomas, and D. S. Simberloff. 1977. The shape of evolution: a comparison of real and random clades. *Paleobiology* 3:23–40.

Gourbière, S., and J. Mallet. 2010. Are species real? The shape of the species boundary with exponential failure, reinforcement, and the "missing snowball." *Evolution* 64:1–24.

Graham, C. H., A. C. Carnaval, C. D. Cadena, K. R. Zamudio, T. E. Roberts, J. L. Parra, C. M. McCain, R. C. Bowie, C. Moritz, and S. B. Baines. 2014. The origin and maintenance of montane diversity: integrating evolutionary and ecological processes. *Ecography* 37:711–19.

Graham, C. H., S. R. Ron, J. C. Santos, C. J. Schneider, C. Moritz, and C. Cunningham. 2004. Integrating phylogenetics and environmental niche models to explore speciation mechanisms in dendrobatid frogs. *Evolution* 58:1781–93.

Grant, P. R., and B. R. Grant. 2008. *How and Why Species Multiply: The Radiation of Darwin's Finches*. Princeton University Press, Princeton, NJ.

Graur, D. 2016. *Molecular and Genome Evolution*. Sinauer Associates, Sunderland, MA.

Graur, D., Y. Zheng, N. Price, R. B. R. Azevedo, R. A. Zufall, and E. Elhaik. 2013. On the immortality of television sets: "function" in the human genome according to the evolution-free gospel of ENCODE. *Genome Biology and Evolution* 5:578–90.

Gravel, D., F. Massol, E. Canard, D. Mouillot, and N. Mouquet. 2011. Trophic theory of island biogeography. *Ecology Letters* 14:1010–16.

Green, R. E., J. Krause, A. W. Briggs, T. Maricic, U. Stenzel, M. Kircher, N. Patterson, H. Li, W. Zhai, M. H. Fritz, N. F. Hansen, E. Y. Durand, A. S. Malaspinas, J. D. Jensen, T. Marques-Bonet, C. Alkan, K. Prüfer, M. Meyer, H. A. Burbano, J. M. Good, R. Schultz, A. Aximu-Petri, A. Butthof, B. Höber, B. Höffner, M. Siegemund, A. Weihmann, C. Nusbaum, E. S. Lander, C. Russ, N. Novod, J. Affourtit, M. Egholm, C. Verna, P. Rudan, D. Brajkovic, Z. Kucan, I. Gusic, V. B. Doronichev, L. V. Golovanova, C. Lalueza-Fox, M. de la Rasilla, J. Fortea, A. Rosas, R. W. Schmitz, P. L. F. Johnson, E. E. Eichler, D. Falush, E. Birney, J. C. Mullikin, M. Slatkin, R. Nielsen, J. Kelso, M. Lachmann, D. Reich, and S. Pääbo. 2010. A draft sequence of the Neandertal genome. *Science* 328:710–22.

Gremer, J. R., and D. L. Venable. 2014. Bet hedging in desert winter annual plants: optimal germination strategies in a variable environment. *Ecology Letters* 17:380–87.

Grene, M., and D. Depew. 2004. *The Philosophy of Biology: An Episodic History*. Cambridge University Press, Cambridge, UK.

Griesemer, J. 1984. Presentations and the status of theories. *PSA: Proceedings of the Biennial Meeting of the Philosophy of Science Association* 1984, no. 1:102–14.

Griesemer, J. 2014. Reproduction and scaffolded developmental processes: an integrated evolutionary perspective. Pp. 183–202 in A. Minelli and T. Pradeu, eds., *Towards a Theory of Development*. Oxford University Press, New York.

Griffing, B. 1969. Selection in reference to biological groups IV: application of selection index theory. *Australian Journal of Biological Sciences* 22:131–42.

Griffing, B. 1977. Selection for populations of interacting phenotypes. Pp. 413–34 in E. Pollak, O. Kempthorne, and T. B. Bailey, eds., *Proceedings of the International Conference on Quantitative Genetics*. Iowa State University Press, Ames.

Griffing, B. 1981. A theory of natural selection incorporating interaction among individuals I: the modeling process. *Journal of Theoretical Biology* 89:635–58.

Griffing, B. 1982. A theory of natural selection incorporating interaction among individuate X: use of groups consisting of a mating pair together with haploid and diploid caste members. *Journal of Theoretical Biology* 95:199–223.

Griffing, B. 1989. Genetic analysis of plant mixtures. *Genetics* 122:943–56.

Grime, J. P. 1977. Evidence for the existence of three primary strategies in plants and its relevance to ecological and evolutionary theory. *American Naturalist* 111:1169–94.

Gupta, M., N. G. Prasad, S. Dey, A. Josi, and T. N. C. Vidya. 2017. Niche construction in evolutionary theory: the construction of an academic niche? *Journal of Genetics* 96:491–504.

Haber, M. H. 2012. Multilevel lineages and multidimensional trees: the levels of lineage and phylogeny reconstruction. *Philosophy of Science* 79.609–23.

Haccou, P., P. Jagers, and V. A. Vatutin. 2005. *Branching Processes: Variation, Growth, and Extinction of Populations*. Cambridge University Press, Cambridge, UK.

Haeckel, E. H. P. A. 1866. *Generelle Morphologie der Organismen: allgemeine Grundzuge der organischen Formen-Wissenschaft, mechanisch begrundet durch die von Charles Darwin reformirte Descendenz-Theorie*. G. Reimer, Berlin.

Haffer, J. 1997. Alternative models of vertebrate speciation in Amazonia: an overview. *Biodiversity and Conservation* 6:451–76.

Haggerty, L. S., P. A. Jachiet, W. P. Hanage, D. A. Fitzpatrick, P. Lopez, M. J. O'Connell, D. Pisani, M. Wilkinson, E. Bapteste, and J. O. McInerney. 2014. A pluralistic account of homology: adapting the models to the data. *Molecular Biology and Evolution* 31:501–16.

Hahn, M. W. 2008. Toward a selection theory of molecular evolution. *Evolution* 62:255–65.

Haig, S. M., and K. Winker. 2010. Avian subspecies: summary and prospectus. *Ornithological Monographs* 67:172–75.

Haigh, J. 1978. The accumulation of deleterious genes in a population—Muller's Ratchet. *Theoretical Population Biology* 14:251–67.

Halary, S., J. W. Leigh, B. Cheaib, P. Lopez, and E. Bapteste. 2010. Network analyses structure genetic diversity in independent genetic worlds. *Proceedings of the National Academy of Sciences* 107:127–32.

Haldane, J. B. S. 1922. Sex ratio and unisexual sterility in hybrid animals. *Journal of Genetics* 12:101–9.

Haldane, J. B. S. 1924. A mathematical theory of natural and artificial selection I. *Transactions of the Cambridge Philosophical Society* 23:19–41.

Haldane, J. B. S. 1930. A mathematical theory of natural and artificial selection 6: isolation. *Proceedings of the Cambridge Philosophical Society* 26:220–30.

Haldane, J. B. S. 1932. *The Causes of Evolution*. Longmans, Green, London.

Haldane, J. B. S. 1953. Forward. *Evolution: Symposia of the Society for Experimental Biology*. Academic Press, New York.

Haldane, J. B. S. 1964. A defense of beanbag genetics. *Perspectives in Biology and Medicine* 7:343–60.

Hall, B. K. 1999. *Evolutionary Developmental Biology*. Kluwer Academic, Dordrecht.

Hall, B. K. 2013. Homology, homoplasy, novelty, and behavior. *Developmental Psychobiology* 55:4–12.

Hall, B. K., and R. Kerney. 2012. Levels of biological organization and the origin of novelty. *Journal of Experimental Zoology Part B: Molecular and Developmental Evolution* 318B:428–37.

Hallgrímsson, B., and B. K. Hall, eds. 2011. *Epigenetics: Linking Genotype and Phenotype in Development and Evolution*. University of California Press, Berkeley.

Hallgrimsson, B., W. Mio, R. S. Marcucio, and R. Spritz. 2014. Let's face it—complex traits are just not that simple. *PLoS Genetics* 10:e1004724.

Hamilton, W. D. 1964a. The genetical evolution of social behaviour I. *Journal of Theoretical Biology* 7:1–16.

Hamilton, W. D. 1964b. The genetical evolution of social behaviour II. *Journal of Theoretical Biology* 7:17–52.

Hamilton, W. D. 1970. Selfish and spiteful behaviour in an evolutionary model. *Nature* 228:1218–20.

Hamilton, W. D. 1975. Innate social aptitudes of man: an approach from evolutionary genetics. Pp. 133–55 in R. Fox, ed., *Biosocial Anthropology*. Wiley, New York.

Hamilton, W. D. 1980. Sex versus non-sex versus parasite. *Oikos* 35:282–90.

Hanken, J. 2015. Is heterochrony still an effective paradigm for contemporary studies of evo-devo? Pp. 97–110 in A. C. Love, ed., *Conceptual Change in Biology: Scientific and Philosophical Perspectives on Evolution and Development*. Springer, Dordrecht.

Hansen, T. F. 1997. Stabilizing selection and the comparative analysis of adaptation. *Evolution* 51:1341–51.

Hansen, T. F. 2006. The evolution of genetic architecture. *Annual Review of Ecology, Evolution, and Systematics* 37:123–57.

Hansen, T. F. 2012. Adaptive landscapes and macroevolutionary dynamics. Pp. 205–26 in E. I. Svensson and R. Calsbeek, eds., *The Adaptive Landscape in Evolutionary Biology*. Oxford University Press, Oxford, UK.

Haq, B. U., and S. R. Schutter. 2008. A chronology of Paleozoic sea-level changes. *Science* 322:64–68.

Haraway, D. 1988. Remodelling the human way of life: Sherwood Washburn and the new physical anthropology, 1950–1980. Pp. 206–59 in G. Stocking Jr., ed., *Bones, Bodies and Behavior: Essays on Biological Anthropology*. University of Wisconsin Press, Madison.

Hardy, N. B., and S. P. Otto. 2014. Specialization and generalization in the diversification of phytophagous insects: tests of the musical chairs and oscillation hypotheses. *Proceedings of the Royal Society of London B: Biological Sciences* 281:20132960.

Harmon, L. J., and S. Harrison. 2015. Species diversity is dynamic and unbounded at local and continental scales. *American Naturalist* 185:584–93.

Harmon, L. J., J. B. Losos, T. Jonathan Davies, R. G. Gillespie, J. L. Gittleman, W. Bryan Jennings, K. H. Kozak, M. A. McPeek, F. Moreno-Roark, T. J. Near, A. Purvis, R. E. Ricklefs, D. Schluter, J. A. Schulte Ii, O. Seehausen, B. L. Sidlauskas, O. Torres-Carvajal, J. T. Weir, and A. Ø. Mooers. 2010. Early bursts of body size and shape evolution are rare in comparative data. *Evolution* 64:2385–96.

Harms, M. J., and J. W. Thornton. 2010. Analyzing protein structure and function using ancestral gene reconstruction. *Current Opinion in Structural Biology* 20:360–66.

Harper, D. A. T. 2010. The Ordovician brachiopod radiation: roles of alpha, beta, and gamma diversity. *Geological Society of America Special Papers* 466:69–83.

Harrison, R. G. 1991. Molecular changes in speciation. *Annual Review of Ecology and Systematics* 22:281–308.

Harrison, R. G. 2012. The language of speciation. *Evolution* 66:3643–57.

Harte, J. 2011. *Maximum Entropy and Ecology: A Theory of Abundance, Distribution, and Energetics*. Oxford University Press, New York.

Harte, J., J. Kitzes, E. A. Newman, A. J. Rominger, J. G. 2013. Taxon categories and the universal species-area relationship. *American Naturalist* 181:282–87.

Hartfield, M. 2016. Evolutionary genetic consequences of facultative sex and outcrossing. *Journal of Evolutionary Biology* 29:5–22.

Hartfield, M., and P. D. Keightley. 2012. Current hypotheses for the evolution of sex and recombination. *Integrative Zoology* 7:192–209.

Harvey, P. H., and M. D. Pagel. 1991. *The Comparative Method in Evolutionary Biology*. Oxford University Press, Oxford, UK.

Hastie, T., R. Tibshirani, and J. Friedman. 2001. *The Elements of Statistical Learning: Data Mining, Inference, and Prediction*. Springer, New York.

Hawksworth, D. L., M. T. Kalin-Arroyo, P. M. Hammond, R. E. Ricklefs, R. M. Cowling, and M. J. Samways. 1995. Magnitude and distribution of biodiversity. Pp. 107–92 in V. H. Heywood and R. T. Watson, eds., *Global Biodiversity Assessment*. Cambridge University Press, Cambridge, UK.

Hawthorne, D. J., and S. Via. 2001. Genetic linkage of ecological specialization and reproductive isolation in pea aphids. *Nature* 412:904–7.

Heath, T. A., J. P. Huelsenbeck, and T. Stadler. 2014. The fossilized birth-death process for coherent calibration of divergence-time estimates. *Proceedings of the National Academy of Sciences* 111:E2957–E2966.

Heffer, A., J. W. Shultz, and L. Pick. 2010. Surprising flexibility in a conserved Hox transcription factor over 550 million years of evolution. *Proceedings of the National Academy of Sciences* 107:18040–45.

Heisler, I. L., and J. Damuth. 1987. A method for analyzing selection in hierarchically structured populations. *American Naturalist* 130:582–602.

Hendry, A. P. 2016. Key questions on the role of phenotypic plasticity in eco-evolutionary dynamics. *Journal of Heredity* 107:25–41.

Hennig, W. 1950. *Grundzüge einer Theorie der phylogenetischen Systematik*. Deutscher Zentralverlag, Berlin.

Hennig, W. 1966. *Phylogenetic Systematics*. University of Illinois Press, Urbana.

Herbers, M. J., and S. V. Banschbach. 1999. Plasticity of social organization in a forest ant species. *Behavioral Ecology and Sociobiology* 45:451–65.

Hereford, J. 2009. A quantitative survey of local adaptation and fitness trade-offs. *American Naturalist* 173:579–88.

Hersch, E. I., and P. C. Phillips. 2004. Power and potential bias in field studies of natural selection. *Evolution* 58:479–85.

Hershberg, R., and D. A. Petrov. 2008. Selection on codon bias. *Annual Review of Genetics* 42:287–99.

Hey, J. 2001. *Genes, Categories, and Species: The Evolutionary and Cognitive Causes of the Species Problem*. Oxford University Press, Oxford, UK.

Hey, J. 2006. On the failure of modern species concepts. *Trends in Ecology and Evolution* 21:447–50.

Hey, J., R. S. Waples, M. L. Arnold, R. K. Butlin, and R. G. Harrison. 2003. Understanding and confronting species uncertainty in biology and conservation. *Trends in Ecology and Evolution* 18:597–603.

Hickman, J. C., and L. F. Pitelka. 1975. Dry weight indicates energy analysis in ecological strategy analysis of plants. *Oecologia* 221:117–21.

Higashi, M., G. Takimoto, and N. Yamamura. 1999. Sympatric speciation by sexual selection. *Nature* 402:523–26.

Hilborn, R., and M. Mangel. 1997. *The Ecological Detective: Confronting Models with Data*. Princeton University Press, Princeton, NJ.

Hill, W. G., and A. Robertson. 1966. The effect of linkage on limits to artificial selection. *Genetics Research* 8:269–94.

HilleRisLambers, J., P. Adler, W. Harpole, J. Levine, and M. Mayfield. 2012. Rethinking community assembly through the lens of coexistence theory. *Annual Review of Ecology, Evolution, and Systematics* 43:227–48.

Hillis, D. M. 1994. Homology in molecular biology. Pp. 339–68 in B. K. Hall, ed., *Homology: The Hierarchical Basis of Comparative Biology*. Academic Press, San Diego, CA.

Hirshfield, M. F., and D. W. Tinkle. 1975. Natural selection and the evolution of reproductive effort. *Proceedings of the National Academy of Sciences* 72:2227–31.

Hirzel, A. H., and G. Le Lay. 2008. Habitat suitability modelling and niche theory. *Journal of Applied Ecology* 45:1372–81.

Hitchcock, C. 1995. The mechanist and the snail. *Philosophical Studies* 84:91–105.

Hitchcock, C., and E. Sober. 2004. Prediction versus accommodation and the risk of overfitting. *British Journal for the Philosophy of Science* 55:1–34.

Hoberg, E. P., and D. R. Brooks. 2015. Evolution in action: climate change, biodiversity dynamics and emerging infectious disease. *Philosophical Transactions of the Royal Society of London B: Biological Sciences* 370:20130553.

Hochachka, W. M., and A. A. Dhondt. 2000. Density-dependent decline of host abundance resulting from a new infectious disease. *Proceedings of the National Academy of Sciences* 97:5303–6.

Hoehn, K. B., P. G. Harnik, and V. L. Roth. 2016. A framework for detecting natural selection on traits above the species level. *Methods in Ecology and Evolution* 7:331–39.

Hoekstra, H. E., and J. A. Coyne. 2007. The locus of evolution: evo-devo and the genetics of adaptation. *Evolution* 61:995–1016.

Hoekstra, H. E., R. J. Hirschmann, R. A. Bundey, P. A. Insel, and J. P. Crossland. 2006. A single amino acid mutation contributes to adaptive beach mouse color pattern. *Science* 313:101–4.

Hoekstra, H. E., J. M. Hoekstra, D. Berrigan, S. N. Vignieri, A. Hoang, C. E. Hill, P. Beerli, and J. G. Kingsolver. 2001. Strength and tempo of directional selection in the wild. *Proceedings of the National Academy of Sciences* 98:9157–60.

Hofbauer, J., and K. Sigmund. 1988. *The Theory of Evolution and Dynamical Systems*. Cambridge University Press, Cambridge, UK.

Höhna, S., T. A. Heath, B. Boussau, M. J. Landis, F. Ronquist, and J. P. Huelsenbeck. 2014. Probabilistic graphical model representation in phylogenetics. *Systematic Biology* 63:753–71.

Holman, E. W. 1985. Evolutionary and psychological effects in pre-evolutionary classifications. *Journal of Classification* 2:29–39.

Holman, E. W. 2007. How comparable are categories in different phyla? *Taxon* 56:179–84.

Holsinger, K. E. 1984. The nature of biological species. *Philosophy of Science* 51:293–307.

Holt, R. D. 2009. Bringing the Hutchinsonian niche into the 21st century: ecological and evolutionary perspectives. *Proceedings of the National Academy of Sciences* 106:19659–65.

Hom, C. L. 1992. Modeling reproductive allocation of dusky salamanders using optimal control theory: pros, cons and caveats. *Evolutionary Ecology* 6:458–81.

Hopkins, R., and M. D. Rausher. 2011. Identification of two genes causing reinforcement in the Texas wildflower *Phlox drummondii*. *Nature* 469:411–14.

Hopkins, R., and M. D. Rausher. 2012. Pollinator-mediated selection on flower color allele drives reinforcement. *Science* 335:1090–92.

Hossfeld, U., and L. Olsson. 2005. The history of the homology concept and the "Phylogenetisches Symposium." *Theory in Biosciences* 124:243–53.

Houle, D. 1991. Genetic covariance of fitness correlates: what genetic correlations are made of and why it matters. *Evolution* 45:630–48.

House, C. H. 2009. The Tree of Life viewed through the contents of genomes. Pp. 141–61 in M. B. Gogarten, J. P. Gogarten, and L. C. Olendzenski, eds., *Horizontal Gene Transfer: Genomes in Flux*. Humana Press, Totowa, NJ.

Huang, S., K. Roy, and D. Jablonski. 2015. Origins, bottlenecks, and present-day diversity: patterns of morphospace occupation in marine bivalves. *Evolution* 69:735–46.

Hubbell, S. P. 2001. *The Unified Theory of Biodiversity and Biogeography*. Princeton University Press, Princeton, NJ.

Hubbell, S. P. 2009. Neutral theory and the theory of island biogeography. Pp. 264–92 in J. B. Losos and R. E. Ricklefs, eds., *The Theory of Island Biogeography Revisited*. Princeton University Press, Princeton, NJ.

Hubby, J. L., and R. C. Lewontin. 1966. A molecular approach to the study of genic heterozygosity in natural populations I: the number of alleles at different loci in *Drosophila pseudoobscura*. *Genetics* 54:577–94.

Huelsenbeck, J. P., F. Ronquist, R. Nielsen, and J. P. Bollback. 2001. Bayesian inference of phylogeny and its impact on evolutionary biology. *Science* 294:2310–14.

Hughes, R. N. 1989. *A Functional Biology of Clonal Animals*. Chapman and Hall, London.

Hull, D. L. 1970. Contemporary systematic philosophies *Annual Review of Ecology and Systematics* 1:19–54.

Hull, D. L. 1978. A matter of individuality. *Philosophy of Science* 45:335–60.

Hull, D. L. 1980. Individuality and selection. *Annual Review of Ecology and Systematics* 11:311–32.

Hull, D. L. 1988. *Science as a Process: An Evolutionary Account of the Social and Conceptual Development of Science*. University of Chicago Press, Chicago, IL.

Hull, D. L. 1989. *The Metaphysics of Evolution*. SUNY Press, Albany, NY.

Hull, D. L. 1997. The ideal species concept—and why we can't get it. Pp. 357–80 in M. F. Claridge, H. Dawah, and M. Wilson, eds., *The Units of Biodiversity*. Chapman and Hall, London.

Hull, D. L. 1999a. On the plurality of species: questioning the party line. Pp. 23–48 in R. A. Wilson, ed., *Species: New Interdisciplinary Essays*. MIT Press, Cambridge, MA.

Hull, D. L. 1999b. The use and abuse of Sir Karl Popper. *Biology and Philosophy* 14:481–504.

Humphreys, A. M., and T. G. Barraclough. 2014. The evolutionary reality of higher taxa in mammals. *Proceedings of the Royal Society of London B: Biological Sciences* 281:20132750.

Hunt, G. 2007a. Evolutionary divergence in directions of high phenotypic variance in the ostracode genus *Poseidonamicus*. *Evolution* 61:1560–76.

Hunt, G. 2007b. The relative importance of directional change, random walks, and stasis in the evolution of fossil lineages. *Proceedings of the National Academy of Sciences* 104:18404–8.

Hunt, G. 2013. Testing the link between phenotypic evolution and speciation: an integrated palaeontological and phylogenetic analysis. *Methods in Ecology and Evolution* 4:714–23.

Hunt, G., M. J. Hopkins, and S. Lidgard. 2015. Simple versus complex models of trait evolution and stasis as a response to environmental change. *Proceedings of the National Academy of Sciences* 112:4885–90.

Hunt, G., and D. L. Rabosky. 2014. Phenotypic evolution in fossil species: pattern and process. *Annual Review of Earth and Planetary Sciences* 42:421–41.

Hunt, G., K. Roy, and D. Jablonski. 2005. Heritability of geographic range sizes revisited. *American Naturalist* 166:129–35.

Hunt, G., and G. Slater. 2016. Integrating paleontological and phylogenetic approaches to macroevolution. *Annual Review of Ecology, Evolution, and Systematics* 47:189–213.

Hurst, L. D., and A. Pomiankowski. 1991. Causes of sex ratio bias may account for unisexual sterility in hybrids: a new explanation of Haldane's rule and related phenomena. *Genetics* 128:841–58.

Huson, D. H., and D. Bryant. 2006. Application of phylogenetic networks in evolutionary studies. *Molecular Biology and Evolution* 23:254–67.

Hutchinson, G. E. 1957. Concluding remarks. *Cold Spring Harbor Symposium on Quantitative Biology* 22:415–27.

Huxley, J. S., ed. 1940. *The New Systematics*. Clarendon Press, Oxford, UK.

Huxley, J. S. 1942. *Evolution: The Modern Synthesis*. George Allen and Unwin, London.

Iles, M. M., K. Walters, and C. Cannings. 2003. Recombination can evolve in large finite populations given selection on sufficient loci. *Genetics* 165:2249–58.

Irwin, D. E., S. Bensch, and T. D. Price. 2001. Speciation in a ring. *Nature* 409:333–37.

Irwin, R. E., S. Y. Strauss, S. Storz, A. Emerson, S. Ecology, N. Jul, and G. Guibert. 2003. The role of herbivores in the maintenance of a flower color polymorphism in wild radish. *Ecology* 84:1733–43.

Iwasa, Y., A. Pomiankowski, and S. Nee. 1991. The evolution of costly mate preferences II: the "handicap" principle. *Evolution* 45:1431–42.

Jablonka, E., M. J. Lamb, and A. Zeligowski. 2014. *Evolution in Four Dimensions: Genetic, Epigenetic, Behavioral, and Symbolic Variation in the History of Life*. Rev. ed. MIT Press, Cambridge, MA.

Jablonski, D. 1987. Heritability at the species level: analysis of geographic ranges of Cretaceous mollusks. *Science* 238:360–63.

Jablonski, D. 1993. The tropics as a source of evolutionary novelty through geological time. *Nature* 364:142–44.

Jablonski, D. 1996. Body size and macroevolution. Pp. 256–89 in D. Jablonski, D. H. Erwin, and J. H. Lipps, eds., *Evolutionary Paleobiology*. University of Chicago Press, Chicago, IL.

Jablonski, D. 1997. Body-size evolution in Cretaceous molluscs and the status of Cope's rule. *Nature* 385:250–52.

Jablonski, D. 1998. Geographic variation in the molluscan recovery from the end-Cretaceous extinction. *Science* 279:1327–30.

Jablonski, D. 2000. Micro- and macroevolution: scale and hierarchy in evolutionary biology and paleobiology. *Paleobiology* 26:15–52.

Jablonski, D. 2002. Survival without recovery after mass extinctions. *Proceedings of the National Academy of Sciences* 99:8139–44.

Jablonski, D. 2005a. Evolutionary innovations in the fossil record: the intersection of ecology, development, and macroevolution. *Journal of Experimental Zoology Part B: Molecular and Developmental Evolution* 304B:504–19.

Jablonski, D. 2005b. Mass extinctions and macroevolution. *Paleobiology* 31:192–210.

Jablonski, D. 2007. Scale and hierarchy in macroevolution. *Palaeontology* 50:87–109.

Jablonski, D. 2008a. Biotic interactions and macroevolution: extensions and mismatches across scales and levels. *Evolution* 62:715–39.

Jablonski, D. 2008b. Extinction and the spatial dynamics of biodiversity. *Proceedings of the National Academy of Sciences* 105:11528–35.

Jablonski, D. 2008c. Species selection: theory and data. *Annual Review of Ecology, Evolution, and Systematics* 39:501–24.

Jablonski, D. 2010a. Macroevolutionary trends in time and space. Pp. 25–43 in P. R. Grant and B. R. Grant, eds., *In Search of the Causes of Evolution: From Field Observations to Mechanisms*. Princeton University Press, Princeton, NJ.

Jablonski, D. 2010b. Origination patterns and multilevel processes in macroevolution. Pp. 335–54 in G. B. Müller and M. Pigliucci, eds., *Evolution: The Extended Synthesis*. MIT Press, Cambridge, MA.

Jablonski, D., and D. J. Bottjer. 1990. The origin and diversification of major groups: environmental patterns and macroevolutionary lags. Pp. 17–57 in P. D. Taylor and G. P. Larwood, eds., *Major Evolutionary Radiations*. Clarendon Press, Oxford, UK.

Jablonski, D., and J. A. Finarelli. 2009. Congruence of morphologically-defined genera with molecular phylogenies. *Proceedings of the National Academy of Sciences* 106: 8262–66.

Jablonski, D., S. Huang, K. Roy, and J. W. Valentine. 2017. Shaping the latitudinal diversity gradient: new perspectives from a synthesis of paleobiology and biogeography. *American Naturalist* 189:1–12.

Jablonski, D., and G. Hunt. 2006. Larval ecology, geographic range, and species survivorship in Cretaceous mollusks: organismic vs. species-level explanations. *American Naturalist* 168:556–64.

Jablonski, D., K. Roy, and J. W. Valentine. 2006. Out of the tropics: evolutionary dynamics of the latitudinal diversity gradient. *Science* 314:102–6.

Jaenike, J. 1978a. An hypothesis to account for the maintenance of sex within populations. *Evolutionary Theory* 3:191–94.

Jaenike, J. 1978b. On optimal oviposition behavior in phytophagous insects. *Theoretical Population Biology* 14:350–56.

Jansson, R., G. Rodríguez-Castañeda, and L. E. Harding. 2013. What can multiple phylogenies say about the latitudinal diversity gradient? A new look at the tropical conservatism, out of the tropics, and diversification rate hypotheses. *Evolution* 67:1741–55.

Janzen, D. H. 1967. Why mountain passes are higher in the tropics. *American Naturalist* 101:233–49.

Jepson, G. L., E. Mayr, and G. G. Simpson, eds. 1949. *Genetics, Paleontology and Evolution*. Princeton University Press, Princeton, NJ.

Jiao, Y., J. Leebens-Mack, S. Ayyampalayam, J. Bowers, M. McKain, J. McNeal, M. Rolf, D. Ruzicka, E. Wafula, N. Wickett, X. Wu, Y. Zhang, J. Wang, Y. Zhang, E. Carpenter, M. Deyholos, T. Kutchan, A. Chanderbali, P. Soltis, D. Stevenson, R. McCombie, J. Pires, G. Wong, D. Soltis, and C. dePamphilis. 2012. A genome triplication associated with early diversification of the core eudicots. *Genome Biology* 13:R3.

Jiao, Y., N. J. Wickett, S. Ayyampalayam, A. S. Chanderbali, L. Landherr, P. E. Ralph, L. P. Tomsho, Y. Hu, H. Liang, P. S. Soltis, D. E. Soltis, S. W. Clifton, S. E. Schlarbaum, S. C.

Schuster, H. Ma, J. Leebens-Mack, and C. W. dePamphilis. 2011. Ancestral polyploidy in seed plants and angiosperms. *Nature* 473:97–100.

Jiggins, C. D., R. E. Naisbit, R. L. Coe, and J. Mallet. 2001. Reproductive isolation caused by colour pattern mimicry. *Nature* 411:302–5.

Johnson, C. K., and S. R. Voss. 2013. Salamander paedomorphosis: linking thyroid hormone to life history and life cycle evolution. *Current Topics in Developmental Biology* 103:229–58.

Johnson, N. A. 2008. Direct selection for reproductive isolation: the Wallace effect and reinforcement. Pp. 114–24 in C. H. Smith and G. Beccaloni, eds., *Natural Selection and Beyond: The Intellectual Legacy of Alfred Russel Wallace*. Oxford University Press, Oxford, UK.

Joron, M., L. Frézal, R. T. Jones, N. L. Chamberlain, S. F. Lee, C. R. Haag, A. Whibley, M. Becuwe, S. W. Baxter, L. Ferguson, P. A. Wilkinson, C. Salazar, C. Davidson, R. Clark, M. A. Quail, H. Beasley, R. Glithero, C. Lloyd, S. Sims, M. C. Jones, and J. Rogers. 2011. Chromosomal rearrangements maintain a polymorphic supergene controlling butterfly mimicry. *Nature* 477:204–8.

Judson, O. P., and B. B. Normark. 1996. Ancient asexual scandals. *Trends in Ecology and Evolution* 11:41–46.

Jukes, T. H., and C. R. Cantor. 1969. Evolution of protein molecules. Pp. 21–123 in H. N. Munro, ed., *Mammalian Protein Metabolism*. Academic Press, New York.

Julliard, R., J. Clavel, V. Devictor, F. Jiguet, and D. Couvet. 2006. Spatial segregation of specialists and generalists in bird communities. *Ecology Letters* 9:1237–44.

Kaci-Chaouch, T., O. Verneau, and Y. Desdevises. 2008. Host specificity is linked to intraspecific variability in the genus *Lamellodiscus* (Monogenea). *Parasitology* 135:607–16.

Kadmon, R., and O. Allouche. 2007. Integrating the effects of area, isolation, and habitat heterogeneity on species diversity: a unification of island biogeography and niche theory. *American Naturalist* 170:443–54.

Kassen, R. 2002. The experimental evolution of specialists, generalists, and the maintenance of diversity. *Journal of Evolutionary Biology* 15:173–90.

Kawecki, T. J. 1994. Accumulation of deleterious mutations and the evolutionary cost of being a generalist. *American Naturalist* 144:833–38.

Kawecki, T. J., R. E. Lenski, D. Ebert, B. Hollis, I. Olivieri, and M. C. Whitlock. 2012. Experimental evolution. *Trends in Ecology and Evolution* 27:547–60.

Kearney, M. 2008. Philosophy and phylogenetics: historical and current connections. Pp. 211–32 in D. Hull and M. Ruse, eds., *The Cambridge Companion to the Philosophy of Biology*. Cambridge University Press, Cambridge, UK.

Kearney, M., and O. Rieppel. 2006. Rejecting "the given" in systematics. *Cladistics* 22: 369–77.

Keeler, S. P., M. S. Dalton, A. M. Cressler, R. D. Berghaus, and D. Stallknecht. 2014. Abiotic factors affecting persistence of avian influenza virus in surface water from waterfowl habitats. *Applied and Environmental Microbiology* 80:2910–17.

Kéfi, S., E. L. Berlow, E. A. Wieters, S. A. Navarrete, O. L. Petchey, S. A. Wood, A. Boit, L. N. Joppa, K. D. Lafferty, R. J. Williams, N. D. Martinez, B. A. Menge, C. A. Blanchette, A. C. Iles, and U. Brose. 2012. More than a meal . . . integrating non-feeding interactions into food webs. *Ecology Letters* 15:291–300.

Keightley, P. D. 1989. Models of quantitative variation of flux in metabolic pathways. *Genetics* 121:869–76.

Keightley, P. D. 1994. The distribution of mutation effects on viability in *Drosophila melanogaster*. *Genetics* 138:1315–22.

Keightley, P. D., and A. Eyre-Walker. 2010. What can we learn about the distribution of fitness effects of new mutations from DNA sequence data? *Philosophical Transactions of the Royal Society of London B: Biological Sciences* 365:1187–93.

Keightley, P. D., and S. P. Otto. 2006. Interference among deleterious mutations favours sex and recombination in finite populations. *Nature* 443:89–92.

Keightley, P. D., U. Trivedi, M. Thomson, F. Oliver, S. Kumar, and M. Blaxter. 2009. Analysis of the genome sequences of three *Drosophila melanogaster* spontaneous mutation accumulation lines. *Genome Research* 19:1195–1201.

Kelly, J. K. 1996. Kin selection in the annual plant *Impatiens capensis*. *American Naturalist* 147:899–918.

Kern, A. D., and M. W. Hahn. 2018. The neutral theory in light of natural selection. *Molecular Biology and Evolution* 35:1366–71.

Kerr, W. E., and S. Wright. 1954a. Experimental studies of the distribution of gene frequencies in very small populations of *Drosophila melanogaster* III: aristapedia and spineless. *Evolution* 8:293–302.

Kerr, W. E., and S. Wright. 1954b. Experimental studies of the distribution of gene frequencies in very small populations of *Drosophila melanogaster* I: forked. *Evolution* 8:172–77.

Kerswell, K. J., and M. Burd. 2012. Frequency-dependent and density-dependent larval competition between life-history strains of a fly, Lucilia cuprina. *Ecological Entomology* 37:109–16.

Khidr, A.-A. A., A. E. Said, O. A. Abu Samak, and S. E. Abu Sheref. 2012. The impacts of ecological factors on prevalence, mean intensity and seasonal changes of the monogenean gill parasite, *Microcotyloides* sp., infesting the *Terapon puta* fish inhabiting coastal region of Mediterranean Sea at Damietta region. *Journal of Basic and Applied Zoology* 65:109–15.

Kiessling, W., C. Simpson, and M. Foote. 2010. Reefs as cradles of evolution and sources of biodiversity in the Phanerozoic. *Science* 327:196–98.

Kimura, M. 1968. Evolutionary rate at the molecular level. *Nature* 217:624–26.

Kimura, M. 1980. A simple method for estimating evolutionary rate of base substitution through comparative studies of nucleotide sequences. *Journal of Molecular Evolution* 16:111–20.

Kimura, M. 1983. *The Neutral Theory of Molecular Evolution*. Cambridge University Press, New York.

King, D., and J. Roughgarden. 1982a. Graded allocation between vegetative and reproductive growth for annual plants in growing seasons of random length. *Theoretical Population Biology* 22:1–16.

King, D., and J. Roughgarden. 1982b. Multiple switches between vegetative and reproductive growth in annual plants. *Theoretical Population Biology* 21:194–204.

King, J. L., and T. H. Jukes. 1969. Non-Darwinian evolution. *Science* 164:788–98.

Kingsolver, J. G., H. E. Hoekstra, J. M. Hoekstra, D. Berrigan, S. N. Vignieri, C. E. Hill, A. Hoang, P. Gibert, and P. Beerli. 2001. The strength of phenotypic selection in natural populations. *American Naturalist* 157:245–61.

Kirkpatrick, M. 2010. How and why chromosome inversions evolve. *PLoS Biology* 8:e1000501.

Kirkpatrick, M., and N. Barton. 2006. Chromosome inversions, local adaptation and speciation. *Genetics* 173:419–34.

Kirkpatrick, M., and S. L. Nuismer. 2004. Sexual selection can constrain sympatric speciation. *Proceedings of the Royal Society of London B: Biological Sciences* 271:687–93.

Kirkpatrick, M., and V. Ravigné. 2002. Speciation by natural and sexual selection. *American Naturalist* 159:S22–S35.

Kirkpatrick, M., and M. J. Ryan. 1991. The evolution of mating preferences and the paradox of the lek. *Nature* 350:33–38.

Kirschner, M. W. 2015. The road to facilitated variation. Pp. 199–217 in A. C. Love, ed., *Conceptual Change in Biology: Scientific and Philosophical Perspectives on Evolution and Development*. Springer, Berlin.

Kirschner, M. W., and J. C. Gerhart. 2010. Facilitated variation. Pp. 253–80 in M. Pigliucci and G. B. Müller, eds., *Evolution: The Extended Synthesis*. MIT Press, Cambrige, MA.

Kisdi, E., and A. H. G. Stefan. 1999. Adaptive dynamics in allele space: evolution of genetic polymorphism by small mutations in a heterogeneous environment. *Evolution* 53:993–1008.

Kisel, Y., and T. G. Barraclough. 2010. Speciation has a spatial scale that depends on levels of gene flow. *American Naturalist* 175:316–34.

Kitcher, P. 1984. 1953 and all that: a tale of two sciences. *Philosophical Review* 93:335–73.

Kitcher, P. 1989. Some puzzles about species. Pp. 183–208 in M. Ruse, ed., *What the Philosophy of Biology Is*. Kluwer Academic, Dordrecht.

Klein, T. W. 1974. Heritability and genetic correlation: statistical power, population comparisons, and sample size. *Behavior Genetics* 4:171–89.

Klimes, L., J. Klimesova, R. Hendriks, and J. van Groenendael. 1997. Clonal plant architecture: a comparative analysis of form and function. Pp. 1–29 in H. De Kroon and J. Van Groenendael, eds., *The Ecology and Evolution of Clonal Plants*. Backhuys, Leiden.

Klingenberg, C. P. 2005. Developmental constraints, modules and evolvability. Pp. 219–47 in B. Hallgrimsson and B. K. Hall, eds., *Variation: A Central Concept in Biology*. Elsevier, Amsterdam.

Klingenberg, C. P. 2014. Studying morphological integration and modularity at multiple levels: concepts and analysis. *Philosophical Transactions of the Royal Society of London B: Biological Sciences* 369:20130249.

Klompmaker, A. A., C. E. Schweitzer, R. M. Feldmann, and M. Kowalewski. 2015. Environmental and scale-dependent evolutionary trends in the body size of crustaceans. *Proceedings of the Royal Society of London B: Biological Sciences* 282:20150440.

Kluge, A. G. 2001. Parsimony with and without scientific justification. *Cladistics* 17:199–210.

Kluge, A. G., and J. S. Farris. 1969. Quantitative phyletics and the evolution of anurans. *Systematic Biology* 18:1–32.

Knowles, L. L., and L. S. Kubatko. 2011. *Estimating Species Trees: Practical and Theoretical Aspects*. John Wiley, Hoboken, NJ.

Koella, J. C. 1988. The tangled bank: the maintenance of sexual reproduction through competitive interactions. *Journal of Evolutionary Biology* 1:95–116.

Kondrashov, A. S. 1984. Deleterious mutations as an evolutionary factor 1: the advantage of recombination. *Genetics Research* 44:199–217.

Kondrashov, A. S. 1988. Deleterious mutations and the evolution of sexual reproduction. *Nature* 336:435–40.

Kondrashov, A. S. 1993. Classification of hypotheses on the advantages of amphimixis. *Journal of Heredity* 84:372–87.

Kondrashov, A. S. 1995. Contamination of the genome by very slightly deleterious mutations: why have we not died 100 times over? *Journal of Theoretical Biology* 175:583–94.

Kondrashov, A. S., and M. Shpak. 1998. On the origin of species by means of assortative mating. *Proceedings of the Royal Society of London B: Biological Sciences* 265:2273–78.

Kooijman, S. A. L. M. 2010. *Dynamic Energy Budget Theory for Metabolic Organisation*. 3rd ed. Cambridge University Press, Cambridge, UK.

Koonin, E. V. 2003. Comparative genomics, minimal gene-sets and the last universal common ancestor. *Nature Reviews Microbiology* 1:127–36.

Koonin, E. V. 2015. Archaeal ancestors of eukaryotes: not so elusive any more. *BMC Biology* 13:1–7.

Koonin, E. V., and Y. I. Wolf. 2009. Is evolution Darwinian or/and Lamarckian? *Biology Direct* 4:1–14.

Kouyos, R. D., S. P. Otto, and S. Bonhoeffer. 2006. Effect of varying epistasis on the evolution of recombination. *Genetics* 173:589–97.

Kouyos, R. D., O. K. Silander, and S. Bonhoeffer. 2007. Epistasis between deleterious mutations and the evolution of recombination. *Trends in Ecology and Evolution* 22: 308–15.

Kozak, K. H., C. H. Graham, and J. J. Wiens. 2008. Integrating GIS-based environmental data into evolutionary biology. *Trends in Ecology and Evolution* 23:141–48.

Kraft, N. J., L. S. Comita, J. M. Chase, N. J. Sanders, N. G. Swenson, T. O. Crist, J. C. Stegen, M. Vellend, B. Boyle, and M. J. Anderson. 2011. Disentangling the drivers of β diversity along latitudinal and elevational gradients. *Science* 333:1755–58.

Kreitman, M. 1983. Nucleotide polymorphism at the alcohol dehydrogenase locus of *Drosophila melanogaster*. *Nature* 304:412–17.

Krug, A. Z., and M. E. Patzkowsky. 2007. Geographic variation in turnover and recovery from the Late Ordovician mass extinction. *Paleobiology* 33:435–54.

Kuijper, B., and R. B. Hoyle. 2015. When to rely on maternal effects and when on phenotypic plasticity? *Evolution* 69:950–68.

Kunte, K., W. Zhang, A. Tenger-Trolander, D. H. Palmer, A. Martin, R. D. Reed, S. P. Mullen, and M. R. Kronforst. 2014. *doublesex* is a mimicry supergene. *Nature* 507:229–32.

Kuparinen, A., and J. Merilä. 2007. Detecting and managing fisheries-induced evolution. *Trends in Ecology and Evolution* 22:652–59.

Küpper, C., M. Stocks, J. E. Risse, N. dos Remedios, L. L. Farrell, S. B. McRae, T. C. Morgan, N. Karlionova, P. Pinchuk, Y. I. Verkuil, A. S. Kitaysky, J. C. Wingfield, T. Piersma, K. Zeng, J. Slate, M. Blaxter, D. B. Lank, and T. Burke. 2016. A supergene determines highly divergent male reproductive morphs in the ruff. *Nature Genetics* 48:79–83.

Kutschera, U., and K. J. Niklas. 2004. The modern theory of biological evolution: an expanded synthesis. *Naturwissenschaften* 91:255–76.

Kutterolf, S., M. Jegen, J. X. Mitrovica, T. Kwasnitschka, A. Freundt, and P. J. Huybers. 2013. A detection of Milankovitch frequencies in global volcanic activity. *Geology* 41:227–30.

Lacey, E. P., and D. Herr. 2005. Phenotypic plasticity, parental effects, and parental care in plants? I. An examination of spike reflectance in *Plantago lanceolata* (Plantaginaceae). *American Journal of Botany* 92:920–30.

Lachapelle, J., J. Reid, and N. Colegrave. 2015. Repeatability of adaptation in experimental populations of different sizes. *Proceedings of the Royal Society of London B: Biological Sciences* 282:20143033.

Laiolo, P., and J. R. Obeso. 2012. Multilevel selection and neighbourhood effects from individual to metapopulation in a wild passerine. *PLoS ONE* 7:e38526.

Lakoff, G., and M. Johnson. 2008. *Metaphors We Live By*. University of Chicago Press, Chicago, IL.

Laland, K. N., T. Uller, M. W. Feldman, K. Sterelny, G. B. Müller, A. Moczek, E. Jablonka, and J. Odling-Smee. 2015. The extended evolutionary synthesis: its structure, assumptions and predictions. *Proceedings of the Royal Society of London B: Biological Sciences* 282:20151019.

Lamarck, J. B. 1809. *Philosophie Zoologique*. Paris.

Lamichhaney, S., G. Fan, F. Widemo, U. Gunnarsson, D. S. Thalmann, M. P. Hoeppner, S. Kerje, U. Gustafson, C. Shi, H. Zhang, W. Chen, X. Liang, L. Huang, J. Wang, E. Liang, Q. Wu, S. M.-Y. Lee, X. Xu, J. Hoglund, X. Liu, and L. Andersson. 2016a. Structural genomic changes underlie alternative reproductive strategies in the ruff (*Philomachus pugnax*). *Nature Genetics* 48:84–88.

Lamichhaney, S., F. Han, J. Berglund, C. Wang, M. S. Almén, M. T. Webster, B. R. Grant, P. R. Grant, and L. Andersson. 2016b. A beak size locus in Darwin's finches facilitated character displacement during a drought. *Science* 352:470–74.

Lande, R. 1977. Statistical tests for natural selection on quantitative characters. *Evolution* 31:442–44.

Lande, R. 1982. A quantitative genetic theory of life-history evolution. *Ecology* 63:607–15.

Lande, R. 1983. Chromosomal rearrangements in speciation. *Heredity* 50:214.

Lande, R. 1984. The expected fixation rate of chromosomal inversions. *Evolution* 38:743–52.

Lande, R. 2009. Adaptation to an extraordinary environment by evolution of phenotypic plasticity and genetic assimilation. *Journal of Evolutionary Biology* 22:1435–46.

Lande, R. 2014. Evolution of phenotypic plasticity and environmental tolerance of a labile quantitative character in a fluctuating environment. *Journal of Evolutionary Biology* 27:866–75.

Lande, R., and S. J. Arnold. 1983. The measurement of selection on correlated characters. *Evolution* 37:1210–26.

Landman, N. H., S. Goolaerts, J. W. M. Jagt, E. A. Jagt-Yazykova, M. Machalski, and M. M. Yacobucci. 2014. Ammonite extinction and nautilid survival at the end of the Cretaceous. *Geology* 42:707–10.

Lankester, E. R. 1870. On the use of the term homology. *Annals and Magazine of Natural History* 6:34–43.

Laubichler, M. D. 2009. Form and function in Evo Devo: historical and conceptual reflections. Pp. 10–46 in M. D. Laubichler and J. Maienschein, eds., *Form and Function in Developmental Evolution*. Cambridge University Press, New York.

Laubichler, M. D. 2010. Evolutionary developmental biology offers a significant challenge to the neo-Darwinian paradigm. Pp. 199–212 in F. J. Ayala and R. Arp, eds., *Contemporary Debates in Philosophy of Biology*. Wiley-Blackwell, Malden, MA.

Laubichler, M. D., and J. Maienschein. 2009. *Form and Function in Developmental Evolution*. Cambridge University Press, Cambridge, UK.

Law, R. 1979. Optimal life histories under age-specific predation. *American Naturalist* 114:399–417.

Lawton, G. 2009. Why Darwin was wrong about the tree of life. *New Scientist* 2692:34–39.

Le Viol, I., F. Jiguet, L. Brotons, S. Herrando, Å. Lindström, J. W. Pearce-Higgins, J. Reif, C. Van Turnhout, and V. Devictor. 2012. More and more generalists: two decades of changes in the European avifauna. *Biology Letters* 8:780–82.

Leclerc, R. D. 2008. Survival of the sparsest: robust gene networks are parsimonious. *Molecular Systems Biology* 4:213.

Leigh, E. G., A. O'Dea, and G. J. Vermeij. 2014. Historical biogeography of the Isthmus of Panama. *Biological Reviews* 89:148–72.

Leimar, O., P. Hammerstein, and J. M. Van Dooren. 2006. A new perspective on developmental plasticity and the principles of adaptive morph development. *American Naturalist* 167:367–76.

Lenormand, T., and S. P. Otto. 2000. The evolution of recombination in a heterogeneous environment. *Genetics* 156:423–38.

Lerat, E., V. Daubin, H. Ochman, and N. A. Moran. 2005. Evolutionary origins of genomic repertoires in bacteria. *PLoS Biology* 3:e130.

Leroi, A. M., M. R. Rose, and G. V. Lauder. 1994. What does the comparative method reveal about adaptation? *American Naturalist* 143:381–402.

Lesoway, M. P. 2016. The future of Evo-Devo: the inaugural meeting of the Pan American Society for evolutionary developmental biology. *Evolution and Development* 18:71–77.

Levine, L. 1995. *The Genetics of Natural Populations: The Continuing Importance of Theodosius Dobzhansky*. Columbia University Press, New York.

Levins, R. 1962. Theory of fitness in a heterogeneous environment I: the fitness set and adaptive function. *American Naturalist* 96:361–78.

Levins, R. 1963. Theory of fitness in a heterogeneous environment II: developmental flexibility and niche selection. *American Naturalist* 97:75–90.

Levins, R. 1965. Theory of fitness in a heterogeneous environment V: optimal genetic systems. *Genetics* 52:891–904.

Levins, R. 1966. The strategy of model building in population biology. *American Scientist* 54:421–31.

Levins, R. 1993. A response to Orzack and Sober: formal analysis and the fluidity of science. *Quarterly Review of Biology* 68:547–55.

Levinson, G., and G. A. Gutman. 1987. Slipped-strand mispairing: a major mechanism for DNA sequence evolution. *Molecular Biology and Evolution* 4:203–21.

Lewontin, R. C. 1965. Selection for colonizing ability. Pp. 77–94 in H. G. Baker and G. L. Stebbins, eds., *The Genetics of Colonizing Species*. Academic Press, New York.

Lewontin, R. C. 1970. The units of selection. *Annual Review of Ecology and Systematics* 1:1–18.

Lewontin, R. C. 1974. *The Genetic Basis of Evolutionary Change*. Columbia University Press, New York.

Lewontin, R. C. 1983. Gene, organism and environment. In D. S. Bendall, ed., *Evolution from Molecules to Men*. Cambridge University Press, Cambridge, UK.

Lewontin, R. C. 1991. Twenty-five years ago in *Genetics*: electrophoresis in the development of evolutionary genetics: milestone or millstone? *Genetics* 128:657–62.

Lewontin, R. C. 2000. The problems of population genetics. Pp. 5–23 in R. S. Singh and C. B. Krimbas, eds., *Evolutionary Genetics from Molecules to Morphology*. Cambridge University Press, Cambridge, UK.

Lewontin, R. C., and D. Cohen. 1969. On population growth in a randomly varying environment. *Proceedings of the National Academy of Sciences* 62:1056–60.

Lewontin, R. C., and J. L. Hubby. 1966. A molecular approach to the study of genic heterozygosity in natural populations II: amount of variation and degree of heterozygosity in natural populations of *Drosophila pseudoobscura*. *Genetics* 54:595–609.

Li, C. C. 1975. *Path Analysis: A Primer*. Boxwood Press, Pacific Grove, CA.

Lieber, M. R. 2010. The mechanism of double-strand DNA break repair by the non-homologous DNA end joining pathway. *Annual Review of Biochemistry* 79:181–211.

Lieberman, D. E., and B. K. Hall. 2007. The evolutionary developmental biology of tinkering: an introduction to the challenge. Pp. 1–19 in G. Bock and J. Goode, eds., *Tinkering: The Microevolution of Development*. John Wiley, Chichester, NY.

Liebold, M. A., and M. A. McPeek. 2006. Coexistence of the niche and neutral perspectives in community ecology. *Ecology* 87:1399–1410.

Lika, K., and S. A. L. M. Kooijman. 2003. Life history implications of allocation to growth versus reproduction in dynamic energy budgets. *Bulletin of Mathematical Biology* 65:809–34.

Lim, J. Y., and C. R. Marshall. 2017. The true tempo of evolutionary radiation and decline revealed on the Hawaiian archipelago. *Nature* 543:710–13.

Linnaeus, C. 1758. *Systema Naturae*. Laurentius Salvius, Stockholm.

Liou, L. W., and T. D. Price. 1994. Speciation by reinforcement of premating isolation. *Evolution* 48:1451–59.

Liow, L. H., T. B. Quental, and C. R. Marshall. 2010. When can decreasing diversification rates be detected with molecular phylogenies and the fossil record? *Systematic Biology* 59:646–59.

Lively, C. M. 1986. Canalization versus developmental conversion in a spatially variable environment. *American Naturalist* 128:561–72.

Lloyd, D. G. 1987. Allocations to pollen, seeds and pollination mechanisms in self-fertilizing plants. *Functional Ecology* 1:83–89.

Lloyd, E. A. 1988. *The Structure and Confirmation of Evolutionary Theory*. Greenwood Press, New York.

Lloyd, E. A. 2005. *The Case of the Female Orgasm: Bias in the Science of Evolution*. Harvard University Press, Cambridge, MA.

Loh, Y. H. E., E. Bezault, F. M. Muenzel, R. B. Roberts, R. Swofford, M. Barluenga, C. E. Kidd, A. E. Howe, F. Di Palma, K. Lindblad-Toh, J. Hey, O. Seehausen, W. Salzburger, T. D. Kocher, and J. T. Streelman. 2013. Origins of shared genetic variation in African cichlids. *Molecular Biology and Evolution* 30:906–17.

Lomolino, M. V., and J. H. Brown. 2009. The reticulating phylogeny of island biogeography theory. *Quarterly Review of Biology* 84:357–90.

Loreau, M. 1998. Biodiversity and ecosystem functioning: a mechanistic model. *Proceedings of the National Academy of Sciences* 95:5632–36.

Loreau, M. 2004. Does functional redundancy exist? *Oikos* 104:606–11.

Losos, J. B. 2008. Phylogenetic niche conservatism, phylogenetic signal and the relationship between phylogenetic relatedness and ecological similarity among species. *Ecology Letters* 11:995–1003.

Losos, J. B. 2009. *Lizards in an Evolutionary Tree: Ecology and Adaptive Radiation of Anoles*. University of California Press, Berkeley.

Losos, J. B. 2010. Adaptive radiation, ecological opportunity, and evolutionary determinism. *American Naturalist* 175:623–39.

Losos, J. B. 2011. Convergence, adaptation, and constraint. *Evolution* 65:1827–40.

Losos, J. B., S. J. Arnold, G. Bejerano, E. D. Brodie III, D. Hibbett, H. E. Hoekstra, D. P. Mindell, A. Monteiro, C. Moritz, H. A. Orr, D. A. Petrov, S. S. Renner, R. E. Ricklefs, P. S. Soltis, and T. L. Turner. 2013. Evolutionary biology for the 21st century. *PLoS Biology* 11:e1001466.

Losos, J. B., D. M. Hillis, and H. W. Greene. 2012. Who speaks with a forked tongue? *Science* 338:1428–29.

Losos, J. B., and C. E. Parent. 2009. The speciation-area relationship. Pp. 415–38 in J. B. Losos and R. E. Ricklefs, eds., *The Theory of Island Biogeography Revisited*. Princeton University Press, Princeton, NJ.

Losos, J. B., and R. E. Ricklefs. 2009. *The Theory of Island Biogeography Revisited*. Princeton University Press, Princeton, NJ.

Love, A. C. 2006. Evolutionary morphology and evo-devo: hierarchy and novelty. *Theory in Biosciences* 124:317–33.

Love, A. C. 2008. Explaining evolutionary innovation and novelty: criteria of explanatory adequacy and epistemological prerequisites. *Philosophy of Science* 75:874–86.

Love, A. C. 2010. Rethinking the structure of evolutionary theory for an extended synthesis. Pp. 403–41 in M. Pigliucci and G. B. Müller, eds., *Evolution: The Extended Synthesis*. MIT Press, Cambridge, MA.

Love, A. C. 2011. Darwin's functional reasoning and homology. Pp. 49–67 in M. Wheeler, ed., *150 Years of Evolution: Darwin's Impact on the Humanities and the Social Sciences*. SDSU Press, San Diego, CA.

Love, A. C. 2013. Theory is as theory does: scientific practice and theory structure in biology. *Biological Theory* 7:325–37.

Love, A. C. 2014. The erotetic organization of developmental biology. Pp. 33–55 in A. Minelli and T. Pradeu, eds., *Towards a Theory of Development*. Oxford University Press, Oxford, UK.

Love, A. C. 2015a. Conceptual change and evolutionary developmental biology. Pp. 1–53 in A. C. Love, ed., *Conceptual Change in Biology: Scientific and Philosophical Perspectives on Evolution and Development*. Springer, Berlin.

Love, A. C., ed. 2015b. *Conceptual Change in Biology: Scientific and Philosophical Perspectives on Evolution and Development*. Springer, Berlin.

Love, A. C. 2015c. Evolutionary developmental biology: philosophical issues. Pp. 265–83 in T. Heams, P. Huneman, L. Lecointre, and M. Silberstein, eds., *Handbook of Evolutionary Thinking in the Sciences*. Springer, Berlin.

Love, A. C. 2017a. Developmental mechanisms. Pp. 332–47 in S. Glennan and P. Illari, eds., *The Routledge Handbook of Mechanisms and Mechanical Philosophy*. Routledge, New York.

Love, A. C. 2017b. Evo-devo and the structure(s) of evolutionary theory: a different kind of challenge. Pp. 159–87 in P. Huneman and D. Walsh, eds., *Challenging the Modern Synthesis: Adaptation, Development, and Inheritance*. Oxford University Press, New York.

Love, A. C., and M. Travisano. 2013. Microbes modeling ontogeny. *Biology and Philosophy* 28:161–88.

Lowe, C. J., and G. A. Wray. 1997. Radical alterations in the roles of homeobox genes during echinoderm evolution. *Nature* 389:718–21.

Lowry, D. B. 2012. Ecotypes and the controversy over stages in the formation of new species. *Biological Journal of the Linnean Society* 106:241–57.

Lowry, D. B., J. L. Modliszewski, K. M. Wright, C. A. Wu, and J. H. Willis. 2008a. The strength and genetic basis of reproductive isolating barriers in flowering plants. *Philosophical Transactions of the Royal Society of London B: Biological Sciences* 363: 3009–21.

Lowry, D. B., C. R. Rockwood, J. H. Willis, and D. Schoen. 2008b. Ecological reproductive isolation of coast and inland races of *Mimulus guttatus*. *Evolution* 62:2196–2214.

Lowry, D. B., and J. H. Willis. 2010. A widespread chromosomal inversion polymorphism contributes to a major life-history transition, local adaptation, and reproductive isolation. *PLoS Biology* 8:e1000500.

Loxdale, H. D., G. Lushai, and J. A. Harvey. 2011. The evolutionary improbability of "generalism" in nature, with special reference to insects. *Biological Journal of the Linnean Society* 103:1–18.

Ludwig, D. 1996. The distribution of population survival times. *American Naturalist* 147:506–26.

Lynch, M. 1988. Design and analysis of experiments on random drift and inbreeding depression. *Genetics* 120:791–807.

Lynch, M. 1990. The rate of morphological evolution in mammals from the standpoint of the neutral expectation. *American Naturalist* 136:727–41.

Lynch, M. 2007. The frailty of adaptive hypotheses for the origins of organismal complexity. *Proceedings of the National Academy of Sciences* 104:8597–8604.

Lynch, M., and J. S. Conery. 2003. The origins of genome complexity. *Science* 302:1401–4.

Lynch, M., and W. Gabriel. 1987. Enviromental tolerance. *American Naturalist* 129:283–303.

Lynch, V., and G. Wagner. 2010. Revisiting a classic example of transcription factor functional equivalence: are Eyeless and Pax6 functionally equivalent or divergent? *Journal of Experimental Zoology Part B: Molecular and Developmental Evolution* 316B:93–98.

Lyons, E. E., and T. W. Mully. 1992. Density effects on flowering phenology and mating potential in Nicotiana alata. *Oecologia* 91:93–100.

MacArthur, R. H., and E. O. Wilson. 1963. An equilibrium theory of insular zoogeography. *Evolution* 17:373–87.

MacArthur, R. H., and E. O. Wilson. 1967. *The Theory of Island Biogeography.* Princeton University Press, Princeton, NJ.

MacCarthy, T., and A. Bergman. 2007. Coevolution of robustness, epistasis, and recombination favors asexual reproduction. *Proceedings of the National Academy of Sciences* 104:12801–6.

Maddison, W. P. 1997. Gene trees in species trees. *Systematic Biology* 46:523–36.

Maddison, W. P., and R. G. FitzJohn. 2015. The unsolved challenge to phylogenetic correlation tests for categorical characters. *Systematic Biology* 64:127–36.

Maddison, W. P., and D. R. Maddison. 2015. Mesquite: a modular system for evolutionary analysis. Version 3.04.

Magiafoglou, A., and A. A. Hoffmann. 2003. Cross-generation effects due to cold exposure in *Drosophila serrata. Functional Ecology* 17:664–72.

Mahall, B. E., and F. H. Bormann. 1978. A quantitative description of the vegetative phenology of herbs in a northern hardwood forest. *Botanical Gazette* 139:467–81.

Malcom, J. W. 2011. Smaller gene networks permit longer persistence in fast-changing environments. *PLoS ONE* 6:e14747.

Mallarino, R., P. R. Grant, B. R. Grant, A. Herrel, W. P. Kuo, and A. Abzhanov. 2011. Two developmental modules establish 3D beak-shape variation in Darwin's finches. *Proceedings of the National Academy of Sciences* 108:4057–62.

Mallet, J. 2010. Why was Darwin's view of species rejected by twentieth century biologists? *Biology and Philosophy* 25:497–527.

Manicardi, G. C., A. Nardelli, and M. Mandrioli. 2015. Fast chromosomal evolution and karyotype instability: recurrent chromosomal rearrangements in the peach potato aphid *Myzus persicae* (Hemiptera: Aphididae). *Biological Journal of the Linnean Society* 116:519–29.

Mao, K., R. I. Milne, L. Zhang, Y. Peng, J. Liu, P. Thomas, R. R. Mill, and S. S. Renner. 2012. Distribution of living Cupressaceae reflects the breakup of Pangea. *Proceedings of the National Academy of Sciences* 109:7793–98.

Maraun, M., M. Heethoff, K. Schneider, S. Scheu, G. Weigmann, J. Cianciolo, R. H. Thomas, and R. A. Norton. 2004. Molecular phylogeny of oribatid mites (Oribatida, Acari): evidence for multiple radiations of parthenogenetic lineages. *Experimental and Applied Acarology* 33:183–201.

Margulis, L. 1991. Symbiogenesis and symbionticism. Pp. 1–14 in L. Margulis and R. Fester, eds., *Symbiosis as a Source of Evolutionary Innovation: Speciation and Morphogenesis.* MIT Press, Cambridge, MA.

Mark Welch, D. B., and M. Meselson. 2000. Evidence for the evolution of bdelloid rotifers without sexual reproduction or genetic exchange. *Science* 288:1211–15.

Marks, J. 2008. Race across the physical-cultural divide in American anthropology. Pp. 242–58 in H. Kuklick, ed., *A New History of Anthropology*. Blackwell, Oxford, UK.

Marquet, P. A., S. A. Navarrete, and J. C. Castilla. 1990. Scaling population density to body size in rocky intertidal communities. *Science* 250:1125–27.

Marriage, T. N., and M. E. Orive. 2012. Mutation-selection balance and mixed mating with asexual reproduction. *Journal of Theoretical Biology* 308:25–35.

Marshall, C. R. 2014. The evolution of morphogenetic fitness landscapes: conceptualising the interplay between the developmental and ecological drivers of morphological innovation. *Australian Journal of Zoology* 62:3–17.

Martin, A., and C. Simon. 1990. Temporal variation in insect life cycles: lessons from periodical cicadas. *Bioscience* 40:359–67.

Martin, G., S. P. Otto, and T. Lenormand. 2006. Selection for recombination in structured populations. *Genetics* 172:593–609.

Martín, H. G., and N. Goldenfeld. 2006. On the origin and robustness of power-law species-area relationships in ecology. *Proceedings of the National Academy of Sciences* 103:10310–15.

Martin, P. R., F. Bonier, and J. J. Tewksbury. 2007. Revisiting Jablonski (1993): cladogenesis and range expansion explain latitudinal variation in taxonomic richness. *Journal of Evolutionary Biology* 20:930–36.

Martin, S. H., K. K. Dasmahapatra, N. J. Nadeau, C. Salazar, J. R. Walters, F. Simpson, M. Blaxter, A. Manica, J. Mallet, and C. D. Jiggins. 2013. Genome-wide evidence for speciation with gene flow in *Heliconius* butterflies. *Genome Research* 23:1817–28.

Maruvka, Y. E., N. M. Shnerb, D. A. Kessler, and R. E. Ricklefs. 2013. Model for macroevolutionary dynamics. *Proceedings of the National Academy of Sciences* 110.E2460–E2469.

Masly, J. P., and D. C. Presgraves. 2007. High-resolution genome-wide dissection of the two rules of speciation in *Drosophila*. *PLoS Biology* 5:e243.

Masta, S. E. 2000. Phylogeography of the jumping spider Habronattus pugillis (Araneae: Salticidae): recent vicariance of sky island populations? *Evolution* 54:1699–1711.

Matute, D. R., I. A. Butler, D. A. Turissini, and J. A. Coyne. 2010. A test of the snowball theory for the rate of evolution of hybrid incompatibilities. *Science* 329:1518–21.

Matzke, N. J. 2014. Model selection in historical biogeography reveals that founder-event speciation is a crucial process in island clades. *Systematic Biology* 63:951–70.

Mayden, R. L. 1997. A hierarchy of species concepts: the denouement in the sage of the species problem. Pp. 381–423 in M. Claridge, H. Dawah, and M. Wilson, eds., *Species: The Units of Biodiversity*. Chapman and Hall, London.

Mayer, W. V. 1986. Biology education in the United States during the twentieth century. *Quarterly Review of Biology* 61:481–507.

Mayfield, M. M., and J. M. Levine. 2010. Opposing effects of competitive exclusion on the phylogenetic structure of communities. *Ecology Letters* 13:1085–93.

Mayhew, P. J. 2007. Why are there so many insect species? Perspectives from fossils and phylogenies. *Biological Reviews* 82:425–54.

Maynard Smith, J. 1964. Group selection and kin selection. *Nature* 201:1145–47.

Maynard Smith, J. 1966. Sympatric speciation. *American Naturalist* 100:637–50.

Maynard Smith, J. 1976a. Group selection. *Quarterly Review of Biology* 51:277–83.

Maynard Smith, J. 1976b. A short-term advantage for sex and recombination through sib-competition. *Journal of Theoretical Biology* 63:245–58.

Maynard Smith, J. 1978a. *The Evolution of Sex*. Cambridge University Press, Cambridge, UK.

Maynard Smith, J. 1978b. Optimization theory in evolution. *Annual Review of Ecology and Systematics* 9:31–56.

Maynard Smith, J. 1986. Evolution: contemplating life without sex. *Nature* 324:300–301.

Maynard Smith, J. 1988a. The evolution of recombination. Pp. 106–25 in R. E. Michod and B. R. Levin, eds., *The Evolution of Sex*. Sinauer Associates, Sunderland, MA.

Maynard Smith, J. 1988b. Selection for recombination in a polygenic model—the mechanism. *Genetics Research* 51:59–63.

Maynard Smith, J., R. M. Burian, S. Kauffman, P. Alberch, J. Campbell, B. Goodwin, R. Lande, D. Raup, and L. Wolpert. 1985. Developmental constraints and evolution: a perspective from the Mountain Lake conference on development and evolution. *Quarterly Review of Biology* 60:265–87.

Mayr, E. 1942. *Systematics and the Origin of Species*. Columbia University Press, New York.

Mayr, E. 1954. Change of genetic environment and evolution. Pp. 157–80 in J. Huxley, A. C. Hardy, and E. B. Ford, eds., *Evolution as Process*. George Allen and Unwin, London.

Mayr, E. 1959. Where are we? *Cold Spring Harbor Symposium on Quantitative Biology* 24:409–40.

Mayr, E. 1960. The emergence of evolutionary novelties. Pp. 349–80 in S. Tax, ed., *Evolution after Darwin*, vol. 1: *The Evolution of Life, Its Origin, History, and Future*. University of Chicago Press, Chicago, IL.

Mayr, E. 1963. *Animal Species and Evolution*. Belknap Press, Harvard University Press, Cambridge, MA.

Mayr, E. 1964. The evolution of living systems. *Proceedings of the National Academy of Sciences* 51:934–41.

Mayr, E. 1969. *Principles of Systematic Zoology*. McGraw-Hill, New York.

Mayr, E. 1970. *Populations, Species, and Evolution*. Harvard University Press, Cambridge, MA.

Mayr, E. 1982. *The Growth of Biological Thought*. Belknap Press, Cambridge, MA.

Mayr, E. 1992. A local flora and the biological species concept. *American Journal of Botany* 79:222–38.

Mayr, E. 1993. What was the evolutionary synthesis? *Trends in Ecology and Evolution* 8:31–34.

Mayr, E. 1996. What is a species, and what is not? *Philosophy of Science* 63:262–77.

Mayr, E. 2002. Interview with Ernst Mayr. *Bioessays* 24:960–73.

Mayr, E., and W. B. Provine. 1980. *The Evolutionary Synthesis: Perspectives on the Unification of Biology*. Harvard University Press, Cambridge, MA.

Mayr, E., and W. B. Provine. 1998. *The Evolutionary Synthesis: Perspectives on the Unification of Biology*. Revised with a new introduction. Harvard University Press, Cambridge, MA.

Mayrose, I., S. H. Zhan, C. J. Rothfels, K. Magnuson-Ford, M. S. Barker, L. H. Rieseberg, and S. P. Otto. 2011. Recently formed polyploid plants diversify at lower rates. *Science* 333:1257.

McDermott, S. R., and M. A. F. Noor. 2010. The role of meiotic drive in hybrid male sterility. *Philosophical Transactions of the Royal Society of London B: Biological Sciences* 365: 1265–72.

McDonald, J. H., and M. Kreitman. 1991. Adaptive protein evolution at the *Adh* locus in *Drosophila*. *Nature* 351:652–54.

McDonald, M. J., D. P. Rice, and M. M. Desai. 2016. Sex speeds adaptation by altering the dynamics of molecular evolution. *Nature* 531:233–36.

McFall-Ngai, M., M. G. Hadfield, T. C. G. Bosch, H. V. Carey, T. Domazet-Lošo, A. E. Douglas, N. Dubilier, G. Eberl, T. Fukami, S. F. Gilbert, U. Hentschel, N. King, S. Kjelleberg, A. H. Knoll, N. Kremer, S. K. Mazmanian, J. L. Metcalf, K. Nealson, N. E. Pierce, J. F. Rawls, A. Reid, E. G. Ruby, M. Rumpho, J. G. Sanders, D. Tautz, and J. J. Wernegreen. 2013. Animals in a bacterial world, a new imperative for the life sciences. *Proceedings of the National Academy of Sciences* 110:3229–36.

McGhee, G. R. 2011. *Convergent Evolution: Limited Forms Most Beautiful*. MIT Press, Cambridge, MA.

McGill, B. J., B. J. Enquist, E. Weiher, and M. Westoby. 2006. Rebuilding community ecology from functional traits. *Trends in Ecology and Evolution* 21:178–85.

McGinley, M. A. 1989. The influence of a positive correlation between clutch size and offspring fitness on the optimal offspring size. *Evolutionary Ecology* 3:150–56.

McGinley, M. A., and E. L. Charnov. 1988. Multiple resources and the optimal balance between size and number of offspring. *Evolutionary Ecology* 2:77–84.

McGinley, M. A., D. H. Temme, and M. A. Geber. 1987. Parental investment in offspring in variable environments: theoretical and empirical. *American Naturalist* 130:370–98.

McInerney, J. O., W. F. Martin, E. V. Koonin, J. F. Allen, M. Y. Galperin, N. Lane, J. M. Archibald, and T. M. Embley. 2011. Planctomycetes and eukaryotes: a case of analogy not homology. *Bioessays* 33:810–17.

McManus, J. F., D. W. Oppo, and J. L. Cullen. 1999. A 0.5-million-year record of millennial-scale climate variability in the North Atlantic. *Science* 283:971–75.

McPheron, B. A., D. C. Smith, and S. H. Berlocher. 1988. Genetic differences between host races of *Rhagoletis pomonella*. *Nature* 336:64–66.

McShea, D. W. 1994. Mechanisms of large-scale evolutionary trends. *Evolution* 48:1747–63.

McShea, D. W. 2000. Trends, tools, and terminology. *Paleobiology* 26:330–33.

McShea, D. W., and R. N. Brandon. 2010. *Biology's First Law: The Tendency for Diversity and Complexity to Increase in Evolutionary Systems*. University of Chicago Press, Chicago, IL.

Medawar, P. B. 1952. *An Unsolved Problem of Biology*. H. K. Lewis, London.

Melián, C. J., O. Seehausen, V. M. Eguíluz, M. A. Fortuna, and K. Deiner. 2015. Diversification and biodiversity dynamics of hot and cold spots. *Ecography* 38:393–401.

Mendelson, T. C., and K. L. Shaw. 2005. Rapid speciation in an arthropod. *Nature* 433:375–76.

Metcalf, C. J. E., and S. Pavard. 2007. Why evolutionary biologists should be demographers. *Trends in Ecology and Evolution* 22:205–12.

Metz, J. A. J. 2011. Thoughts on the geometry of meso-evolution: collecting mathematical elements for a post-modern synthesis. Pp. 193–231 in F. A. C. C. Chalub and J. F. Rodrigues, eds., *The Mathematics of Darwin's Legacy: Mathematics and Biosciences in Interaction*. Springer, Basel.

Metz, J. A. J., and O. Diekmann. 1986. *The Dynamics of Physiologically Structured Populations*. Springer, Berlin.

Metzker, M. L., D. P. Mindell, X.-M. Liu, R. G. Ptak, R. A. Gibbs, and D. M. Hillis. 2002.

Molecular evidence of HIV-1 transmission in a criminal case. *Proceedings of the National Academy of Sciences* 99:14292–97.

Michod, R. E., and B. R. Levin, eds., 1988. *The Evolution of Sex*. Sinauer Associates, Sunderland, MA.

Miller, S. L., and H. C. Urey. 1959. Organic compound synthesis on the primitive earth. *Science* 130:245–51.

Millstein, R. L. 2007. Distinguishing drift and selection empirically: "the great snail debate" of the 1950s. *Journal of the History of Biology* 41:339–67.

Millstein, R. L. 2009. Populations as individuals. *Biological Theory* 4:267–73.

Mindell, D. P. 1992. Phylogenetic consequences of symbioses: Eukarya and Eubacteria are not monophyletic taxa. *Biosystems* 27:53–62.

Mindell, D. P. 2006. *The Evolving World: Evolution in Everyday Life*. Harvard University Press, Cambridge, MA.

Minelli, A. 2010. Evolutionary developmental biology does not offer a significant challenge to the neo-Darwinian paradigm. Pp. 213–26 in F. J. Ayala and R. Arp, eds., *Contemporary Debates in Philosophy of Biology*. Wiley-Blackwell, Malden, MA.

Mishler, B. D. 1999. Getting rid of species? Pp. 307–15 in R. Wilson, ed., *Species: New Interdisciplinary Essays*. MIT Press, Cambridge, MA.

Mishler, B. D. 2010. Species are not uniquely real biological entities. Pp. 110–22 in F. J. Ayala and R. Arp, eds., *Contemporary Debates in Philosophy of Biology*. Wiley-Blackwell, West Sussex, UK.

Mishler, B. D., and M. J. Donoghue. 1982. Species concepts: a case for pluralism. *Systematic Biology* 31:491–503.

Mishler, B. D., and E. Theriot. 2000. Monophyly, apomorphy, and phylogenetic species concepts. Pp. 44–54 in Q. D. Wheeler and R. Meier, eds., *Species Concepts and Phylogenetic Theory: A Debate*. Columbia University Press, New York.

Mittelbach, G. G., and D. W. Schemske. 2015. Ecological and evolutionary perspectives on community assembly. *Trends in Ecology and Evolution* 30:241–47.

Moczek, A. P. 2008. On the origins of novelty in development and evolution. *Bioessays* 30:432–47.

Moczek, A. P., K. E. Sears, A. Stollewerk, P. J. Wittkopp, P. Diggle, I. Dworkin, C. Ledon-Rettig, D. Q. Matus, S. Roth, E. Abouheif, F. D. Brown, C.-H. Chiu, C. S. Cohen, A. W. D. Tomaso, S. F. Gilbert, B. Hall, A. C. Love, D. C. Lyons, T. J. Sanger, J. Smith, C. Specht, M. Vallejo-Marin, and C. G. Extavour. 2015. The significance and scope of evolutionary developmental biology: a vision for the 21st century. *Evolution and Development* 17:198–219.

Moczek, A. P., S. Sultan, S. Foster, C. Ledón-Rettig, I. Dworkin, H. F. Nijhout, E. Abouheif, and D. W. Pfennig. 2011. The role of developmental plasticity in evolutionary innovation. *Proceedings of the Royal Society of London B: Biological Sciences* 278:2705–13.

Moorad, J. A. 2013. Multi-level sexual selection: individual and family-level selection for mating success in a historical human population. *Evolution* 67:1635–48.

Moore, B., and M. Donoghue. 2007. Correlates of diversification in the plant clade Dipsacales: geographic movement and evolutionary innovations. *American Naturalist* 170:S28–S55.

Mora, C., D. P. Tittensor, S. Adl, A. G. B. Simpson, and B. Worm. 2011. How many species are there on Earth and in the ocean? *PLoS Biology* 9:e1001127.

Moran, N. A., and D. B. Sloan. 2015. The hologenome concept: helpful or hollow? *PLoS Biology* 13:e1002311.

Morand, S., A. Simková, I. Matejusová, L. Plaisance, O. Verneau, and Y. Desdevises. 2002.

Investigating patterns may reveal processes: evolutionary ecology of ectoparasitic monogeneans. *International Journal for Parasitology* 32:111–19.

Morgan, C. L. 1896. On modification and variation. *Science* 4:733–40.

Moritz, C., C. J. Schneider, and D. B. Wake. 1992. Evolutionary relationships within the *Ensatina eschscholtzii* complex confirm the ring species interpretation. *Systematic Biology* 41:273–91.

Morlon, H. 2014. Phylogenetic approaches for studying diversification. *Ecology Letters* 17:508–25.

Morlon, H., S. Kefi, and N. D. Martinez. 2014. Effects of trophic similarity on community composition. *Ecology Letters* 17:1495–1506.

Morlon, H., T. L. Parsons, and J. B. Plotkin. 2011. Reconciling molecular phylogenies with the fossil record. *Proceedings of the National Academy of Sciences* 108:16327–32.

Morris, D. W. 1996. Coexistence of specialist and generalist rodents via habitat selection. *Ecology* 77:2352–64.

Morrison, D. A. 2014. Is the Tree of Life the best metaphor, model, or heuristic for phylogenetics? *Systematic Biology* 63:628–38.

Morrow, E. H., T. E. Pitcher, and G. Arnqvist. 2003. No evidence that sexual selection is an "engine of speciation" in birds. *Ecology Letters* 6:228–34.

Mouquet, N., V. Devictor, C. N. Meynard, F. Munoz, L. F. Bersier, J. Chave, P. Couteron, A. Dalecky, C. Fontaine, and D. Gravel. 2012. Ecophylogenetics: advances and perspectives. *Biological Reviews* 87:769–85.

Mousseau, T. A., and D. A. Roff. 1987. Natural selection and the heritability of fitness components. *Heredity* 59:181–97.

Moyle, L. C., and T. Nakazato. 2010. Hybrid incompatibility "snowballs" between *Solanum* species. *Science* 329:1521–23.

Muir, W. M. 1996. Group selection for adaptation to multiple-hen cages: selection program and direct responses. *Poultry Science* 75:447–58.

Muir, W. M. 2005. Incorporation of Competitive Effects in Forest Tree or Animal Breeding Programs. *Genetics* 170:1247–59.

Muir, W. M., P. Bijma, and A. Schinckel. 2013. Multilevel selection with kin and nonkin groups, experimental results with Japanese quail (*Coturnix japonica*). *Evolution* 67:1598–1606.

Muirhead, C. A., and R. Lande. 1997. Inbreeding depression under joint selfing, outcrossing and asexuality. *Evolution* 51:1409–15.

Müller, G. B. 2003. Homology: the evolution of morphological organization. Pp. 52–69 in G. B. Müller and S. A. Newman, eds., *Origination of Organismal Form: Beyond the Gene in Developmental and Evolutionary Biology*. MIT Press, Cambridge, MA.

Müller, G. B. 2007. Evo-devo: extending the evolutionary synthesis. *Nature Reviews Genetics* 8:943–49.

Müller, G. B., and S. A. Newman. 2003. Origination of organismal form: the forgotten cause in evolutionary theory. Pp. 3–10 in G. B. Müller and S. A. Newman, eds., *Origination of Organismal Form: Beyond the Gene in Developmental and Evolutionary Biology*. MIT Press, Cambridge, MA.

Müller, G. B., and G. P. Wagner. 1991. Novelty in evolution: restructuring the concept. *Annual Review of Ecology and Systematics* 22:229–56.

Muller, H. J. 1932. Some genetic aspects of sex. *American Naturalist* 66:118–38.

Muller, H. J. 1942. Isolating mechanisms, evolution and temperature. *Biological Symposium* 6:71–125.

Muller, H. J. 1949. Redintegration of the symposium on genetics, paleontology, and evolution. Pp. 421–45 in G. L. Jepson, G. G. Simpson, and E. Mayr, eds., *Genetics, Paleontology and Evolution*. Princeton University Press, Princeton, NJ.

Muller, H. J. 1964. The relation of recombination to mutational advance. *Mutation Research/Fundamental and Molecular Mechanisms of Mutagenesis* 1:2–9.

Murren, C. J., J. R. Auld, H. Callahan, C. K. Ghalambor, C. A. Handelsman, M. A. Heskel, J. G. Kingsolver, H. J. Maclean, J. Masel, H. Maughan, D. W. Pfennig, R. A. Relyea, S. Seiter, E. Snell-Rood, U. K. Steiner, and C. D. Schlichting. 2015. Constraints on the evolution of phenotypic plasticity: limits and costs of phenotype and plasticity. *Heredity* 115:293–301.

Murren, C. J., H. J. Maclean, S. E. Diamond, U. K. Steiner, M. A. Heskel, C. A. Handelsman, K. G. Cameron, J. R. Auld, H. S. Callahan, D. W. Pfennig, R. A. Relyea, C. D. Schlichting, and J. Kingsolver. 2014. Evolutionary change in continuous reaction norms. *American Naturalist* 183:453–67.

Muschick, M., A. Indermaur, and W. Salzburger. 2012. Convergent evolution within an adaptive radiation of cichlid fishes. *Current Biology* 22:2362–68.

Myers, C. E., and E. E. Saupe. 2013. A macroevolutionary expansion of the modern synthesis and the importance of extrinsic abiotic factors. *Palaeontology* 56:1179–98.

Nagy, A., L. Banyai, and L. Patthy. 2011. Reassessing domain architecture evolution of metazoan proteins: major impact of errors caused by confusing paralogs and epaktologs. *Genes* 2:516–61.

Nakhleh, L. 2013. Computational approaches to species phylogeny inference and gene tree reconciliation. *Trends in Ecology and Evolution* 28:719–28.

Nance, R. D., and J. B. Murphy. 2013. Origins of the supercontinent cycle. *Geoscience Frontiers* 4:439–48.

Nance, R. D., T. R. Worsley, and J. B. Moody. 1986. Post-Archean biogeochemical cycles and long-term episodicity in tectonic processes. *Geology* 14:514–18.

Nance, R. D., T. R. Worsley, and J. B. Moody. 1988. The supercontinent cycle. *Scientific American* 256:72–79.

Nathan, R. 2006. Long-distance dispersal of plants. *Science* 313:786–88.

Nathan, R., and D. Shohami. 2013. Dispersal. *Oxford Bibliographies Online: Ecology*. Oxford University Press, Oxford, UK.

Nee, S. 1989. Antagonistic co-evolution and the evolution of genotypic randomization. *Journal of Theoretical Biology* 140:499–518.

Nee, S. 2006. Birth-death models in macroevolution. *Annual Review of Ecology, Evolution, and Systematics* 37:1–17.

Nei, M. 1967. Modifications of linkage intensity by natural selection. *Genetics* 57:625–41.

Nelson, G. J., and N. I. Platnick. 1981. *Systematics and Biogeography: Cladistics and Vicariance*. Columbia University Press, New York.

Newman, S. A. 2012. Physico-genetic determinants in the evolution of development. *Science* 338:217–19.

Newman, S. A., and R. Bhat. 2008. Dynamical patterning modules: physico-genetic determinants of morphological development and evolution. *Physical Biology* 5:1–14.

Nicoglou, A. 2015. The evolution of phenotypic plasticity: genealogy of a debate in genetics. *Studies in History and Philosophy of Science Part C: Studies in History and Philosophy of Biological and Biomedical Sciences* 50:67–76.

Nielsen, R. 2005. Molecular signatures of natural selection. *Annual Review of Genetics* 39:197–218.

Niklas, K. J. 2009. Deducing plant function from organic form: challenges and pitfalls.

Pp. 47–82 in M. D. Laubichler and J. Maienschein, eds., *Form and Function in Developmental Evolution*. Cambridge University Press, Cambridge, UK.

Nisbet, R. M., E. B. Muller, K. Lika, and S. A. L. M. Kooijman. 2000. From molecules to ecosystems through dynamic energy budget models. *Journal of Animal Ecology* 69: 913–26.

Nonaka, E., R. Svanbäck, X. Thibert-Plante, G. Englund, and Å. Brännström. 2015. Mechanisms by which phenotypic plasticity affects adaptive divergence and ecological speciation. *American Naturalist* 186:E126–E143.

Noonburg, E. G., R. M. Nisbet, E. McCauley, W. S. C. Gurney, W. W. Murdoch, and A. M. De Roos. 1998. Experimental testing of dynamic energy budget models. *Functional Ecology* 12:211–22.

Noor, M. A. F. 1995. Speciation driven by natural selection in *Drosophila*. *Nature* 375: 674–75.

Noor, M. A. F., K. L. Grams, L. A. Bertucci, and J. Reiland. 2001. Chromosomal inversions and the reproductive isolation of species. *Proceedings of the National Academy of Sciences* 98:12084–88.

Nosil, P. 2012. *Ecological Speciation*. Oxford University Press, Oxford, UK.

Nowak, S., J. Neidhart, I. G. Szendro, and J. Krug. 2014. Multidimensional epistasis and the transitory advantage of sex. *PLoS Computational Biology* 10:e1003836.

Nuismer, S. L., and S. Gandon. 2008. Moving beyond common-garden and transplant designs: insight into the causes of local adaptation in species interactions. *American Naturalist* 171:658–68.

O'Hara, R. J. 1988. Homage to Clio, or, toward an historical philosophy for evolutionary biology. *Systematic Biology* 37:142–55.

O'Hara, R. J. 1997. Population thinking and tree thinking in systematics. *Zoologica Scripta* 26:323–29.

O'Malley, M. A., and Y. Boucher, eds. 2011. Special thematic series: beyond the Tree of Life. *Biology Direct* 6.

O'Malley, M. A., W. Martin, and J. Dupré. 2010. The tree of life. introduction to an evolutionary debate. *Biology and Philosophy* 25:441–53.

Uchman, H., E. Lerat, and V. Daubin. 2005. Examining bacterial species under the specter of gene transfer and exchange. *Proceedings of the National Academy of Sciences* 102:6595–99.

Odenbaugh, J. 2006. The strategy of model building in population biology. *Biology and Philosophy* 21:607–21.

Odenbaugh, J. 2011. A general, unifying theory of ecology? Pp. 51–61 in S. M. Scheiner and M. R. Willig, eds., *The Theory of Ecology*. University of Chicago Press, Chicago, IL.

Odling-Smee, F. J., K. N. Laland, and M. W. Feldman. 2003. *Niche Construction: The Neglected Process in Evolution*. Princeton University Press, Princeton, NJ.

Okasha, S. 2006. *Evolution and the Levels of Selection*. Oxford University Press, New York.

Olson, E. C. 1960. Morphology, paleontology, and evolution. Pp. 523–45 in S. Tax, ed., *Evolution after Darwin*. University of Chicago Press, Chicago, IL.

Oneal, E., D. Otte, and L. L. Knowles. 2010. Testing for biogeographic mechanisms promoting divergence in Caribbean crickets (genus Amphiacusta). *Journal of Biogeography* 37:530–40.

Oparin, A. I. 1938. *The Origin of Life*. Macmillan, New York.

Ord, T. J., and T. C. Summers. 2015. Repeated evolution and the impact of evolutionary history on adaptation. *BMC Evolutionary Biology* 15:1–12.

Orgel, L. E. 1998. The origin of life—a review of facts and speculations. *Trends in Biochemical Sciences* 23:491–95.

Orr, H. A. 1993. Haldane's rule has multiple genetic causes. *Nature* 361:532–33.

Orr, H. A. 1995. The population genetics of speciation: the evolution of hybrid incompatibilities. *Genetics* 139:1805–13.

Orr, H. A. 2005. The genetic basis of reproductive isolation: insights from *Drosophila*. *Proceedings of the National Academy of Sciences* 102:6522–26.

Orr, H. A., and M. Turelli. 1996. Dominance and Haldane's rule. *Genetics* 143:613–16.

Orr, H. A., and M. Turelli. 2001. The evolution of postzygotic isolation: accumulating Dobzhansky-Muller incompatibilities. *Evolution* 55:1085–94.

Orr, M. R., and T. B. Smith. 1998. Ecology and speciation. *Trends in Ecology and Evolution* 13:502–6.

Orzack, S. H., and E. Sober. 2001. *Adaptationism and Optimality*. Cambridge University Press, Cambridge, UK.

Osborn, H. F. 1896. A mode of evolution requiring neither natural selection nor the inheritance of acquired characters. *Transactions of the New York Academy of Sciences* 15:141–42.

Osborn, H. F. 1897. The limits of organic selection. *American Naturalist* 31:944–51.

Ossowski, S., K. Schneeberger, J. I. Lucas-Lledó, N. Warthmann, R. M. Clark, R. G. Shaw, D. Weigel, and M. Lynch. 2010. The rate and molecular spectrum of spontaneous mutations in *Arabidopsis thaliana*. *Science* 327:92–94.

Otto, S. P., and N. H. Barton. 2001. Selection for recombination in small populations. *Evolution* 55:1921–31.

Otto, S. P., and T. Day. 2007. *A Biologist's Guide to Mathematical Modeling in Ecology and Evolution*. Princeton University Press, Princeton, NJ.

Otto, S. P., and M. W. Feldman. 1997. Deleterious mutations, variable epistatic interactions, and the evolution of recombination. *Theoretical Population Biology* 51:134–47.

Otto, S. P., and T. Lenormand. 2002. Resolving the paradox of sex and recombination. *Nature Reviews Genetics* 3:252–61.

Otto, S. P., and J. Whitton. 2000. Polyploid incidence and evolution. *Annual Review of Genetics* 34:401–37.

Owen, R. 1848. *On the Archetype and Homologies of the Vertebrate Skeleton*. Richard and John E. Taylor, London.

Pace, N. R. 1991. Origin of life—facing up to the physical setting. *Cell* 65:531–33.

Pace, N. R. 1997. A molecular view of microbial diversity and the biosphere. *Science* 276:734–40.

Pace, N. R. 2004. The early branches in the tree of life. Pp. 76–85 in J. Cracraft and M. J. Donoghue, eds., *Assembling the Tree of Life*. Oxford University Press, Oxford, UK.

Palacio-López, K., B. Beckage, S. Scheiner, and J. Molofsky. 2015. The ubiquity of phenotypic plasticity in plants: a synthesis. *Ecology and Evolution* 5:3389–3400.

Palmer, A. R. 2012. Developmental plasticity and the origin of novel forms: unveiling cryptic genetic variation via "use and disuse." *Journal of Experimental Zoology Part B: Molecular and Developmental Evolution* 318B:466–79.

Pálsson, S. 2001. The effects of deleterious mutations on cyclically parthenogenetic organisms. *Journal of Theoretical Biology* 208:201–14.

Papadopoulou, A., and L. L. Knowles. 2015. Species-specific responses to island connectivity cycles: refined models for testing phylogeographic concordance across a Mediterranean Pleistocene Aggregate Island complex. *Molecular Ecology* 24:4252–68.

Papaïx, J., J. J. Burdon, C. Lannou, and P. H. Thrall. 2014. Evolution of pathogen specialisation in a host metapopulation: joint effects of host and pathogen dispersal. *PLoS Computational Biology* 10:e1003633.

Pariselle, A., W. A. Boeger, J. Snoeks, C. F. Bilong Bilong, S. Morand, and M. P. M. Vanhove. 2011. The monogenean parasite fauna of cichlids: a potential tool for host biogeography. *International Journal of Evolutionary Biology* 2011:471480.

Parker, G. A., R. R. Baker, and V. G. F. Smith. 1972. The origin and evolution of gamete dimorphism and the male-female phenomenon. *Journal of Theoretical Biology* 36: 529–53.

Parker, G. A., and J. Maynard Smith. 1990. Optimality theory in evolutionary biology. *Nature* 348:27–33.

Parvinen, K., U. Dieckmann, M. Gyllenberg, and J. A. J. Metz. 2003. Evolution of dispersal in metapopulations with local density dependence and demographic stochasticity. *Journal of Evolutionary Biology* 16:143–53.

Paterson, S., T. Vogwill, A. Buckling, R. Benmayor, A. J. Spiers, N. R. Thomson, M. Quail, F. Smith, D. Walker, B. Libberton, A. Fenton, N. Hall, and M. A. Brockhurst. 2010. Antagonistic coevolution accelerates molecular evolution. *Nature* 464:275–78.

Patterson, C. 1988. Homology in classical and molecular biology. *Molecular Biology and Evolution* 2:603–25.

Patzkowsky, M. E. 2016. A hierarchical branching model of evolutionary radiations. *Paleobiology* 21:440–60.

Pavlicev, M., and G. P. Wagner. 2012. A model of developmental evolution: selection, pleiotropy and compensation. *Trends in Ecology and Evolution* 27:316–22.

Pavlicev, M., and S. Widder. 2015. Wiring for independence: positive feedback motifs facilitate individuation of traits in development and evolution. *Journal of Experimental Zoology Part B: Molecular and Developmental Evolution* 324:104–13.

Pearse, W. D., A. Purvis, J. Cavender-Bares, and M. R. Helmus. 2014. Metrics and models of community phylogenetics. Pp. 451–64 in L. Z. Garamszegi, ed., *Modern Phylogenetic Comparative Methods and Their Application in Evolutionary Biology*. Springer, Heidelberg.

Pearson, K. 1886. Notes on regression and inheritance in the case of two parents. *Proceedings of the Royal Society of London* 58:240–42.

Pearson, K. 1904. Mathematical contributions to the theory of evolution XII: on a generalised theory of alternative inheritance, with special reference to Mendel's laws. *Proceedings of the Royal Society of London* 203:53–86.

Pease, J. B., D. C. Haak, M. W. Hahn, and L. C. Moyle. 2016. Phylogenomics reveals three sources of adaptive variation during a rapid radiation. *PLoS Biology* 14:e1002379.

Peckham, M. 1959. Darwinism and Darwinisticism. *Victorian Studies* 3:19–40.

Peirson, B. R. E. 2015. Plasticity, stability, and yield: the origins of Anthony David Bradshaw's model of adaptive phenotypic plasticity. *Studies in History and Philosophy of Science Part C: Studies in History and Philosophy of Biological and Biomedical Sciences* 50:51–66.

Pennell, M. W., R. G. FitzJohn, W. K. Cornwell, and L. J. Harmon. 2015. Model adequacy and the macroevolution of angiosperm functional traits. *American Naturalist* 186: E33–E50.

Pennell, M. W., and L. J. Harmon. 2013. An integrative view of phylogenetic comparative methods: connections to population genetics, community ecology, and paleobiology. *Annals of the New York Academy of Sciences* 1289:90–105.

Pennisi, E. 1999. Is it time to uproot the Tree of Life? *Science* 284:1305–7.

Penny, D., and A. Poole. 1999. The nature of the last universal common ancestor. *Current Opinion in Genetics and Development* 9:672–77.

Peter, I., and E. H. Davidson. 2015. *Genomic Control Process: Development and Evolution*. Academic Press, San Diego, CA.

Phadnis, N., and H. A. Orr. 2009. A single gene causes both male sterility and segregation distortion in *Drosophila* hybrids. *Science* 323:376–79.

Philippi, T., and J. Seger. 1989. Hedging one's evolutionary bets, revisited. *Trends in Ecology and Evolution* 4:41–44.

Phillimore, A. B. 2010. Subspecies origination and extinction in birds. *Ornithological Monographs* 67:42–53.

Phillips, P. C. 2005. Testing hypotheses regarding the genetics of adaptation. *Genetica* 123:15–24.

Phillips, P. C. 2008. Epistasis—the essential role of gene interactions in the structure and evolution of genetic systems. *Nature Reviews Genetics* 9:855–67.

Phillips, P. C., and N. A. Johnson. 1998. The population genetics of synthetic lethals. *Genetics* 150:449–58.

Pianka, E. R. 1966. Latitudinal gradients in species diversity: a review of concepts. *American Naturalist* 100:33–46.

Pianka, E. R. 1970. On r- and K-selection. *American Naturalist* 104:592–97.

Pickett, S. T. A., J. Kolasa, and C. G. Jones. 2007. *Ecological Understanding: The Nature of Theory and the Theory of Nature*. 2nd ed. Elsevier, New York.

Pickrell, J. K., and D. Reich. 2014. Toward a new history and geography of human genes informed by ancient DNA. *Trends in Genetics* 30:377–89.

Pigliucci, M. 2009. An extended synthesis for evolutionary biology. *Annals of the New York Academy of Sciences* 1168:218–28.

Pigliucci, M., and G. B. Müller. 2010a. Elements of an extended evolutionary synthesis. Pp. 3–18 in G. B. Müller and M. Pigliucci, eds., *Evolution: The Extended Synthesis*. MIT Press, Cambridge, MA.

Pigliucci, M., and G. B. Müller. 2010b. *Evolution: The Extended Synthesis*. MIT Press, Cambridge, MA.

Platt, J. R. 1964. Strong inference. *Science* 146:347–53.

Poisot, T., J. D. Bever, A. Nemri, P. H. Thrall, and M. E. Hochberg. 2011. A conceptual framework for the evolution of ecological specialisation. *Ecology Letters* 14:841–51.

Poisot, T., E. Canard, N. Mouquet, and M. E. Hochberg. 2012. A comparative study of ecological specialization estimators. *Methods in Ecology and Evolution* 3:537–44.

Poisot, T., S. Kéfi, S. Morand, M. Stanko, P. A. Marquet, and M. E. Hochberg. 2015. A continuum of specialists and generalists in empirical communities. *PLoS ONE* 10:e0114674.

Poisot, T., M. Lounnas, and M. E. Hochberg. 2013a. The structure of natural microbial enemy-victim networks. *Ecological Processes* 2:1–9.

Poisot, T., M. Stanko, D. Miklisová, and S. Morand. 2013b. Facultative and obligate parasite communities exhibit different network properties. *Parasitology* 140:1340–45.

Poisot, T., O. Verneau, and Y. Desdevises. 2011. Morphological and molecular evolution are not linked in *Lamellodiscus* (Plathyhelminthes, Monogenea). *PLoS ONE* 6:e26252.

Polechová, J., and N. H. Barton. 2005. Speciation through competition: a critical review. *Evolution* 59:1194–1210.

Polly, P. D. 2004. On the simulation of the evolution of morphological shape: multivariate shape under selection and drift. *Palaeontologia Electronica* 7:1–28.

Pontarp, M., L. Bunnefeld, J. S. Cabral, R. S. Etienne, S. A. Fritz, R. Gillespie, C. H. Graham, O. Hagen, F. Hartig, S. Huang, R. Jansson, O. Maliet, T. Münkemüller, L. Pellissier, T. F. Rangel, D. Storch, T. Wiegand, and A. H. Hurlbert. 2019. The latitudinal diversity gradient: novel understanding through mechanistic eco-evolutionary models. *Trends in Ecology and Evolution* 34:211–23.

Popper, K. R. 1959. *The Logic of Scientific Discovery*. Hutchinson, London.

Popper, K. R. 1963. *Conjectures and Refutations: The Growth of Scientific Knowledge*. Routledge and Kegan Paul, London.

Posada, D. 2012. *Selection of Phylogenetic Models of Molecular Evolution*. Encyclopedia of Life Sciences. John Wiley and Sons.

Pottin, K., C. Hyacinthe, and S. Rétaux. 2010. Conservation, development, and function of a cement gland-like structure in the fish *Astyanax mexicanus*. *Proceedings of the National Academy of Sciences* 107:17256–61.

Presgraves, D. C., L. Balagopalan, S. M. Abmayr, and H. A. Orr. 2003. Adaptive evolution drives divergence of a hybrid incompatibility gene between two species of *Drosophila*. *Nature* 423:715–19.

Presgraves, D. C., and H. A. Orr. 1998. Haldane's rule in taxa lacking a hemizygous X. *Science* 282:952–54.

Price, G. R. 1970. Selection and covariance. *Nature* 227:520–21.

Price, G. R. 1972a. Extension of covariance selection mathematics. *Annals of Human Genetics* 35:485–90.

Price, G. R. 1972b. Fisher's "fundamental theorem" made clear. *Annals of Human Genetics* 36:129–40.

Price, P. W. 1980. *Evolutionary Biology of Parasites*. Princeton University Press, Princeton, NJ.

Price, P. W. 2003. *Macroevolutionary Theory on Macroecological Patterns*. Cambridge University Press, Cambridge, UK.

Price, T. 1998. Sexual selection and natural selection in bird speciation. *Philosophical Transactions of the Royal Society of London B: Biological Sciences* 353:251–60.

Price, T. 2008. *Speciation in Birds*. Roberts, Greenwood Village, CO.

Price, T., and D. Schluter. 1991. On the low heritability of life-history traits. *Evolution* 45:853–61.

Promislow, D. E. L., and P. H. Harvey. 1990. Living fast and dying young: a comparative analysis of life-history variation among mammals. *Journal of Zoology* 220:417–37.

Provine, W. B. 1971. *The Origins of Theoretical Population Genetics*. University of Chicago Press, Chicago, IL.

Provine, W. B. 1981. Origins of the genetics of natural populations series. Pp. 5–83 in R. C. Lewontin, A. J. Moore, W. B. Provine, and B. Wallace, eds., *Dobzhansky's Genetics of Natural Populations* I–XLIII. Columbia University Press, New York.

Provine, W. B. 1986. *Sewall Wright and Evolutionary Biology*. University of Chicago Press, Chicago, IL.

Provine, W. B. 1988. Progress in evolution and meaning in life. Pp. 49–74 in M. H. Nitecki, ed., *Evolutionary Progress*. University of Chicago Press, Chicago, IL.

Pruitt, J. N., and C. J. Goodnight. 2014. Site-specific group selection drives locally adapted group compositions. *Nature* 514:359–62.

Pyron, R. A., and F. T. Burbrink. 2013. Phylogenetic estimates of speciation and extinction rates for testing ecological and evolutionary hypotheses. *Trends in Ecology and Evolution* 28:729–36.

Quammen, D. 1996. *The Song of the Dodo: Island Biogeography in an Age of Extinctions*. Random House, London.

Quek, S.-P., S. J. Davies, P. S. Ashton, T. Itino, and N. E. Pierce. 2007. The geography of diversification in mutualistic ants: a gene's-eye view into the Neogene history of Sundaland rain forests. *Molecular Ecology* 16:2045–62.

Queller, D. C. 1984. Pollen-ovule ratios and hermaphrodite sexual allocation strategies. *Evolution* 38:1148–51.

Queller, D. C. 1992. A general model for kin selection. *Evolution* 46:376–80.

Quental, T. B., and C. R. Marshall. 2010. Diversity dynamics: molecular phylogenies need the fossil record. *Trends in Ecology and Evolution* 25:434–41.

Quine, W. V. O. 1951. Two dogmas of empiricism. *Philosophical Review* 60:20–46.

Rabosky, D. L. 2010. Extinction rates should not be estimated from molecular phylogenies. *Evolution* 64:1816–24.

Rabosky, D. L. 2014. Automatic detection of key innovations, rate shifts, and diversity-dependence on phylogenetic trees. *PLoS ONE* 9:e89543.

Rabosky, D. L., and E. E. Goldberg. 2015. Model inadequacy and mistaken inferences of trait-dependent speciation. *Systematic Biology* 64:340–55.

Rabosky, D. L., A. H. Hurlbert, A. President, D. Moderator, and T. Price. 2015. Species richness at continental scales is dominated by ecological limits. *American Naturalist* 185:572–83.

Rabosky, D. L., F. Santini, J. Eastman, S. A. Smith, B. Sidlauskas, J. Chang, and M. E. Alfaro. 2013. Rates of speciation and morphological evolution are correlated across the largest vertebrate radiation. *Nature Communications* 4:1958.

Raff, R. A. 1996. *The Shape of Life: Genes, Development, and the Evolution of Animal Form.* University of Chicago Press, Chicago, IL.

Raff, R. A. 2000. Evo-devo: the evolution of a new discipline. *Nature Reviews Genetics* 1:74–79.

Raff, R. A. 2007. Written in stone: fossils, genes, and evo-devo. *Nature Reviews Genetics* 8:911–20.

Raj, A., M. Stephens, and J. K. Pritchard. 2014. fastSTRUCTURE: variational inference of population structure in large SNP data sets. *Genetics* 197:573–89.

Rajon, E., and J. B. Plotkin. 2013. The evolution of genetic architectures underlying quantitative traits. *Proceedings of the Royal Society of London B: Biological Sciences* 280:20131552.

Ramsey, J., H. D. Bradshaw, and D. W. Schemske. 2003. Components of reproductive isolation between the monkeyflowers *Mimulus lewisii* and *M. cardinalis* (Phrymaceae). *Evolution* 57:1520–34.

Ramsey, J., and D. W. Schemske. 2002. Neopolyploidy in flowering plants. *Annual Review of Ecology and Systematics* 33:589–639.

Randerson, J. P., and L. D. Hurst. 2001. The uncertain evolution of the sexes. *Trends in Ecology and Evolution* 16:571–79.

Rankin, B. D., J. W. Fox, C. R. Barrón-Ortiz, A. E. Chew, P. A. Holroyd, J. A. Ludtke, X. Yang, and J. M. Theodor. 2015. The extended Price equation quantifies species selection on mammalian body size across the Palaeocene/Eocene Thermal Maximum. *Proceedings of the Royal Society of London B: Biological Sciences* 282:20151097.

Rao, V., and V. Nanjundiah. 2011. J. B. S. Haldane, Ernst Mayr and the beanbag genetics dispute. *Journal of the History of Biology* 44:233–81.

Raup, D. M. 1985. Mathematical models of cladogenesis. *Paleobiology* 11:42–52.

Ravigné, V., U. Dieckmann, and I. Olivieri. 2009. Live where you thrive: joint evolution of habitat choice and local adaptation facilitates specialization and promotes diversity. *American Naturalist* 174:E141–E169.

Rebeiz, M., N. H. Patel, and V. F. Hinman. 2015. Unraveling the tangled skein: the evolution of transcriptional regulatory networks in development. *Annual Review of Genomics and Human Genetics* 16:12.11–12.29.

Redfield, R. J. 1993. Evolution of natural transformation: testing the DNA repair hypothesis in *Bacillus subtilis* and *Haemophilus influenzae*. *Genetics* 133:755–61.

Redfield, R. J. 2001. Do bacteria have sex? *Nature Reviews Genetics* 2:634–39.

Redish, A. D., E. Kummerfeld, R. L. Morris, and A. C. Love. 2018. Opinion: reproducibility failures are essential to scientific inquiry. *Proceedings of the National Academy of Sciences* 115:5042–46.

Reeck, G. R., C. de Haën, D. C. Teller, R. F. Doolittle, W. M. Fitch, R. E. Dickerson, P. Chambon, A. D. McLachlan, E. Margoliash, and T. H. Jukes. 1987. "Homology" in proteins and nucleic acids: a terminology muddle and a way out of it. *Cell* 50:667.

Reed, T. E., R. S. Waples, D. E. Schindler, J. J. Hard, and M. T. Kinnison. 2010. Phenotypic plasticity and population viability: the importance of environmental predictability. *Proceedings of the Royal Society of London B: Biological Sciences* 277:3391–3400.

Reese, A. T., H. C. Lanier, and E. J. Sargis. 2013. Skeletal indicators of ecological specialization in pika (Mammalia, Ochotonidae). *Journal of Morphology* 274:585–602.

Relyea, R. A., P. R. Stephens, L. N. Barrow, A. R. Blaustein, P. W. Bradley, J. C. Buck, A. Chang, J. P. Collins, B. Crother, J. Earl, S. S. Gervasi, J. T. Hoverman, O. Hyman, E. M. Lemmon, T. M. Luhring, M. Michelson, C. Murray, S. Price, R. D. Semlitsch, A. Sih, A. B. Stoler, N. VandenBroek, A. Warwick, G. Wengert, and J. I. Hammond. 2018. Phylogenetic patterns of trait and trait plasticity evolution: insights from amphibian embryos. *Evolution* 72:663–78.

Rensch, B. 1947. *Neuere Probleme der Abstammungslehre*. Ferdinand Encke Verlag, Stuttgart.

Reznick, D., M. J. Bryant, and F. Bashey. 2002. r- and K-selection revisited: the role of population regulation in life-history evolution. *Ecology* 83:1509–20.

Rheindt, F. E., and S. V. Edwards. 2011. Genetic introgression: an integral but neglected component of speciation in birds. *Auk* 128:620–32.

Rice, S. H. 1998. The evolution of canalization and the breaking of Von Baer's laws: modeling the evolution of development with epistasis. *Evolution* 52:647–56.

Rice, S. H. 2004. *Evolutionary Theory: Mathematical and Conceptual Foundations*. Sinauer Associates, Sunderland, MA.

Rice, S. H. 2012. The place of development in mathematical evolutionary theory. *Journal of Experimental Zoology Part B: Molecular and Developmental Evolution* 318:480–88.

Ricklefs, R. E. 2006. Time, species, and the generation of trait variance in clades. *Systematic Biology* 55:151–59.

Riddle, B. R., M. N. Dawson, E. A. Hadly, D. J. Hafner, M. J. Hickerson, S. J. Mantooth, and A. D. Yoder. 2008. The role of molecular genetics in sculpting the future of integrative biogeography. *Progress in Physical Geography* 32:173–202.

Ridley, M. 1993. *Evolution*. Blackwell Science, Cambridge, MA.

Rieppel, O. 2007. Species: kinds of individuals or individuals of a kind. *Cladistics* 23:373–84.

Rieppel, O., and M. Kearney. 2002. Similarity. *Biological Journal of the Linnean Society* 75:59–82.

Rieppel, O., and M. Kearney. 2007. The poverty of taxonomic characters. *Biology and Philosophy* 22:95–113.

Rissler, L. J. 2016. Union of phylogeography and landscape genetics. *Proceedings of the National Academy of Sciences* 113:8079–86.

Ritchie, M. G. 2007. Sexual selection and speciation. *Annual Review of Ecology Evolution and Systematics* 38:79–102.

Rius, M., and J. A. Darling. 2014. How important is intraspecific genetic admixture to the success of colonising populations? *Trends in Ecology and Evolution* 29:233–42.

Robertson, A. 1966. A mathematical model of the culling process in dairy cattle. *Animal Production* 8:95–108.

Roff, D. A. 1992. *Evolution of Life Histories: Theory and Analysis*. Sinauer Associates, Sunderland, MA.

Roff, D. A., and D. J. Fairbairn. 2007. The evolution of trade-offs: where are we? *Journal of Evolutionary Biology* 20:433–47.

Rominger, A. J., K. R. Goodman, J. Y. Lim, F. S. Valdovinos, E. Armstrong, G. M. Bennett, M. S. Brewer, D. D. Cotoras, C. P. Ewing, J. Harte, N. Martinez, P. O'Grady, D. Percy, D. Price, G. K. Roderick, K. Shaw, D. S. Gruner, and R. G. Gillespie. 2015. Community assembly on isolated islands: macroecology meets evolution. *Global Ecology and Biogeography* 25:769–80.

Rominger, A. J., T. E. Miller, and S. L. Collins. 2009. Relative contributions of neutral and niche-based processes to the structure of a desert grassland grasshopper community. *Oecologia* 161:791–800.

Rose, M. R., and T. H. Oakley. 2007. The new biology: beyond the Modern Synthesis. *Biology Direct* 2:30.

Rosenberg, E., and I. Zilber-Rosenberg. 2013. *The Hologenome Concept: Human, Animal and Plant Microbiota*. Springer, Berlin.

Rosenberg, N. A., J. K. Pritchard, J. L. Weber, H. M. Cann, K. K. Kidd, L. A. Zhivotovsky, and M. W. Feldman. 2002. Genetic structure of human populations. *Science* 298: 2381–85.

Rosenzweig, M. L. 1995. *Species Diversity in Space and Time*. Cambridge University Press, Cambridge, UK.

Rosindell, J., and S. J. Cornell. 2007. Species-area relationships from a spatially explicit neutral model in an infinite landscape. *Ecology Letters* 10:586–95.

Rosindell, J., and A. B. Phillimore. 2011. A unified model of island biogeography sheds light on the zone of radiation. *Ecology Letters* 14:552–60.

Roth, V. L. 1988. The biological basis of homology. Page 236 in C. J. Humphries, ed., *Ontogeny and Systematics*. Columbia University Press, New York.

Roy, K. 1994. Effects of the Mesozoic Marine Revolution on the taxonomic, morphologic, and biogeographic evolution of a group: aporrhaid gastropods during the Mesozoic. *Paleobiology* 20:274–96.

Roze, D. 2014. Selection for sex in finite populations. *Journal of Evolutionary Biology* 27: 1304–22.

Ruddiman, W. F., M. Raymo, and A. McIntyre. 1986. Matuyama 41,000-year cycles: North Atlantic Ocean and northern hemisphere ice sheets. *Earth and Planetary Science Letters* 80:117–29.

Rudolph, J. L. 2002. *Scientists in the Classroom: The Cold War Reconstruction in American Science Education*. Palgrave, New York.

Rundell, R. J., and T. D. Price. 2009. Adaptive radiation, nonadaptive radiation, ecological speciation and nonecological speciation. *Trends in Ecology and Evolution* 24:394–99.

Rundle, H. D., L. Nagel, J. W. Boughman, and D. Schluter. 2000. Natural selection and parallel speciation in sympatric sticklebacks. *Science* 287:306–8.

Rundle, H. D., and P. Nosil. 2005. Ecological speciation. *Ecology Letters* 8:336–52.

Rutter, M. T., A. Roles, J. K. Conner, R. G. Shaw, F. H. Shaw, K. Schneeberger, S. Ossowski,

D. Weigel, and C. B. Fenster. 2012. Fitness of *Arabidopsis thaliana* mutation accumulation lines whose spontaneous mutations are known. *Evolution* 66:2335–39.

Saenko, S. V., V. French, P. M. Brakefield, and P. Beldade. 2008. Conserved developmental processes and the formation of evolutionary novelties: examples from butterfly wings. *Philosophical Transactions of the Royal Society of London B: Biological Sciences* 363: 1549–56.

Safran, R. J., E. S. C. Scordato, L. B. Symes, R. L. Rodríguez, and T. C. Mendelson. 2013. Contributions of natural and sexual selection to the evolution of premating reproductive isolation: a research agenda. *Trends in Ecology and Evolution* 28:643–50.

Sakai, S., and Y. Harada. 2001. Why do large mothers produce large offspring? Theory and a test. *American Naturalist* 157:348–59.

Salazar-Ciudad, I. 2008. Making evolutionary predictions about the structure of development and morphology: beyond the neo-Darwinian and constraints paradigms. Pp. 31–49 in A. Minelli and G. Fusco, eds., *Evolving Pathways*. Cambridge University Press, Cambridge, UK.

Salazar-Ciudad, I., and J. Jernvall. 2010. A computational model of teeth and the developmental origins of morphological variation. *Nature* 464:583–86.

Salguero-Gómez, R., O. R. Jones, E. Jongejans, S. P. Blomberg, D. J. Hodgson, C. Mbeau-Ache, P. A. Zuidema, H. de Kroon, and Y. M. Buckley. 2016. Fast-slow continuum and reproductive strategies structure plant life-history variation worldwide. *Proceedings of the National Academy of Sciences* 113:230–35.

Salthe, S. N. 2013. *Evolving Hierarchical Systems: Their Structure and Representation*. Columbia University Press, New York.

Salzburger, W., B. Van Bocxlaer, and A. S. Cohen. 2014. Ecology and evolution of the African Great Lakes and their faunas. *Annual Review of Ecology, Evolution, and Systematics* 45:519–45.

Sandage, A., L. Brown, and P. P. Craig. 2004. *Centennial History of the Carnegie Institution of Washington*, vol. 4: *The Department of Plant Biology*. Cambridge University Press, New York.

Sandel, B., L. Arge, B. Dalsgaard, R. G. Davies, K. J. Gaston, W. J. Sutherland, and J-C Svenning. 2011. The influence of Late Quaternary climate-change velocity on species endemism. *Science* 334:660–64.

Sanders, N. J., and C. Rahbek. 2012. The patterns and causes of elevational diversity gradients. *Ecography* 35:1–3.

Sanderson, M. J. 2002. Estimating absolute rates of molecular evolution and divergence times: a penalized likelihood approach. *Molecular Biology and Evolution* 19:101–9.

Sarkar, S. 2004. From the *Reaktionsnorm* to the evolution of adaptive plasticity: a historical sketch, 1909–1999. Pp. 10–30 in T. J. DeWitt and S. M. Scheiner, eds., *Phenotypic Plasticity: Functional and Conceptual Approaches*. Oxford University Press, New York.

Savage, V. M., A. P. Allen, J. H. Brown, J. F. Gillooly, A. B. Herman, W. H. Woodruff, and G. B. West. 2007. Scaling of number, size, and metabolic rate of cells with body size in mammals. *Proceedings of the National Academy of Sciences* 104:4718–23.

Schaefer, I., K. Domes, M. Heethoff, K. Schneider, I. Schön, R. A. Norton, S. Scheu, and M. Maraun. 2006. No evidence for the "Meselson effect" in parthenogenetic oribatid mites (Oribatida, Acari). *Journal of Evolutionary Biology* 19:184–93.

Schaffer, W. M. 1974. Selection for optimal life histories: the effects of age structure. *Ecology* 55:291–303.

Schaffer, W. M. 1981. On reproductive value and fitness. *Ecology* 62:1683–83.

Schaffer, W. M., R. S. Inouye, and T. S. Whittam. 1982. Energy allocation by an annual plant when the effects of seasonality on growth and reproduction are decoupled. *American Naturalist* 120:787–815.

Schaffer, W. M., and M. L. Rosenzweig. 1977. Selection for optimal life histories II: multiple equilibria and the evolution of alternative reproductive strategies. *Ecology* 58: 60–72.

Scheiner, S. M. 1993a. Genetics and evolution of phenotypic plasticity. *Annual Review of Ecology and Systematics* 24:35–68.

Scheiner, S. M. 1993b. Plasticity as a selectable trait: reply to Via. *American Naturalist* 142: 372–74.

Scheiner, S. M. 2009. The evolutionary ecology of development. *Evolution* 63:1666–70.

Scheiner, S. M. 2010. Toward a conceptual framework for biology. *Quarterly Review of Biology* 85:293–318.

Scheiner, S. M. 2013a. The ecological literature, an idea-free distribution. *Ecology Letters* 16:1421–23.

Scheiner, S. M. 2013b. The genetics of phenotypic plasticity XII: temporal and spatial heterogeneity. *Ecology and Evolution* 3:4596–4609.

Scheiner, S. M. 2014. The Baldwin Effect: neglected and misunderstood. *American Naturalist* 184:ii–iii.

Scheiner, S. M., M. Barfield, and R. D. Holt. 2012. The genetics of phenotypic plasticity XI: joint evolution of plasticity and dispersal rate. *Ecology and Evolution* 2:2027–39.

Scheiner, S. M., M. Barfield, and R. D. Holt. 2017. The genetics of phenotypic plasticity XV: genetic assimilation, the Baldwin effect, and evolutionary rescue. *Ecology and Evolution* 7:8788–8803.

Scheiner, S. M., and D. Berrigan. 1998. The genetics of phenotypic plasticity VIII: the cost of plasticity in *Daphnia pulex*. *Evolution* 52:368–78.

Scheiner, S. M., and T. J. DeWitt. 2004. Future research directions. Pp. 201–6 in T. J. DeWitt and S. M. Scheiner, eds., *Phenotypic Plasticity: Functional and Conceptual Approaches*. Oxford University Press, New York.

Scheiner, S. M., R. Gomulkiewicz, and R. D. Holt. 2015. The genetics of phenotypic plasticity XIV: coevolution. *American Naturalist* 185:594–609.

Scheiner, S. M., and R. D. Holt. 2012. The genetics of phenotypic plasticity X: variation versus uncertainty. *Ecology and Evolution* 2:751–67.

Scheiner, S. M., and R. F. Lyman. 1991. The genetics of phenotypic plasticity II: response to selection. *Journal of Evolutionary Biology* 4:23–50.

Scheiner, S. M., R. J. Mitchell, and H. S. Callahan. 2000. Using path analysis to measure natural selection. *Journal of Evolutionary Biology* 13:423–33.

Scheiner, S. M., and M. R. Willig. 2008. A general theory of ecology. *Theoretical Ecology* 1:21–28.

Scheiner, S. M., and M. R. Willig. 2011. A general theory of ecology. Pp. 3–19 in S. M. Scheiner and M. R. Willig, eds., *The Theory of Ecology*. University of Chicago Press, Chicago, IL.

Schemske, D. 2010. Adaptation and the origin of species. *American Naturalist* 176:S4–S25.

Schilthuizen, M., M. C. W. G. Giesbers, and L. W. Beukeboom. 2011. Haldane's rule in the 21st century. *Heredity* 107:95–102.

Schlichting, C. D. 1986. The evolution of phenotypic plasticity in plants. *Annual Review of Ecology and Systematics* 17:667–93.

Schlichting, C. D. 2004. The role of phenotypic plasticity in diversification. Pp. 191–200 in T. J. DeWitt and S. M. Scheiner, eds., *Phenotypic Plasticity: Functional and Conceptual Approaches*. Oxford University Press, New York.

Schlichting, C. D., and M. Pigliucci. 1993. Control of phenotypic plasticity via regulatory genes. *American Naturalist* 142:366–70.

Schluter, D. 1996. Adaptive radiation along genetic lines of least resistance. *Evolution* 50:1766–74.

Schluter, D. 2000. *The Ecology of Adaptive Radiation*. Oxford University Press, New York.

Schmalhausen, I. I. 1949. *Factors of Evolution*. Blakiston, Philadelphia, PA.

Schnable, P. S., and R. P. Wise. 1998. The molecular basis of cytoplasmic male sterility and fertility restoration. *Trends in Plant Science* 3:175–89.

Schön, I., and K. Martens. 2003. No slave to sex. *Proceedings of the Royal Society of London B: Biological Sciences* 270:827–33.

Schwenk, K., and G. P. Wagner. 2004. The relativism of constraints on phenotypic evolution. Pp. 390–408 in M. Pigliucci and K. Preston, eds., *Phenotypic Integration*. Oxford University Press, Oxford, UK.

Secord, J. A. 2000. *Victorian Sensation: The Extraordinary Publication, Reception, and Secret Authorship of Vestiges of the Natural History of Creation*. University of Chicago Press, Chicago, IL.

Seehausen, O., R. K. Butlin, I. Keller, C. E. Wagner, J. W. Boughman, P. A. Hohenlohe, C. L. Peichel, G.-P. Saetre, C. Bank, A. Brannstrom, A. Brelsford, C. S. Clarkson, F. Eroukhmanoff, J. L. Feder, M. C. Fischer, A. D. Foote, P. Franchini, C. D. Jiggins, F. C. Jones, A. K. Lindholm, K. Lucek, M. E. Maan, D. A. Marques, S. H. Martin, B. Matthews, J. I. Meier, M. Most, M. W. Nachman, E. Nonaka, D. J. Rennison, J. Schwarzer, E. T. Watson, A. M. Westram, and A. Widmer. 2014. Genomics and the origin of species. *Nature Reviews Genetics* 15:176–92.

Seehausen, O., P. J. Mayhew, and J. J. M. van Alphen. 1999. Evolution of colour patterns in East African cichlid fish. *Journal of Evolutionary Biology* 12:514–34.

Seehausen, O., and J. J. M. van Alphen. 1999. Can sympatric speciation by disruptive sexual selection explain rapid evolution of cichlid diversity in Lake Victoria? *Ecology Letters* 2:262–71.

Sepkoski, D. 2012. *Rereading the Fossil Record: The Growth of Paleobiology as an Evolutionary Discipline*. University of Chicago Press, Chicago, IL.

Sepkoski, D., and M. Ruse, eds. 2009. *The Paleobiological Revolution: Essays on the Growth of Modern Paleontology*. University of Chicago Press, Chicago, IL.

Sepkoski, J. J., Jr. 1996. Competition in macroevolution: the double wedge revisited. Pp. 211–55 in D. Jablonski, D. H. Erwin, and J. H. Lipps, eds., *Evolutionary Paleobiology*. University of Chicago Press, Chicago, IL.

Servedio, M. R., Y. Brandvain, S. Dhole, C. L. Fitzpatrick, E. E. Goldberg, C. A. Stern, J. Van Cleve, and D. J. Yeh. 2014. Not just a theory—the utility of mathematical models in evolutionary biology. *PLoS Biology* 12:e1002017.

Servedio, M. R., and R. Bürger. 2015. The effects of sexual selection on trait divergence in a peripheral population with gene flow. *Evolution* 69:2648–61.

Servedio, M. R., and M. Kirkpatrick. 1997. The effects of gene flow on reinforcement. *Evolution* 51:1764–72.

Servedio, M. R., G. S. Van Doorn, M. Kopp, A. M. Frame, and P. Nosil. 2011. Magic traits in speciation: "magic" but not rare? *Trends in Ecology and Evolution* 26:389–97.

Sexton, J. P., S. B. Hangartner, and A. A. Hoffmann. 2014. Genetic isolation by environment or distance: which pattern of gene flow is most common? *Evolution* 68:1–15.

Shapira, M. 2016. Gut microbiotas and host evolution: scaling up symbiosis. *Trends in Ecology and Evolution* 31:539–49.

Shaw, K. L., and S. P. Mullen. 2011. Genes versus phenotypes in the study of speciation. *Genetica* 139:649–61.

Shaw, R. G., and C. J. Geyer. 2010. Inferring fitness landscapes. *Evolution* 64:2510–20.

Shermer, M., and F. J. Sulloway. 2000. The grand old man of evolution: an interview with evolutionary biologist Ernst Mayr. *Skeptic* 8:76–82.

Shirai, L. T., S. V. Saenko, R. A. Keller, M. A. Jerónimo, P. M. Brakefield, H. Descimon, N. Wahlberg, and P. Beldade. 2012. Evolutionary history of the recruitment of conserved developmental genes in association to the formation and diversification of a novel trait. *BMC Evolutionary Biology* 21:21.

Shubin, N. H., C. Tabin, and S. Carroll. 2009. Deep homology and the origins of evolutionary novelty. *Nature* 457:818–23.

Shubin, N. H., and D. Wake. 1996. Phytogeny, variation, and morphological integration. *American Zoologist* 36:51–60.

Sibly, R., P. Calow, and N. Nichols. 1985. Are patterns of growth adaptive? *Journal of Theoretical Biology* 112:553–74.

Siddall, M. E., and A. G. Kluge. 1997. Probabilism and phylogenetic inference. *Cladistics* 13:313–36.

Sidlauskas, B. 2008. Continuous and arrested morphological diversification in sister clades of characiform fishes: a phylomorphospace approach. *Evolution* 62:3135–56.

Simberloff, D. S. 1974. Equilibrium theory of island biogeography and ecology. *Annual Review of Ecology and Systematics* 5:161–82.

Simberloff, D. S., and E. O. Wilson. 1969. Experimental zoogeography of islands: the colonization of empty islands. *Ecology* 50:278–96.

Šimková, A., O. Verneau, M. Gelnar, and S. Morand. 2006. Specificity and specialization of congeneric monogeneans parasitizing cyprinid fish. *Evolution* 60:1023–37.

Simons, A. M., and M. O. Johnston. 1997. Developmental instability as a bet-hedging strategy. *Oikos* 80:401–6.

Simpson, C. 2010. Species selection and driven mechanisms jointly generate a large-scale morphological trend in monobathrid crinoids. *Paleobiology* 36:481–96.

Simpson, C. 2013. Species selection and the macroevolution of coral coloniality and photosymbiosis. *Evolution* 67:1607–21.

Simpson, C., and J. Müller. 2012. Species selection in the molecular age. Pp. 116–34 in J. Müller and R. Asher, eds., *From Clone to Bone: The Synergy of Morphological and Molecular Tools in Paleobiology*. Cambridge University Press, New York.

Simpson, G. G. 1944a. "Introductory remarks." Bulletin, no. 4, 14 November, William B. Provine Collection, Cornell University Archives.

Simpson, G. G. 1944b. *Tempo and Mode in Evolution*. Columbia University Press, New York.

Simpson, G. G. 1951. The species concept. *Evolution* 5:285–98.

Simpson, G. G. 1953a. The Baldwin Effect. *Evolution* 7:110–17.

Simpson, G. G. 1953b. *The Major Features of Evolution*. Columbia University Press, New York.

Simpson, G. G. 1961. *Principles of Animal Taxonomy*. Columbia University Press, New York.

Sinervo, B., K. Zamudio, P. Doughty, and R. B. Huey. 1992. Allometric engineering: a causal analysis of natural selection on offspring size. *Science* 258:1927–30.

Slater, G. J. 2013. Phylogenetic evidence for a shift in the mode of mammalian body size evolution at the Cretaceous-Paleogene boundary. *Methods in Ecology and Evolution* 4:734–44.

Slater, G. J. 2015. Iterative adaptive radiations of fossil canids show no evidence for diversity-dependent trait evolution. *Proceedings of the National Academy of Sciences* 112:4897–902.

Slater, G. J., and L. J. Harmon. 2013. Unifying fossils and phylogenies for comparative analyses of diversification and trait evolution. *Methods in Ecology and Evolution* 4:699–702.

Slater, G. J., L. J. Harmon, and M. E. Alfaro. 2012. Integrating fossils with molecular phylogenies improves inference of trait evolution. *Evolution* 66:3931–44.

Slater, G. J., and M. W. Pennell. 2014. Robust regression and posterior predictive simulation increase power to detect early bursts of trait evolution. *Systematic Biology* 63:293–308.

Slater, M. H. 2013. *Are Species Real?* Palgrave Macmillan, London.

Slater, M. H. 2015. Natural kindness. *British Journal for the Philosophy of Science* 66:375–411.

Slatkin, M. 1974. Hedging one's evolutionary bets. *Nature* 250:704–5.

Slatkin, M. 1981. A diffusion model of species selection. *Paleobiology* 7:421–25.

Slatyer, R. A., M. Hirst, and J. P. Sexton. 2013. Niche breadth predicts geographical range size: a general ecological pattern. *Ecology Letters* 16:1104–14.

Slotten, R. A. 2004. *The Heretic in Darwin's Court: The Life of Alfred Russel Wallace.* Columbia University Press, New York.

Smith, C. C., and S. D. Fretwell. 1974. The optimal balance between size and number of offspring. *American Naturalist* 108:499–506.

Smith, J. A., L. B. Chenoweth, S. M. Tierney, and M. P. Schwarz. 2013. Repeated origins of social parasitism in allodapine bees indicate that the weak form of Emery's rule is widespread, yet sympatric speciation remains highly problematic. *Biological Journal of the Linnean Society* 109:320–31.

Smith, S. A., and B. C. O'Meara. 2012. treePL: divergence time estimation using penalized likelihood for large phylogenies. *Bioinformatics* 28:2689–90.

Smocovitis, V. B. 1992. Unifying biology: the evolutionary synthesis and evolutionary biology. *Journal of the History of Biology* 25:1–65.

Smocovitis, V. B. 1994. Organizing evolution: founding the society for the study of evolution (1939–1950). *Journal of the History of Biology* 27:241–309.

Smocovitis, V. B. 1996. *Unifying Biology: The Evolutionary Synthesis and Evolutionary Biology.* Princeton University Press, Princeton, NJ.

Smocovitis, V. B. 1999. The 1959 Darwin Centennial celebration in America. *Osiris* 14:274–323.

Smocovitis, V. B. 2005. "It ain't over 'til it's over": rethinking the Darwinian revolution. *Journal of the History of Biology* 38:33–49.

Smocovitis, V. B. 2009. Darwin's botany in the *Origin of Species*. Pp. 216–36 in M. Ruse and R. J. Richards, eds., *The Cambridge Companion to the "Origin of Species."* Cambridge University Press, Cambridge, UK.

Smocovitis, V. B. 2012. Humanizing evolution: anthropology, the evolutionary synthesis, and the prehistory of biological anthropology, 1927–1962. *Current Anthropology* 53:S108–S125.

Smolin, L. 2006. *The Trouble with Physics: The Rise of String Theory, the Fall of a Science, and What Comes Next.* Houghton Mifflin Harcourt, New York.

Smorkatcheva, A. V., and V. A. Lukhtanov. 2014. Evolutionary association between subterranean lifestyle and female sociality in rodents. *Mammalian Biology—Zeitschrift für Säugetierkunde* 79:101–9.

Sneath, P. H. A., and R. R. Sokal. 1973. *Numerical Taxonomy: The Principles and Practice of Numerical Classification.* Freeman, San Francisco, CA.

Snell-Rood, E. C. 2012. Selective processes in development: implications for the costs and benefits of phenotypic plasticity. *Integrative and Comparative Biology* 52:31–42.

Snell-Rood, E. C., J. D. Van Dyken, T. Cruickshank, M. J. Wade, and A. P. Moczek. 2010. Toward a population genetic framework of developmental evolution: the costs, limits, and consequences of phenotypic plasticity. *Bioessays* 32:71–81.

Sobel, J. M., and G. F. Chen. 2014. Unification of methods for estimating the strength of reproductive isolation. *Evolution* 68:1511–22.

Sober, E. 1984. *The Nature of Selection*. University of Chicago Press, Chicago, IL.

Sober, E. 1994. *Conceptual Issues in Evolutionary Biology*. MIT Press, Cambridge, MA.

Sober, E. 2008. *Evidence and Evolution: The Logic behind the Science*. Cambridge University Press, Cambridge, UK.

Sober, E. 2011. *Did Darwin Write the Origin Backwards? Philosophical Essays on Darwin's Theory*. Prometheus Books, New York.

Sober, E., and M. Steel. 2011. Entropy increase and information loss in Markov models of evolution. *Biology and Philosophy* 26:223–50.

Sober, E., and M. Steel. 2014. Time and knowability in evolutionary processes. *Philosophy of Science* 81:558–79.

Soltis, P. S., and D. E. Soltis. 2009. The role of hybridization in plant speciation. *Annual Review of Plant Biology* 60:561–88.

Song, N., J. M. Joseph, G. B. Davis, and D. Durand. 2008. Sequence similarity network reveals common ancestry of multidomain proteins. *PLoS Computational Biology* 4:e1000063.

Sonnhammer, E. L., and E. V. Koonin. 2002. Orthology, paralogy and proposed classification for paralog subtypes. *Trends in Genetics* 18:619–20.

Soul, L. C., and M. Friedman. 2015. Taxonomy and phylogeny can yield comparable results in comparative paleontological analyses. *Systematic Biology* 64:608–20.

Soule, T., and J. A. Foster. 1998. Effects of code growth and parsimony pressure on populations in genetic programming. *Evolutionary Computation* 6:293–309.

Soule, T., and R. B. Heckendorn. 2013. A practical platform for on-line genetic programming for robotics. Pp. 15–29 in R. Riolo, E. Vladislavleva, M. Ritchie, and J. H. Moore, eds., *Genetic Programming Theory and Practice X*. Springer, Ann Arbor, MI.

Spemann, H. 1915. Zur Geschichte und Kritik des Begriffs der Homologie. Pp. 163–85 in C. Chun and W. Johannsen, eds., *Allgemeine Biologie*. B. G. Teubner, Leipzig.

Spencer, H. G., B. H. McArdle, and D. M. Lambert. 1986. A theoretical investigation of speciation by reinforcement. *American Naturalist* 128:241–62.

Stamatakis, A. 2014. RAxML version 8: a tool for phylogenetic analysis and post-analysis of large phylogenies. *Bioinformatics* 30:1312–13.

Stanley, S. M. 1973. An explanation for Cope's Rule. *Evolution* 27:1–26.

Stanley, S. M. 1979. *Macroevolution*. W. H. Freeman, San Francisco, CA.

Stanley, S. M. 1990. The general correlation between rate of speciation and rate of extinction: fortuitous causal linkages. Pp. 103–27 in R. M. Ross and W. D. Allmon, eds., *Causes of Evolution: A Paleontological Perspective*. University of Chicago Press, Chicago, IL.

Stanley, S. M., P. W. Signor, S. Lidgard, and A. F. Karr. 1981. Natural clades differ from "random" clades: simulations and analyses. *Paleobiology* 7:115–27.

Stanley, S. O., and R. Y. Morita. 1968. Salinity effect on the maximal growth temperature of some bacteria isolated from marine environments. *Journal of Bacteriology* 95:169–73.

Stanton, M. L., and C. Galen. 1997. Life on the edge: adaptation versus environmentally mediated gene flow in the snow buttercup, *Ranunculus adoneus. American Naturalist* 150:143–78.

Starrfelt, J., and H. Kokko. 2012. Bet-hedging: a triple trade-off between means, variances and correlations. *Biological Reviews* 87:742–55.

Stearns, S. C. 1977. The evolution of life history traits: a critique of the theory and a review of the data. *Annual Review of Ecology and Systematics* 8:145–71.

Stearns, S. C. 1992. *The Evolution of Life Histories.* Oxford University Press, Oxford, UK.

Stebbins, G. L. 1950. *Variation and Evolution in Plants.* Columbia University Press, New York.

Stebbins, G. L. 1965. From gene to character in higher plants. *American Scientist* 53: 104–26.

Steele, E. J. 1981. *Somatic Selection and Adaptive Evolution: On the Inheritance of Acquired Characters.* 2nd ed. University of Chicago Press, Chicago, IL.

Sterelny, K. 2005. Another view of life [review of Conway Morris 2003]. *Studies in History and Philosophy of Science Part C: Studies in History and Philosophy of Biological and Biomedical Sciences* 36:585–93.

Sterelny, K. 2011. Evolvability reconsidered. Pp. 83–100 in B. Calcott and K. Sterelny, eds., *The Major Transitions in Evolution Revisited.* MIT Press, Cambridge, MA.

Stevens, G. C. 1992. The elevational gradient in altitudinal range: an extension of Rapoport's latitudinal rule to altitude. *American Naturalist* 140:893–911.

Stevens, L., C. J. Goodnight, and S. Kalisz. 1995. Multilevel selection in natural populations of *Impatiens capensis. American Naturalist* 145:513–26.

Storch, D., P. Keil, and W. Jetz. 2012. Universal species-area and endemics-area relationships at continental scales. *Nature* 488:78–81.

Stukenbrock, E. H. 2013. Evolution, selection and isolation: a genomic view of speciation in fungal plant pathogens. *New Phytologist* 199:895–907.

Sulloway, F. J. 1982. Darwin and his finches: the evolution of a legend. *Journal of the History of Biology* 15:1–53.

Sultan, S. E. 2007. Development in context: the timely emergence of eco-devo. *Trends in Ecology and Evolution* 22:575–82.

Sultan, S. E., and H. G. Spencer. 2002. Metapopulation structure favors plasticity over local adaptation. *American Naturalist* 160:271–83.

Svensson, E. I. 2018. On reciprocal causation in the evolutionary process. *Biology and Philosophy* 45:1–14.

Sweigart, A. L., and L. E. Flagel. 2015. Evidence of natural selection acting on a polymorphic hybrid incompatibility locus in *Mimulus. Genetics* 199:543–54.

Szathmáry, E. 1993. Do deleterious mutations act synergistically? Metabolic control theory provides a partial answer. *Genetics* 133:127–32.

Szathmáry, E. 2015. Toward major evolutionary transitions theory 2.0. *Proceedings of the National Academy of Sciences* 112:10104–11.

Tajima, F. 1989. Statistical method for testing the neutral mutation hypothesis by DNA polymorphism. *Genetics* 123:585–95.

Tamura, K., D. Peterson, N. Peterson, G. Stecher, M. Nei, and S. Kumar. 2011. MEGA5: molecular evolutionary genetics analysis using maximum likelihood, evolutionary distance, and maximum parsimony methods. *Molecular Biology and Evolution* 28:2731–39.

Tang, S., and D. C. Presgraves. 2009. Evolution of the *Drosophila* nuclear pore complex results in multiple hybrid incompatibilities. *Science* 323:779–82.

Tang, S., and D. C. Presgraves. 2015. Lineage-specific evolution of the complex *Nup160* hybrid incompatibility between *Drosophila melanogaster* and its sister species. *Genetics* 200:1245–54.

Tank, D. C., J. M. Eastman, M. W. Pennell, P. S. Soltis, D. E. Soltis, C. E. Hinchliff, J. W. Brown, E. B. Sessa, and L. J. Harmon. 2015. Nested radiations and the pulse of angiosperm diversification: increased diversification rates often follow whole genome duplications. *New Phytologist* 207:454–67.

Tattersall, I. 2000. Paleoanthropology: the last half-century. *Evolutionary Anthropology* 9:2–16.

Tatusov, R. L., M. Y. Galperin, D. A. Natale, and E. V. Koonin. 2000. The COG database: a tool for genome-scale analysis of protein functions and evolution. *Nucleic Acids Research* 28:33–36.

Tautz, D. 1998. Debatable homologies. *Nature* 395:17–19.

Tax, S., ed. 1960a. *Evolution after Darwin*, vol. 1: *The Evolution of Life, Its Origin, History, and Future*. University of Chicago Press, Chicago. IL.

Tax, S., ed. 1960b. *Evolution after Darwin*, vol. 2: *The Evolution of Man*. University of Chicago Press, Chicago, IL.

Tax, S., and C. Callender, eds., 1960. *Evolution after Darwin*, vol. 3: *Issues in Evolution*. University of Chicago Press, Chicago, IL.

Taylor, H. M., R. S. Gourley, C. E. Lawrence, and R. S. Kaplan. 1974. Natural selection of life history attributes: an analytical approach. *Theoretical Population Biology* 122: 104–22.

Telford, M. J., and G. E. Budd. 2003. The place of phylogeny and cladistics in evo-devo research. *International Journal of Developmental Biology* 47:479–90.

Tëmkin, I., and N. Eldredge. 2015. Networks and hierarchies: approaching complexity in evolutionary theory. Pp. 183–226 in E. Serrelli and N. Gontier, eds., *Macroevolution: Explanation, Interpretation and Evidence*. Springer International, Cham, Switzerland.

Tenaillon, O. 2014. The utility of Fisher's geometric model in evolutionary genetics. *Annual Review of Ecology, Evolution, and Systematics* 45:179–201.

Teotónio, H., S. Estes, P. C. Phillips, and C. F. Baer. 2017. Experimental evolution with *Caenorhabditis* nematodes. *Genetics* 206:691–716.

Theis, K. R., N. M. Dheilly, J. L. Klassen, R. M. Brucker, J. F. Baines, T. C. G. Bosch, J. F. Cryan, S. F. Gilbert, C. J. Goodnight, E. A. Lloyd, J. Sapp, P. Vandenkoornhuyse, I. Zilber-Rosenberg, E. Rosenberg, and S. R. Bordenstein. 2016. Getting the hologenome concept right: an eco-evolutionary framework for hosts and their microbiomes. *mSystems* 1:e00028-00016.

Thibert-Plante, X., and A. P. Hendry. 2011. The consequences of phenotypic plasticity for ecological speciation. *Journal of Evolutionary Biology* 24:326–42.

Thoday, J. M. 1953. Components of fitness. *Symposium of the Society of Experimental Biology* 7:96–113.

Thompson, J. N. 2005. *The Geographic Mosaic of Coevolution*. University of Chicago Press, Chicago, IL.

Thompson, K., and A. J. A. Stewart. 1981. The measurement and meaning of reproductive effort in plants. *American Naturalist* 117:205–11.

Thornton, J. W. 2004. Resurrecting ancient genes: experimental analysis of extinct molecules. *Nature Reviews Genetics* 5:366–75.

Thrall, P. H., M. E. Hochberg, J. J. Burdon, and J. D. Bever. 2007. Coevolution of symbiotic mutualists and parasites in a community context. *Trends in Ecology and Evolution* 22:120–26.

Tonsor, S. J., T. W. Elnaccash, and S. M. Scheiner. 2013. Developmental instability is genetically correlated with phenotypic plasticity, constraining heritability, and fitness. *Evolution* 67:2923–35.

Triantis, K. A., and S. A. Bhagwat. 2011. Applied island biogeography. Pp. 190–223 in R. J. Ladle and R. J. Whittaker, eds., *Conservation Biogeography*. Wiley-Blackwell, Oxford, UK.

Trivers, R. L., and D. E. Willard. 1973. Natural selection of parental ability to vary the sex ratio of offspring. *Science* 179:90–92.

Troost, T. A., B. W. Kooi, and S. A. L. M. Kooijman. 2005. When do mixotrophs specialize? Adaptive dynamics theory applied to a dynamic energy budget model. *Mathematical Biosciences* 193:159–82.

True, J. R., and S. B. Carroll. 2002. Gene co-option in physiological and morphological evolution. *Annual Review of Cell and Developmental Biology* 18:53–80.

True, J. R., and E. S. Haag. 2001. Developmental system drift and flexibility in evolutionary trajectories. *Evolution and Development* 3:109–19.

True, J. R., J. M. Mercer, and C. C. Laurie. 1996. Differences in crossover frequency and distribution among three sibling species of *Drosophila*. *Genetics* 142:507–23.

Tsuji, K. 1995. Reproductive conflicts and levels of selection in the ant *Pristomyrmex pungens*: contextual analysis and partitioning of covariance. *American Naturalist* 146: 586–607.

Tufto, J. 2015. Genetic evolution, plasticity, and bet-hedging as adaptive responses to temporally autocorrelated fluctuating selection: a quantitative genetic model. *Evolution* 69:2034–49.

Tuljapurkar, S. 1990. *Population Dynamics in Variable Environments*. Springer-Verlag, New York.

Tuljapurkar, S. D., and S. H. Orzack. 1980. Population dynamics in variable environments I: long-run growth rates and extinction. *Theoretical Population Biology* 18:314–42.

Turelli, M. 1984. Heritable genetic variation via mutation-selection balance: Lerch's zeta meets the abdominal bristle. *Theoretical Population Biology* 25:138–93.

Turelli, M., N. H. Barton, and J. A. Coyne. 2001. Theory and speciation. *Trends in Ecology and Evolution* 16:330–43.

Turelli, M., J. R. Lipkowitz, and Y. Brandvain. 2014. On the Coyne and Orr-igin of species: effects of intrinsic postzygotic isolation, ecological differentiation, X chromosome size, and sympatry on *Drosophila* speciation. *Evolution* 68:1176–87.

Turelli, M., and L. C. Moyle. 2007. Asymmetric postmating isolation: Darwin's corollary to Haldane's rule. *Genetics* 176:1059–88.

Turelli, M., and H. A. Orr. 1995. The dominance theory of Haldane's rule. *Genetics* 140: 389–402.

Turesson, G. 1922. The genotypical response of the plant species to the habitat. *Hereditas* 3:341–47.

Turner, D. 2015. Historical contingency and the explanation of evolutionary trends. Pp. 73–90 in P.-A. Braillard and C. Malaterre, eds., *Explanation in Biology: An Enquiry into the Diversity of Explanatory Patterns in the Life Sciences*. Springer Netherlands, Dordrecht.

Turney, P., D. Whitley, and R. W. Anderson. 1996. Evolution, learning, and instinct: 100 years of the Baldwin effect. *Evolutionary Computation* 4:iv–viii.

Tuttle, Elaina M., Alan O. Bergland, Marisa L. Korody, Michael S. Brewer, Daniel J. Newhouse, P. Minx, M. Stager, A. Betuel, Zachary A. Cheviron, Wesley C. Warren, Rusty A. Gonser, and Christopher N. Balakrishnan. 2016. Divergence and functional degradation of a sex chromosome-like supergene. *Current Biology* 26:344–50.

Upchurch, P. 2008. Gondwanan break-up: legacies of a lost world? *Trends in Ecology and Evolution* 23:229–36.

Valente, L. M., R. S. Etienne, and A. B. Phillimore. 2014. The effects of island ontogeny on species diversity and phylogeny. *Proceedings of the Royal Society of London B: Biological Sciences* 281:20133227.

Valente, L. M., A. B. Phillimore, and R. S. Etienne. 2015. Equilibrium and non-equilibrium dynamics simultaneously operate in the Galápagos islands. *Ecology Letters* 18:844–52.

Valentine, J. W. 1973. *Evolutionary Paleoecology of the Marine Biosphere*. Prentice-Hall, Englewood Cliffs, NJ.

Valentine, J. W. 1980. Determinants of diversity in higher taxonomic categories. *Paleobiology* 6:444–50.

Valentine, J. W. 2004. *On the Origin of Phyla*. University of Chicago Press, Chicago, IL.

Valentine, J. W. ,and C. L. May. 1996. Hierarchies in biology and paleontology. *Paleobiology* 22:23–33.

Vamosi, J. C., W. S. Armbruster, and S. S. Renner. 2014. Evolutionary ecology of specialization: insights from phylogenetic analysis. *Proceedings of the Royal Society of London B: Biological Sciences* 281:20142004.

Van Buskirk, J., and U. K. Steiner. 2009. The fitness costs of developmental canalization and plasticity. *Journal of Evolutionary Biology* 22:852–60.

Van Dyken, J. D., and M. J. Wade. 2010. The genetic signature of conditional expression. *Genetics* 184:557–70.

Van Noordwijk, A. J., and G. De Jong. 1986. Acquisition and allocation of resources: their influence on variation in life history tactics. *American Naturalist* 128:137–42.

Van Tienderen, P. H. 1991. Evolution of generalists and specialists in spatially heterogeneous environments. *Evolution* 45:1317–31.

Van Tienderen, P. H., and J. Antonovics. 1994. Constraints in evolution: on the baby and the bath water. *Functional Ecology* 8:139–40.

Van Valen, L. 1973. A new evolutionary law. *Evolutionary Theory* 1:1–30.

Van Valkenburgh, B., X. Wang, and J. Damuth. 2004. Cope's rule, hypercarnivory, and extinction in North American canids. *Science* 306:101–4.

Vasek, F. 1980. Creosote bush: long-lived clones in the Mojave Desert. *American Journal of Botany* 67:246–55.

Vaupel, J. W. 1988. Inherited frailty and longevity. *Demography* 25:277–87.

Velasco, J. D. 2008. Species concepts should not conflict with evolutionary history, but often do. *Studies in History and Philosophy of Science Part C: Studies in History and Philosophy of Biological and Biomedical Sciences* 39:407–14.

Velasco, J. D. 2012. The future of systematics: tree-thinking without the tree. *Philosophy of Science* 79:624–36.

Vellend, M. 2010. Conceptual synthesis in community ecology. *Quarterly Review of Biology* 85:183–206.

Venable, D. L. 1992. Size-number trade-offs and the variation of seed size with plant resource status. *American Naturalist* 140:287–87.

Venable, D. L., and L. Lawlor. 1980. Delayed germination and dispersal in desert annuals—escape in space and time. *Oecologia* 46:272–82.

Venail, P. A., R. C. MacLean, T. Bouvier, M. A. Brockhurst, M. E. Hochberg, and N. Mouquet. 2008. Diversity and productivity peak at intermediate dispersal rate in evolving metacommunities. *Nature* 452:210–14.

Vermeij, G. J. 2006. Historical contingency and the purported uniqueness of evolutionary innovations. *Proceedings of the National Academy of Sciences* 103:1804–9.

Vermeij, G. J. 2012. Crucibles of creativity: the geographic origins of tropical molluscan innovations. *Evolutionary Ecology* 26:357–73.

Vermeij, G. J. 2015. Forbidden phenotypes and the limits of evolution. *Interface Focus* 5:20150028.

Vermeij, G. J., and P. D. Roopnarine. 2013. Reining in the Red Queen: the dynamics of adaptation and extinction reexamined. *Paleobiology* 39:560–75.

Verzijden, M. N., C. ten Cate, M. R. Servedio, G. M. Kozak, J. W. Boughman, and E. I. Svensson. 2012. The impact of learning on sexual selection and speciation. *Trends in Ecology and Evolution* 27:511–19.

Via, S. 1993. Adaptive phenotypic plasticity: target or byproduct of selection in a variable environment. *American Naturalist* 142:352–65.

Via, S., R. Gomulkiewicz, G. De Jong, S. M. Scheiner, C. D. Schlichting, and P. Van Tienderen. 1995. Adaptive phenotypic plasticity: consensus and controversy. *Trends in Ecology and Evolution* 10:212–17.

Via, S., and R. Lande. 1985. Genotype-environment interaction and the evolution of phenotypic plasticity. *Evolution* 39:505–22.

Via, S., and J. West. 2008. The genetic mosaic suggests a new role for hitchhiking in ecological speciation. *Molecular Ecology* 17:4334–45.

Vincent, T. L., and H. R. Pulliam. 1980. Evolution of life history strategies for an asexual annual plant model. *Theoretical Population Biology* 17:215–31.

Violle, C., D. R. Nemergut, Z. Pu, and L. Jiang. 2011. Phylogenetic limiting similarity and competitive exclusion. *Ecology Letters* 14:782–87.

Visscher, P. M., M. A. Brown, M. I. McCarthy, and J. Yang. 2012. Five years of GWAS discovery. *American Journal of Human Genetics* 90:7–24.

Vitti, J. J., S. R. Grossman, and P. C. Sabeti. 2013. Detecting natural selection in genomic data. *Annual Review of Genetics* 47:97–120.

Voigt, W. 1973. *Homologie und Typus in der Biologie.* Gustav Fischer, Jena.

Voje, K. L., Ø. H. Holen, L. H. Liow, and N. C. Stenseth. 2015. The role of biotic forces in driving macroevolution: beyond the Red Queen. *Proceedings of the Royal Society of London B: Biological Sciences* 282:20150186.

Vos, M. 2009. Why do bacteria engage in homologous recombination? *Trends in Microbiology* 17:226–32.

Vrba, E. S., and N. Eldredge. 1984. Individuals, hierarchies and processes: towards a more complete evolutionary theory. *Paleobiology* 10:146–71.

Vrba, E. S., and S. J. Gould. 1986. The hierarchical expansion of sorting and selection: sorting and selection cannot be equated. *Paleobiology* 12:217–28.

Waddington, C. H. 1942. Canalization of development and the inheritance of acquired characters. *Nature* 150:563–65.

Waddington, C. H. 1953a. The "Baldwin effect," "genetic assimilation" and "homeostasis." *Evolution* 7:386–87.

Waddington, C. H. 1953b. Epigenetics and evolution. Pp. 186–99 in *Symposium Society for Experimental Biology VII: Evolution.* Cambridge University Press, Cambridge.

Waddington, C. H. 1953c. Genetic assimilation of an acquired character. *Evolution* 7:118–26.

Wade, M. J. 1977. An experimental study of group selection. *Evolution* 31:134–53.

Wade, M. J. 1978. A critical review of the models of group selection. *Quarterly Review of Biology* 53:101–14.

Wade, M. J. 1985. Soft selection, hard selection, kin selection, and group selection. *American Naturalist* 125:61–73.

Wade, M. J., P. Bijma, E. D. Ellen, and W. Muir. 2010. Group selection and social evolution in domesticated animals. *Evolutionary Applications* 3:453–65.

Wade, M. J., and C. J. Goodnight. 1991. Wright's shifting balance theory: an experimental study. *Science* 253:1015–18.

Wade, M. J., N. A. Johnson, and Y. Toquenaga. 1999. Temperature effects and genotype-by-environment interactions in hybrids: Haldane's rule in flour beetles. *Evolution* 53:855–65.

Wade, M. J., N. A. Johnson, and G. Wardle. 1994. Analysis of autosomal polygenic variation for the expression of Haldane's rule in flour beetles. *Genetics* 138:791–99.

Wade, M. J., and S. Kalisz. 1990. The causes of natural selection. *Evolution* 44:1947–55.

Wade, M. J., and D. E. McCauley. 1984. Group selection: the interaction of local deme size and migration in the differentiation of small populations. *Evolution* 38:1047–58.

Wagner, G. P. 1989. The biological homology concept. *Annual Review of Ecology and Systematics* 20:51–69.

Wagner, G. P. 2000. What is the promise of developmental evolution? Part I: why is developmental biology necessary to explain evolutionary innovations? *Journal of Experimental Zoology Part B: Molecular and Developmental Evolution* 288:95–98.

Wagner, G. P. 2007. The developmental genetics of homology. *Nature Reviews Genetics* 8:473–79.

Wagner, G. P. 2010. Evolvability: the missing piece in the neo-Darwinian synthesis. Pp. 197–213 in M. A. Bell, D. J. Futuyma, W. F. Eanes, and J. S. Levinton, eds., *Evolution since Darwin: The First 150 Years*. Sinauer Associates, Sunderland, MA.

Wagner, G. P. 2014. *Homology, Genes, and Evolutionary Innovation*. Princeton University Press, Princeton, NJ.

Wagner, G. P. 2015. Evolutionary innovations and novelties: let us get down to business! *Zoologischer Anzeiger* 256:75–81.

Wagner, G. P., C.-H. Chiu, and M. Laubichler. 2000. Developmental evolution as a mechanistic science: the inference from developmental mechanisms to evolutionary processes. *American Zoologist* 40:819–31.

Wagner, G. P., and J. Draghi. 2010. Evolution of evolvability. Pp. 379–99 in M. Pigliucci and G. B. Müller, eds., *Evolution: The Extended Synthesis*. MIT Press, Cambridge, MA.

Wagner, G. P., and V. J. Lynch. 2010. Evolutionary novelties. *Current Biology* 20:R48–R52.

Wagner, G. P., and J. Zhang. 2011. The pleiotropic structure of the genotype-phenotype map: the evolvability of complex organisms. *Nature Reviews Genetics* 12:204–13.

Wagner, P. J. 1996. Contrasting the underlying patterns of active trends in morphologic evolution. *Evolution* 50:990–1007.

Wagner, P. J. 2010. Paleontological perspectives on morphological evolution. Pp. 451–78 in M. A. Bell, D. J. Futuyma, W. F. Eanes, and J. S. Levinton, eds., *Evolution since Darwin: The First 150 Years*. Sinauer Associates, Sunderland, MA.

Wagner, P. J., and G. F. Estabrook. 2014. Trait-based diversification shifts reflect differential extinction among fossil taxa. *Proceedings of the National Academy of Sciences* 111:16419–24.

Wagner, S. M., A. J. Martinez, Y.-M. Ruan, K. L. Kim, P. A. Lenhart, A. C. Dehnel, K. M. Oliver, and J. A. White. 2015. Facultative endosymbionts mediate dietary breadth in a polyphagous herbivore. *Functional Ecology* 29:1402–10.

Wainwright, P. C. 2007. Functional versus morphological diversity in macroevolution. *Annual Review of Ecology, Evolution, and Systematics* 38:381–401.

Wake, D. 2003. Homology and homoplasy. Pp. 191–200 in B. K. Hall and W. M. Olson, eds., *Keywords and Concepts in Evolutionary Developmental Biology*. Harvard University Press, Cambridge, MA.

Waldron, A. 2007. Null models of geographic range size evolution reaffirm its heritability. *American Naturalist* 170:221–31.

Wallace, A. R. 1858. On the tendency of varieties to depart indefinitely from the original type. In C. Darwin and A. R. Wallace, On the tendency of species to form varieties; and on the perpetuation of varieties and species by natural means of selection. *Journal of the Linnean Society (Zoology)* 3:45–62.

Wallace, A. R. 1889. *Darwinism. An Exposition of the Theory of Natural Selection with Some of Its Applications*. Macmillan, London.

Walters, C. J. 1986. *Adaptive Management of Renewable Resources*. Macmillan, New York.

Wang, I. J., R. E. Glor, and J. B. Losos. 2013. Quantifying the roles of ecology and geography in spatial genetic divergence. *Ecology Letters* 16:175–82.

Warren, D. L. 2012. In defense of "niche modeling." *Trends in Ecology and Evolution* 27:497–500.

Warren, D. L., M. Cardillo, D. F. Rosauer, and D. I. Bolnick. 2014. Mistaking geography for biology: inferring processes from species distributions. *Trends in Ecology and Evolution* 29:572–80.

Washburn, S. L. 1951. The new physical anthropology. *Transactions of the New York Academy of Sciences* 13:298–304.

Watson, R. A., D. M. Weinreich, and J. Wakeley. 2011. Genome structure and the benefit of sex. *Evolution* 65:523–36.

Waxman, D., and S. Gavrilets. 2005. 20 questions on adaptive dynamics. *Journal of Evolutionary Biology* 18:1139–54.

Webb, C. O., D. D. Ackerly, M. A. McPeek, and M. J. Donoghue. 2002. Phylogenies and community ecology. *Annual Review of Ecology and Systematics*:475–505.

Webb, C. O., J. B. Losos, and A. A. Agrawal. 2006. Integrating phylogenies into community ecology. *Ecology* 87:S1–S2.

Weinig, C., J. A. Johnston, C. G. Willis, and J. N. Maloof. 2007. Antagonistic multilevel selection on size and architecture in variable density settings. *Evolution* 61:58–67.

Weinreich, D. M., R. A. Watson, and L. Chao. 2005. Perspective: sign epistasis and genetic constraints on evolutionary trajectories. *Evolution* 59:1165–74.

Weismann, A. 1891. *Essays upon Heredity and Kindred Biological Problems*. Trans. E. B. Poulton, S. Schonland, and A.E. Shipley. Clarendon Press, Oxford, UK.

Weissman, D. B., and N. H. Barton. 2012. Limits to the rate of adaptive substitution in sexual populations. *PLoS Genetics* 8:e1002740.

West, D. A. 2016. *Darwin's Man in Brazil: The Evolving Science of Fritz Muller*. University of Florida Press, Gainesville.

West, G. B., J. H. Brown, and B. J. Enquist. 1999. The fourth dimension of life: fractal geometry and allometric scaling of organisms. *Science* 284:1677–79.

West, G. B., J. H. Brown, and B. J. Enquist. 2001. A general model for ontogenetic growth. *Nature* 413:628–31.

West, S. 2009. *Sex Allocation*. Princeton University Press, Princeton, NJ.

West-Eberhard, M. J. 1989. Phenotypic plasticity and the origins of diversity. *Annual Review of Ecology and Systematics* 20:249–78.

West-Eberhard, M. J. 2003. *Developmental Plasticity and Evolution*. Oxford University Press, New York.

West-Eberhard, M. J. 2005. Phenotypic accommodation: adaptive innovation due to developmental plasticity. *Journal of Experimental Zoology Part B: Molecular and Developmental Evolution* 304B:610–18.

Wheeler, Q. D., and R. Meier, eds., 2000. *Species Concepts and Phylogenetic Theory: A Debate*. Columbia University Press, New York.

Wheeler, Q. D., and N. I. Platnick. 2000. The phylogenetic species concept (*sensu* Wheeler and Platnick). Pp. 55–69 in Q. D. Wheeler and R. Meier, eds., *Species Concepts and Phylogenetic Theory: A Debate*. Columbia University Press, New York.

White, M. J. D. 1973. *Animal Cytology and Evolution*. Cambridge University Press, Cambridge, UK.

White, M. J. D. 1978. *Modes of Speciation*. W. H. Freeman, San Francisco, CA.

Whitlock, M., and P. Phillips. 2000. The exquisite corpse: a shifting view of the shifting balance. *Trends in Ecology and Evolution* 15:347–48.

Whittaker, R. W., K. A. Triantis, and R. J. Ladle. 2008. A general dynamic theory for oceanic island biogeography. *Journal of Biogeography* 35:977–94.

Whitton, J., C. J. Sears, E. J. Baack, and S. P. Otto. 2008. The dynamic nature of apomixes in the angiosperms. *International Journal of Plant Sciences* 169:169–82.

Wiens, J. J. 2004. The role of morphological data in phylogeny reconstruction. *Systematic Biology* 53:653–61.

Wiens, J. J. 2011. The niche, biogeography and species interactions. *Philosophical Transactions of the Royal Society of London B: Biological Sciences* 366:2336–50.

Wiens, J. J., D. D. Ackerly, A. P. Allen, B. L. Anacker, L. B. Buckley, H. V. Cornell, E. I. Damschen, T. Jonathan Davies, J.-A. Grytnes, S. P. Harrison, B. A. Hawkins, R. D. Holt, C. M. McCain, and P. R. Stephens. 2010. Niche conservatism as an emerging principle in ecology and conservation biology. *Ecology Letters* 13:1310–24.

Wiens, J. J., and C. H. Graham. 2005. Niche conservatism: integrating evolution, ecology, and conservation biology. *Annual Review of Ecology, Evolution, and Systematics* 36:519–39.

Wiens, J. J., C. R. Hutter, D. G. Mulcahy, B. P. Noonan, T. M. Townsend, J. W. Sites, and T. W. Reeder. 2012. Resolving the phylogeny of lizards and snakes (Squamata) with extensive sampling of genes and species. *Biology Letters* 8:1043–46.

Wiley, E. O. 1978. The evolutionary species concept reconsidered. *Systematic Biology* 27:17–26.

Wiley, E. O. 1981. *Phylogenetics: The Theory and Practice of Phylogenetic Systematics*. John Wiley, New York.

Wiley, E. O., and D. R. Brooks. 1982. Victims of history—a nonequilibrium approach to evolution. *Systematic Biology* 31:1–24.

Wiley, E. O., and D. R. Brooks. 1986. *Evolution as Entropy: Toward a Unified Theory of Biology*. University of Chicago Press, Chicago, IL.

Wiley, E. O., and R. L. Mayden. 2000a. A defense of the evolutionary species concept. Pp. 198–208 in Q. D. Wheeler and R. Meier, eds., *Species Concepts and Phylogenetic Theory: A Debate*. Columbia University Press, New York.

Wiley, E. O., and R. L. Mayden. 2000b. The evolutionary species concept. Pp. 70–89 in Q. D. Wheeler and R. Meier, eds., *Species Concepts and Phylogenetic Theory: A Debate*. Columbia University Press, New York.

Wilkins, J. S. 2003. How to be a chaste species pluralist-realist: the origins of species modes and the synapomorphic species concept. *Biology and Philosophy* 18:621–38.

Wilkins, J. S., and M. C. Ebach. 2013. *The Nature of Classification: Relationships and Kinds in the Natural Sciences*. Palgrave Macmillan, London.

Williams, G. C. 1966a. *Adaptation and Natural Selection*. Princeton University Press, Princeton, NJ.

Williams, G. C. 1966b. Natural selection, the costs of reproduction, and a refinement of Lack's principle. *American Naturalist* 161:153–67.

Williams, G. C. 1975. *Sex and Evolution*. Princeton University Press, Princeton, NJ.

Williams, M. B. 1989. Evolvers are individuals: extension of the species as individual claim. Pp. 301–8 in M. Ruse, ed., *What the Philosophy of Biology Is*. Kluwer Academic, Dordrecht.

Wilson, A. B., K. Noack-Kunnmann, and A. Meyer. 2000. Incipient speciation in sympatric Nicaraguan crater lake cichlid fishes: sexual selection versus ecological diversification. *Proceedings of the Royal Society of London B: Biological Sciences* 267:2133–41.

Wilson, A. C., G. L. Bush, S. M. Case, and M. C. King. 1975. Social structuring of mammalian populations and rate of chromosomal evolution. *Proceedings of the National Academy of Sciences* 72:5061–65.

Wilson, D. S. 1980. *The Natural Selection of Populations and Communities*. Benjamin/Cummings, Menlo Park, CA.

Wilson, D. S., and J. Yoshimura. 1994. On the coexistence of specialists and generalists. *American Naturalist* 144:692–707.

Wilson, E. O. 1975. *Sociobiology: The New Synthesis*. Belknap Press, Cambridge, MA.

Wilson, E. O. 1996. *Naturalist*. Shearwater Press, New York.

Wilson, J. T. 1963. A possible origin of the Hawaiian Islands. *Canadian Journal of Physics* 41:863–70.

Wilson, J. T. 1966. Did the Atlantic Ocean close and then re-open? *Nature* 211:676–81.

Wilson, M. 2006. *Wandering Significance: An Essay on Conceptual Behavior*. Oxford University Press, New York.

Wilson, R. A. 1999. Realism, essence, and kind: resuscitating species essentialism? Pp. 187–207 in R. Wilson, ed., *Species: New Interdisciplinary Essays*. MIT Press, Cambridge, MA.

Wimsatt, W. C. 2001. Generative entrenchment and the developmental systems approach to evolutionary processes. Pp. 219–37 in S. Oyama, P. E. Griffiths, and R. D. Gray, eds., *Cycles of Contingency*. MIT Press, Cambridge, MA.

Wimsatt, W. C. 2007. *Re-engineering Philosophy for Limited Beings: Piecewise Approximations to Reality*. Harvard University Press, Cambridge, MA.

Winsor, M. P. 2006. Linnaeus's biology was not essentialist. *Annals of the Missouri Botanical Garden* 93:2–7.

Winther, R. G. 2015. The structure of scientific theories. In E. N. Zalta, ed., *The Stanford Encyclopedia of Philosophy*, http://plato.stanford.edu/archives/spr2015/entries/structure-scientific-theories/.

Wittgenstein, L. 1922. *Tractatus Logico-Philosophicus*. Routledge and Kegan Paul, London.

Woese, C. R. 2004. A new biology for a new century. *Microbiology and Molecular Biology Reviews* 68:173–86.

Woese, C. R., and G. E. Fox. 1977. Phylogenetic structure of the prokaryotic domain: the primary kingdoms. *Proceedings of the National Academy of Sciences* 74:5088–90.

Woese, C. R., O. Kandler, and M. L. Wheelis. 1990. Towards a natural system of organisms: proposal for the domains Archaea, Bacteria, and Eucarya. *Proceedings of the National Academy of Sciences* 87:4576–79.

Wolf, J. B., E. D. Brodie III, J. M. Cheverud, A. J. Moore, and M. J. Wade. 1998. Evolutionary consequences of indirect genetic effects. *Trends in Ecology and Evolution* 13:64–69.

Wolfe, K. H. 2001. Yesterday's polyploids and the mystery of diploidization. *Nature Reviews Genetics* 2:333–41.

Wolfe, K. H., C. W. Morden, and J. D. Palmer. 1992. Function and evolution of a minimal plastid genome from a nonphotosynthetic parasitic plant. *Proceedings of the National Academy of Sciences* 89:10648–52.

Woltereck, R. 1909. Weitere experimentelle Untersuchungen über Artveränderung, speziell über das Wesen quantitativer Artunterschiede bei Daphniden. *Verhandlungen der Deutschen Zooligischen Gesellschaft* 19:110–72.

Wood, H. M., N. J. Matzke, R. G. Gillespie, and C. E. Griswold. 2012. Treating fossils as terminal taxa in divergence time estimation reveals ancient vicariance patterns in the palpimanoid spiders. *Systematic Biology* 62:264–84.

Wood, T. E., N. Takebayashi, M. S. Barker, I. Mayrose, P. B. Greenspoon, and L. H. Rieseberg. 2009. The frequency of polyploid speciation in vascular plants. *Proceedings of the National Academy of Sciences* 106:13875–79.

Wootton, J. T. 2005. Field parameterization and experimental test of the neutral theory of biodiversity. *Nature* 433:309–12.

Worley, A. C., D. Houle, and S. C. H. Barrett. 2003. Consequences of hierarchical allocation for the evolution of life-history traits. *American Naturalist* 161:153–67.

Worsley, T. R., R. D. Nance, and J. B. Moody. 1984. Global tectonics and eustasy for the past 2 billion years. *Marine Geology* 58:373–400.

Worsley, T. R., R. D. Nance, and J. B. Moody. 1991. Tectonics, life and climate for the last three billion years: a unified system? Pp. 200–210 in S. H. Schneider and P. J. Boston, eds., *Scientists on Gaia*. MIT Press, Cambridge, MA.

Wray, G. A., H. E. Hoekstra, D. J. Futuyma, R. E. Lenski, T. F. C. Mackay, D. Schulter, and J. E. Strassman. 2014. Does evolutionary theory need a rethink? *Nature* 514:161–64.

Wright, S. 1921. Correlation and causation. *Journal of Agricultural Research* 20:557–85.

Wright, S. 1931. Evolution in Mendelian populations. *Genetics* 16:97–159.

Wright, S. 1932. The roles of mutation, inbreeding, crossbreeding, and selection in evolution. *Proceedings of the Sixth International Congress of Genetics* 1:356–66.

Wright, S. 1942. Statistical genetics and evolution. *Bulletin of the American Mathematical Society* 48:223–46.

Wright, S. 1951. The genetical structure of populations. *Annals of Eugenics* 15:323–54.

Wright, S. 1968. *Evolution and the Genetics of Populations*, vol. 1: *Genetic and Biometric Foundations*. University of Chicago Press, Chicago, IL.

Wright, S. 1982. The shifting balance theory and macroevolution. *Annual Review of Genetics* 16:1–19.

Wright, S., and W. E. Kerr. 1954. Experimental studies of the distribution of gene frequencies in very small populations of *Drosophila melanogaster*. II. Bar. *Evolution* 8:225–40.

Wund, M. A. 2012. Assessing the impacts of phenotypic plasticity on evolution. *Integrative and Comparative Biology* 52:5–15.

Wynne-Edwards, V. C. 1962. *Animal Dispersion in Relation to Social Behavior*. Hafner, New York.

Yachi, S., and M. Loreau. 1999. Biodiversity and ecosystem productivity in a fluctuating environment: the insurance hypothesis. *Proceedings of the National Academy of Sciences* 96:1463–68.

Yang, Z. 2006. *Computational Molecular Evolution*. Oxford University Press, New York.

Yang, Z., and B. Rannala. 2012. Molecular phylogenetics: principles and practice. *Nature Reviews Genetics* 13:303–14.

Yesson, C., and A. Culham. 2006. Phyloclimatic modeling: combining phylogenetics and bioclimatic modeling. *Systematic Biology* 55:785–802.

Zachos, J., M. Pagani, L. Sloan, E. Thomas, and K. Billups. 2001. Trends, rhythms, and aberrations in global climate 65 Ma to present. *Science* 292:686–93.

Zamer, W. E., and S. M. Scheiner. 2014. A conceptual framework for organismal biology: linking theories, models, and data. *Integrative and Comparative Biology* 54:736–56.

Zelditch, M. L., and W. L. Fink. 1996. Heterochrony and heterotopy: stability and innovation in the evolution of form. *Paleobiology* 22:241–54.

Zhang, C., T. Stadler, S. Klopfstein, T. A. Heath, and F. Ronquist. 2016. Total-evidence dating under the fossilized birth-death process. *Systematic Biology* 65:228–49.

Zhao, Y., and S. S. Potter. 2002. Functional comparison of the *Hoxa4*, *Hoxa10*, and *Hoxa11* homeoboxes. *Developmental Biology* 244:21–36.

Zhaxybayeva, O., P. Lapierre, and J. P. Gogarten. 2004. Genome mosaicism and organismal lineages. *Trends in Genetics* 20:254–60.

Zilber-Rosenberg, I., and E. Rosenberg. 2008. Role of microorganisms in the evolution of animals and plants: the hologenome theory of evolution. *FEMS Microbiology Reviews* 32:723–35.

Zimmer, C. 2016. The biologists who want to overhaul evolution. *Atlantic Monthly*. https://www.theatlantic.com/science/archive/2016/11/the-biologists-who-want-to-overhaul-evolution/508712/.

Zuckerkandl, E., and L. Pauling. 1965. Evolutionary divergence and convergence in proteins. Pp. 97–166 in V. Bryson and H. J. Vogel, eds., *Evolving Genes and Proteins*. Academic Press, New York.

Zufall, R. A. 2016. Mating systems and reproductive strategies in *Tetrahymena*. Pp. 221–33 in G. Witzany and M. Nowacki, eds., *Biocommunication of Ciliates*. Springer, New York

INDEX

adaptive landscape, 10, 289, 347. *See also* fitness landscape

allocation: hierarchical, 215, 221, 227; resource, 216–17, 225

anagenesis, 112, 331, 336, 350, 355

assimilation, genetic, 256, 266

Baldwin effect, 255–56, 266

biogeography: island, 28, 244, 322, 325, 327; patterns in, 319–20; vicariance, 320

Brownian motion model, 351, 355, 362

Cambrian explosion, 353–54, 357, 364

causation: downward, 339, 340, 358, 359, 360; upward, 339, 352, 358, 359, 360

cis-regulatory element, 145, 157–58, 159, 163

clonal interference, 290, 294

coevolution, 19, 264, 270, 285

common descent, 2, 13, 21, 93, 124, 153

competition, 75, 203, 223, 228, 242, 247, 282, 290, 312

concrete practice, 147, 150, 162, 167

constitutive theory, definition of, 2, 6, 15, 21, 46, 63, 65, 152, 158, 168, 214, 264, 336, 340

contingency, 9, 155, 339, 342, 361

convergence, 9, 20, 89, 246, 343

Darwin, 3, 11, 26, 46, 66, 103, 121, 123, 141, 171, 228, 255, 297, 301, 309, 312, 338; Darwin centennial, 26, 35–37, 38, 42, 44

deductive theory, 172

demography, 54, 57, 76, 211, 215, 217, 231, 312

detection method, definition of, 48, 55, 60

developmental theory, 155

disparity, 339, 341, 354

dispersal, 222, 235, 239, 251, 282, 321, 328, 330, 352

dispersal theory, 321

diversification, 5, 11, 20, 131, 153, 166, 274, 315, 318, 321, 325, 335, 339, 342, 354, 363

diversity: functional, 339, 341; taxonomic, 114, 339, 356, 358

Dobzhansky-Muller incompatibilities, 301

drift: genetic, 30, 39, 50, 67, 71, 74, 275, 278, 285, 315; phylogenetic, 361; species, 302, 326, 331, 360, 364

early burst model, 355

ecological specialization theory: domain, 239; evolution of, 233; propositions, 239

emergent property, 126, 328, 338; community, 239, 247

environmental heterogeneity, 11, 239, 259, 275, 282

epigenetics, 5, 163, 346, 353

epistasis, 71, 259, 265, 275, 278, 283, 285, 287, 299, 345